NEW DIRECTIONS IN LINEAR ACOUSTICS AND VIBRATION

The field of acoustics is of immense industrial and scientific importance. The subject is built on the foundations of linear acoustics, which is widely regarded as so mature that it is fully encapsulated in the physics texts of the 1950s. This view was changed by developments in physics such as the study of quantum chaos. Developments in physics throughout the last four decades, often equally applicable to both quantum and linear acoustic problems but overwhelmingly more often expressed in the language of the former, have explored this. There is a significant new amount of theory that can be used to address problems in linear acoustics and vibration, but only a small amount of reported work does so. This book is an attempt to bridge the gap between theoreticians and practitioners, as well as the gap between quantum and acoustic. Tutorial chapters provide introductions to each of the major aspects of the physical theory and are written using the appropriate terminology of the acoustical community. The book will act as a quick-start guide to the new methods while providing a wide-ranging introduction to the physical concepts.

Matthew Wright is a senior lecturer in Acoustics at the Institute of Sound and Vibration Research (ISVR). His B.Eng. was in engineering acoustics and vibration, and his Ph.D. was in Volterra series characterization and identification of nonlinear bioacoustic systems, both from the University of Southampton. Since then he has worked on flow control for drag and noise reduction, turbofan inlet design, aeroacoustic theory, violin acoustics, and quantum chaos in acoustics, for the study of which he was awarded an EPSRC Advanced Research Fellowship. His current interests include wind farm noise and the neuroscience of hearing. He is a Fellow of the Institute of Acoustics, a Fellow of the Institute of Mathematics and Its Applications, a Senior Member of the American Institute of Aeronautics and Astronautics, a Member of the Acoustical Society of America, and the book reviews editor of the *Journal of Sound and Vibration*. He teaches musical instrument acoustics and acoustical design.

Richard Weaver received an A.B. degree in physics from Washington University in St. Louis in 1971 and a Ph.D. in astrophysics from Cornell University in 1977. He has been at the University of Illinois since 1981, after a research associateship in theoretical elastic wave propagation and ultrasonics at Cornell. He was elected a Fellow of the Acoustical Society of America in 1996 and received the Hetényi Award from the Society for Experimental Mechanics in 2004. He is associate editor of the *Journal of the Acoustical Society of America*.

New Directions in Linear Acoustics and Vibration

QUANTUM CHAOS, RANDOM MATRIX THEORY, AND COMPLEXITY

Edited by

Matthew Wright
University of Southampton

Richard Weaver
University of Illinois

CAMBRIDGE
UNIVERSITY PRESS

32 Avenue of the Americas, New York NY 10013-2473, USA

Cambridge University Press is part of the University of Cambridge.

It furthers the University's mission by disseminating knowledge in the pursuit of education, learning and research at the highest international levels of excellence.

www.cambridge.org
Information on this title: www.cambridge.org/9781107513457

First published 2010
First paperback edition 2015

A catalogue record for this publication is available from the British Library

Library of Congress Cataloguing in Publication data

New directions in linear acoustics and vibration : quantum chaos, random matrix theory, and complexity / [edited by] Matthew Wright, Richard Weaver.
p. cm.
Includes bibliographical references and index.
ISBN 978-0-521-88508-9
1. Sound-waves. 2. Vibration. 3. Acoustical engineering. I. Wright, Matthew, 1966– II. Weaver, Richard, 1949– III. Title.
QC243.N49 2010
620.2–dc22 2010019825

ISBN 978-0-521-88508-9 Hardback
ISBN 978-1-107-51345-7 Paperback

Contents

Foreword

Michael Berry

H. H. Wills Physics Laboratory, University of Bristol, Bristol, UK

In the early 1970s, Martin Gutzwiller and Roger Balian and Claude Bloch described quantum spectra in terms of classical periodic orbits, and in the mid 1970s it became clear that the random matrix theory devised for nuclear physics would also describe the statistics of quantum energy levels in classically chaotic systems. It seemed obvious even then that these two great ideas would find application in acoustics, but it has taken more than three decades for this insight to be fully implemented. The chapters in this fine collection provide abundant demonstration of the continuing fertility, in the understanding of acoustic spectra, of periodic orbit theory and the statistical approach. The editors' kind invitation to me to write this foreword provides an opportunity to make a remark about each of these two themes.

First, here is a simple argument for periodic orbit theory being the uniquely appropriate tool for describing the acoustics of rooms. The reason for confining music and speech within auditoriums – at least in climates where there is no need to protect listeners from the weather – is to prevent sound from being attentuated by radiating into the open air. But if the confinement were perfect, that is, if the walls of the room were completely reflecting, sounds would reverberate forever and get confused. To avoid these extremes, the walls in a real room must be partially absorbing. This has the effect of converting the discrete eigenvalues with perfectly reflecting walls into resonances. I will argue that for real rooms the width of resonances usually exceeds their spacing. This is important because it casts doubt on the usefulness of the concept of an individual mode in assessing the acoustic response of rooms; a smoothed description of the spectrum seems preferable. But smoothing is precisely what periodic orbit theory naturally describes. When there is no absorption, the contributions from the long periodic orbits make the convergence of the sum problematic, frustrating the direct calculation of individual eigenvalues, for example, in quantum chaology. Absorption attentuates the long orbits, and the oscillatory contributions from few shortest orbits are sufficient to describe the acoustic response. But these few orbits are important: the crudest smoothing, based simply on the average spectral density, obliterates all the spectral oscillations and fails to capture the characteristics of most real rooms.

To assess the significance of absorption, start from the Weyl counting formula for the number N of modes with frequencies less than f, for a room of volume L^3:

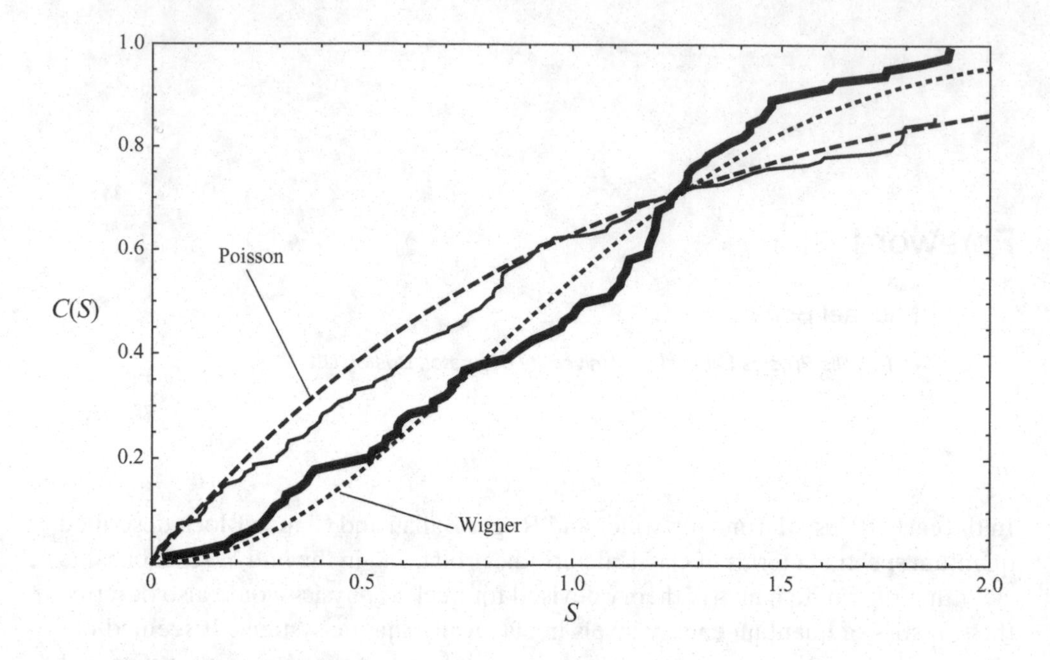

if the speed of sound is $c = 330\,\mathrm{ms}^{-1}$,

$$N = \frac{4\pi L^3 f^3}{3c^3}.$$

In the presence of absorption, modeled approximately by an exponential amplitude decay time T, that is, intensity $\sim \exp(-2t/T)$, the resonance width corresponds to a frequency broadening,

$$\Delta f = \frac{1}{2\pi T}.$$

Thus, incorporating the reverberation time T_{60}, corresponding to 60-dB intensity reduction, that is, $T = T_{60}/3\log_e 10$, the number ΔN of modes smoothed over by the broadening is

$$\Delta N = 6\log_e 10\frac{L^3 f^2}{c^3 T_{60}}.$$

For estimates, we can choose the frequency middle A ($f = 440\,\mathrm{Hz}$). Then, for a small auditorium with $L = 6\,\mathrm{m}$, and a reverberation time $T_{60} = 0.7\,\mathrm{s}$, $\Delta N \sim 23$, which is unexpectedly large for such a small room. For the Albert Hall in London, where the effective $L \sim 60\,\mathrm{m}$, and taking $T_{60} = 2\,\mathrm{s}$, $\Delta N \sim$ 8,200. These estimates strongly suggest that there is little sense in studying individual modes.

Second, here is an unusual application of spectral statistics from 1993, inspired by a visit to Loughborough University, where I talked about quantum chaos and mentioned that the ideas could be usefully applied in acoustics. Afterward, Robert Perrin showed me his measurements (Perrin et al. 1983) of eigenfrequencies of one English church bell, ranging from 292.72 Hz – the lowest mode, called the hum, through the first few harmonics, with their traditional names Fundamental, Tierce, Quint, Nominal, Twister, Superquint – up to the 134th frequency of 9,285 Hz. This provided sufficient data to make a first attempt to understand the frequency spacings distribution.

I did this in two ways. First, taking the whole set of 134 frequencies, unfolding them by fitting the counting function (spectral staircase) to a cubic function, and then calculating the 133 spacings, normalized to unit mean. The resulting cumulative spacings distribution $C(S) =$ fraction of spacings less than S, fits the Poisson distribution $1 - \exp(-S)$ reasonably well (the thin and dashed curves in the figure). This is not surprising because the bell has approximate rotation symmetry, and the whole set of frequencies conflates subsets with different numbers l of nodal meridians ("angular momentum quantum number"). Fortunately the value of l for each frequency was given; l ranged from 0 to 28, but only the subsets with $0 \leq l \leq 10$ included sufficient frequencies to generate sensible statistics. In the second procedure, I unfolded these subsets separately and conflated the spacings afterwards, thereby generating the heavy curve in the figure. This is better fitted to the Wigner cumulative distribution $1 - \exp(-S^2/4)$ (the dotted curve in the figure), indicating strong repulsion of neighboring frequencies in each l-subset. The precise fit is not important because the Wigner distribution should apply when the ray geodesics on the bell – "classical paths" – are chaotic, whereas the vibrations of the bell, regarded as a thin elastic sheet, are probably integrable, with frequencies given by the modes of a one-dimensional "radial" equation, albeit of fourth order.

Reference

Perrin, R., Charnley, T. & DePont, L. (1983), "Normal modes of the modern English church bell," *J. Sound. Vib.* **90**, 29–49.

Introduction

Matthew Wright and Richard Weaver

This book has some of its genesis in the, possibly apocryphal, story that at an acoustics conference in the late 1980s a certain distinguished professor, tiring of the proceedings, turned to the assembled researchers and announced

> Listen! If what you're doing isn't nonlinear or transonic, then don't bother! It's all been done!

Certainly it has become easy to think of linear acoustics as essentially completed. After all, classic texts such as Morse and Feshbach (1953) give admirably thorough expositions of very general techniques, particularly those based on Green's functions. Cases described by coordinate systems in which the governing equations are separable are extensively tabulated and admit analytic solutions. The alternative is to employ numerical methods, many of them also based on Green's functions, which work in arbitrarily complex geometries. There is perhaps a perception that notwithstanding a host of important applied problems, there are no fundamental issues remaining in linear acoustics. Increased understanding of the richness and complexity of nonlinear problems with the explosion of interest in chaos only serves to make linear systems seem "done and dusted" in comparison.

And yet this picture is overly dismissive. A solution of a linear differential equation depends nonlinearly on its coefficients and the shape of the boundary. The dependence is all the richer if those coefficients are random or if boundary reflections are defocusing. Developments in physics throughout the last four decades, often equally applicable to both quantum and linear acoustic problems, but overwhelmingly more often expressed in the language of the former, have explored this. More than that they have provided a new way of thinking about such things. We have been impressed at the significant new body of theory that can be used to address problems in linear acoustics and vibration, although also disappointed at the small amount of reported work that does so. This book is an attempt to bridge the gap between theoreticians and practitioners, as well as the gap between quantum and acoustic, a gap that is mostly terminological but should nevertheless not be underestimated. Our hope is that acousticians and vibration engineers who wish to see what can be done with these new tools will find in this book a comprehensible introduction and that physicists may also learn what problems might usefully be addressed.

So what is on offer? We would like to take the reader on a short guided tour of the terrain. We begin with what is known as the *semiclassical trace formula* (Chapter 1), which expresses the modal density of a closed, lossless enclosure (membrane or cavity) in terms of its *periodic orbits*, closed internal ray paths that repeat indefinitely. As a way to determine eigenvalues (let alone response to arbitrary excitations) it cannot compete with the numerical techniques that have been refined for use in engineering (such as finite elements) or physics (such as plane-wave decomposition); its significance lies in the fact that it provides an explicit link between the shape of an enclosure and its acoustic characteristics, both in an average sense (via the Weyl series) and at the level of individual eigenvalues, and in a way that doesn't depend on separability.

This connection is important because for many shapes the periodic orbits are unstable and the ray paths are *chaotic*, the implications of which are explored in Chapter 2. It can be disconcerting to find chaos having such a profound influence on linear systems. This is due to the nonlinearity of ray motion in the high-frequency limit, and the study of the effects of this on the finite-frequency wave motion has come to be known as quantum chaology or (despite linguistic objections) *quantum chaos*. It used to be easy to imagine that almost all ordinary differential equations had well-behaved, predictable solutions because almost all the ones in books did. That misapprehension was shattered by the explosion of awareness about chaos. In the same way it is easy to fall into the trap of thinking that modeshapes and natural frequencies are as simple and regular in arbitrary shapes as those of the simple textbook examples used to teach the subject. They are not, and for very similar reasons.

One of the consequences of chaotic ray motion is that eigenfunctions often resemble superpositions of Gaussian random waves, the properties of which are explored in more detail in Chapter 4. Those that do not are referred to as "scarred modes"; Chapter 5 presents an ingenious formulation that allows the eigenfunctions to be represented with impressive efficiency in a basis built out of deliberately constructed scar functions. Of course acousticians rarely encounter truly lossless systems in practice; so some of the implications of opening the enclosure are explored in Chapter 6. And in Chapter 7 the central result of the periodic orbit theory is re-derived in a form suitable for elasticity so as to expand the range of possible applications.

Before that, however, we introduce the second major theme of this book: *random matrix theory*. The study of the statistics of the eigenvalues of ensembles of matrices whose elements are random variables and exhibit a particular symmetry began in nuclear physics as an exploration of the conjecture that a sufficiently complex system might have properties statistically similar to those of a random Hamiltonian. Modern computational capabilities have made it easier to test conjectures and confirm analytic results. For example, the fact that the normalized spacings of the eigenvalues of a large Gaussian Orthogonal matrix are close to the Rayleigh distribution (obeyed exactly by an ensemble of pairs of eigenvalues of 2×2 Gaussian orthogonal matrices) can be shown using less than 10 lines of

MATLAB[†] and can be computed in a few seconds. Chapter 3 introduces the theory that allows such predictions and, as its name implies, explores why such an approach should be so effective in describing the behavior of the wave-bearing and vibrating systems we are considering here.

Our third theme, *complexity* does not get a chapter to itself or even an index entry. Instead it is embedded throughout the book in the richness of the behavior of simple systems and the diversity of applications in the later chapters. Each reader will make their own connections between the various topics here, but one striking example is worth noting here: how in a multitude of contexts "the part contains the whole." Just as each cell of an organism contains the DNA of the whole being, a few short periodic orbits contain information about a large part of the eigenstructure; in seismology and underwater acoustics a short part of a time history reveals information about the whole system.

Subsequent chapters survey several applied topics related in varying degrees to the earlier chapters. Inasmuch as multiple scattering plays such a recurrent and important role in mesoscopics (the subject of Chapter 8), we also include a review of the, often too obscure to the non-initiate, diagrammatic methods for the theory of randomly scattered acoustics in Chapter 9. The surprising and highly applicable results of the theory of time-reversed waves are explored in Chapter 10 with particular reference to the themes of this book, which have led to important applications in ultrasonics.

Chapter 11 shows the relevance of ray chaos for long-range propagation in the ocean, whereas Chapter 12 demonstrates applications in seismology. Chapter 13 shows how random matrix theory can be applied to structural acoustics and vibrations, whereas Chapter 14 explains an alternative random matrix theory approach to the problem of estimating the likely variation in response that results from the inevitable small variations that arise in manufacturing.

It is impossible in a book of practical length to cover all the modern applications of these ideas that we might have, and we apologize to those who have noted holes in our coverage. Perhaps there will be a need for another book.

As editors we wish to thank the authors and the publishers for their patience during the unfortunately long time it has taken to turn their contributions into this book. We express our gratitude to all the publishers who granted permission for the chapter authors to reuse figures from their published articles without payment, and our greater gratitude to those who provided it as a matter of policy without being asked. We have tried to attribute all reused figures; if we have inadvertently failed to do so we would be grateful to be informed and will endeavor to correct the

[†] For the avoidance of doubt they are as follows:

```
n = 2000;
A = randn(n);
E = eig((A + A')/2);
s = diff(E).*real(sqrt(2*n - E(1:n-1).^2)/pi);
[N,x] = hist(s,40);
bar(x,N/n/(x(2) - x(1)))
hold on
plot(x,(pi/2)*x.*exp((-pi/4)*x.^2),'r','LineWidth',2)
hold off
```

oversight in future editions if there are any. We also wish to thank the organizers of the 2005 Summer School on Chaotic and Random Wave Scattering at the Centro International de Ciencias A. C. in Cuernavaca, at which the idea for this book was born when we accidentally got separated from the rest of our party while exploring the pyramids of Xochicalco and their notable acoustics.

Matthew Wright acknowledges the support provided by an EPSRC Advanced Research Fellowship during this project and thanks his colleagues at ISVR and in particular Chris Howls of the School of Mathematics for useful discussions and Carolyn and David for their tolerance and understanding. Richard Weaver thanks the US National Science Foundation for support from grant 28096. Olivier Legrand and Fabrice Mortessagne thank Valérie Doya for useful criticism and careful reading of the manuscript. Niels Søndergaard thanks Gregor Tanner for valuable discussions. Joseph Turner and Goutam Ghoshal gratefully acknowledge the financial support of the US Department of Energy, the National Science Foundation, and the Federal Railroad Administration. Steven Tomsovic and Michael Brown thank Javier Beron-Vera, Nicholas Cerruti, Katherine Hegewisch, Irina Rypina, and Ilya Udovydchenkov for the benefit of many discussions relating to the material presented and gratefully acknowledge support from the US National Science Foundation, grants PHY-0555301 (ST) and CMG-0417425 (MB), and Code 321 of the Office of Naval Research (MB).

1 The Semiclassical Trace Formula

Matthew Wright

Institute of Sound and Vibration Research, University of Southampton, UK

1.1 Introduction

For a two-dimensional enclosure, such as a membrane or the cross section of an infinitely long duct, those with the very simplest shapes (circles, rectangles, spheres, boxes, etc.) with simple uniform boundary conditions, the modes and natural frequencies can be determined analytically. For any other shape they may be determined numerically by a range of mature numerical techniques of which finite element and boundary element analyses are the best known and the most widely studied. Knowing how to calculate the modes and natural frequencies for any particular shape, however, is not the same as understanding how those modes and natural frequencies depend on the shape. Suppose, for example, that we wish to improve the design of a component by optimizing some quantity such as weight, while leaving its natural frequencies unchanged. In the course of such an optimization changes will be made to the shape, whereupon the process of calculating the modes and natural frequencies must begin all over again; at best, part of the mesh can be re-used. Such an analysis cannot tell us where effort can be most or least profitably concentrated.

It turns out that the shapes that can be analyzed are (for good reason) quite untypical compared with arbitrary shapes. The situation mirrors the one that used to prevail in the study of dynamical systems, where linear differential equations were most widely studied because of their solubility, and the fact that other systems showed radically different qualitative behavior was, for a time, ignored. In both cases the overlooked feature is chaos, but in the case of acoustic morphology the phenomenon is known as quantum chaos. Despite its name, this phenomenon can be exhibited by large-scale systems such as acoustical resonators, whose governing equations are entirely linear. It arises when a ray path is unstable to small perturbations and displays strong sensitivity to initial conditions.

Several surveys (Berry 1987, Guhr et al. 1998, Galdi et al. 2005, Kuhl et al. 2005) and books (Gutzwiller 1990, Ott 1993, Brack & Bhaduri 1997, Stöckmann 1999, Richter 2000, Haake 2001, Nakamura & Harayama 2004, Reichl 2004, Cvitanović et al. 2005) on aspects of this subject have become available in recent years, but these are variously intended for physicists, mathematicians, and electronic engineers. The theory of periodic orbits, and of quantum chaos, is applicable to a far greater range of areas than just acoustics, and naturally these texts span that range.

1.2 Introductory Examples

1.2.1 Modes in a Rectangular Enclosure

The rectangle is perhaps the simplest case to study because an explicit formula exists for its natural frequencies. From here on we shall work with wavenumber rather than frequency, and so we shall use the equation for the eigenwavenumbers of a rectangle with sides a_1, a_2:

$$k_{n,m} = \pi \left(\frac{n^2}{a_1^2} + \frac{m^2}{a_2^2} \right)^{1/2}, \tag{1.1}$$

where the indices n and m run $0, 1, 2, \ldots$ for Neumann boundary conditions and $1, 2, 3, \ldots$ for Dirichlet conditions. The spectral density of this system is defined as

$$\rho(k) = \sum_{n,m} \delta(k - k_{n,m}) \tag{1.2}$$

and the modecount as

$$N(k) = \int_0^k \rho(k')\,\mathrm{d}k' = \sum_{n,m} \mathrm{H}(k - k_{n,m}), \tag{1.3}$$

where H is the Heaviside function. We shall now show how alternative, series-form expressions for $\rho(k)$ and $N(k)$ can be obtained.

The delta functions in (1.2) can be written as the limit of a Gaussian function

$$\delta(k - k_{n,m}) = \lim_{t\to 0} \frac{1}{2\sqrt{\pi t}} \mathrm{e}^{-(k-k_{n,m})^2/4t}. \tag{1.4}$$

We can therefore write the spectral density function in the form

$$\rho(k) = \sum_{n=1}^{\infty} \sum_{m=1}^{\infty} \lim_{t\to 0} \frac{1}{2\sqrt{\pi t}} \mathrm{e}^{-\left(k-\pi\sqrt{n^2/a_1^2+m^2/a_2^2}\right)^2/4t}. \tag{1.5}$$

The Poisson formula for a double sum,

$$\begin{aligned}\sum_{n=0}^{\infty} \sum_{m=0}^{\infty} f(n,m) = &\sum_{M_1=-\infty}^{\infty} \sum_{M_2=-\infty}^{\infty} \iint_0^{\infty} f(n_1,n_2)\mathrm{e}^{2\pi\mathrm{i}(M_1 n_1 + M_2 n_2)}\,\mathrm{d}n_1\,\mathrm{d}n_2 \\ &+ \frac{1}{2} \sum_{M_1=-\infty}^{\infty} \int_0^{\infty} f(n_1,0)\mathrm{e}^{2\pi\mathrm{i}M_1 n_1}\,\mathrm{d}n_1 \\ &+ \frac{1}{2} \sum_{M_2=-\infty}^{\infty} \int_0^{\infty} f(0,n_2)\mathrm{e}^{2\pi\mathrm{i}M_2 n_2}\,\mathrm{d}n_2 \\ &+ \frac{1}{4} f(0,0),\end{aligned} \tag{1.6}$$

can be applied to (1.5). We shall take each term separately, denoting them F_1, F_2, F_3, F_4.

The expression for F_1 can be integrated by making the substitutions

$$n_1 = \frac{a_1 r}{\pi}\cos\theta, \qquad n_2 = \frac{a_2 r}{\pi}\sin\theta, \qquad \mathrm{d}n_1\,\mathrm{d}n_2 = \frac{a_1 a_2}{\pi^2} r\,\mathrm{d}r\,\mathrm{d}\theta, \tag{1.7}$$

giving

$$F_1 = \sum_{M_1=-\infty}^{\infty} \sum_{M_2=-\infty}^{\infty} \int_0^{\infty} \int_0^{\pi/2} \lim_{t\to 0} \frac{a_1 b_1}{\pi^2} \frac{1}{2\sqrt{\pi t}} e^{-(k-r)^2/4t + 2i(M_1 a_1 \cos\theta + M_2 a_2 \sin\theta) r} r \, dr \, d\theta. \tag{1.8}$$

After some manipulation this gives

$$F_1 = \frac{a_1 a_2 k}{2\pi} \sum_{M_1=-\infty}^{\infty} \sum_{M_2=-\infty}^{\infty} J_0(k L_{M_1,M_2}), \tag{1.9}$$

where $L_{M_1,M_2} = 2\sqrt{M_1^2 a_2^2 + M_2^2 a_2^2}$ and J_0 is a Bessel function of zero order.

For F_2 we have

$$\begin{aligned} F_2 &= -\frac{1}{2} \sum_{M_1=-\infty}^{\infty} \lim_{t\to 0} \frac{1}{2\sqrt{\pi t}} \int_0^{\infty} e^{-(k-\pi n_1/a_1)^2/4t + 2\pi i M_1 n_1} dn_1 \\ &= -\frac{1}{2} \sum_{M_1=-\infty}^{\infty} \lim_{t\to 0} \frac{a_1}{2\pi} e^{2M_1 a_1 (ik - 2M_1 a_1 t)} \left[1 + \operatorname{erf}\left(\frac{k + 4i M_1 a_1 t}{2\sqrt{t}} \right) \right] \\ &= -\frac{a_1}{2\pi} \sum_{M_1=-\infty}^{\infty} e^{2ik M_1 a_1} \\ &= -\frac{a_1}{2\pi} \sum_{M_1=-\infty}^{\infty} \cos(2k M_1 a_1), \end{aligned} \tag{1.10}$$

and F_3 is the same with all subscripts 1 changed to 2 throughout. It can be shown that taking the sums on the left-hand side of (1.6) from 1 instead of 0, which would correspond to Neumann, rather than Dirichlet, boundary conditions, would reverse the sign of F_2 and F_3.

We therefore have

$$\rho(k) = \frac{a_1 a_2 k}{2\pi} \sum_{M_1, M_2=-\infty}^{\infty} J_0(k L_{M_1,M_2}) \pm \sum_{i=1,2} \sum_{M=-\infty}^{\infty} \frac{a_i}{2\pi} \cos(2k M a_i) + \frac{\delta(k)}{4} \tag{1.11}$$

for Dirichlet (Neumann), conditions. Figure 1.1 shows a series of ray paths drawn in the rectangular domain, which reflect M_1 and M_2 times from the left and bottom walls, respectively, before returning to their origin with the initial heading so as to be able to repeat indefinitely. Such closed paths are called *periodic orbits*. Their length is given by L_{M_1,M_2}. This is no coincidence, as will be seen. The term $2kMa_i$ that forms the argument of the cosine in the second term can also be interpreted as the length of a ray path traveling between two parallel sides,

Because $\rho(k)$ is singular for all $k = k_n$, it must be smoothed before evaluation. In practice, we find it more convenient to work with $N(k)$, its integral with respect to k. Before evaluating this, however, we shall separate out the terms corresponding to zero-length orbits as

$$\overline{\rho}(k) = \frac{a_1 a_2}{2\pi} k \pm \frac{a_1 + a_2}{2\pi} + \frac{\delta(k)}{4}, \tag{1.12}$$

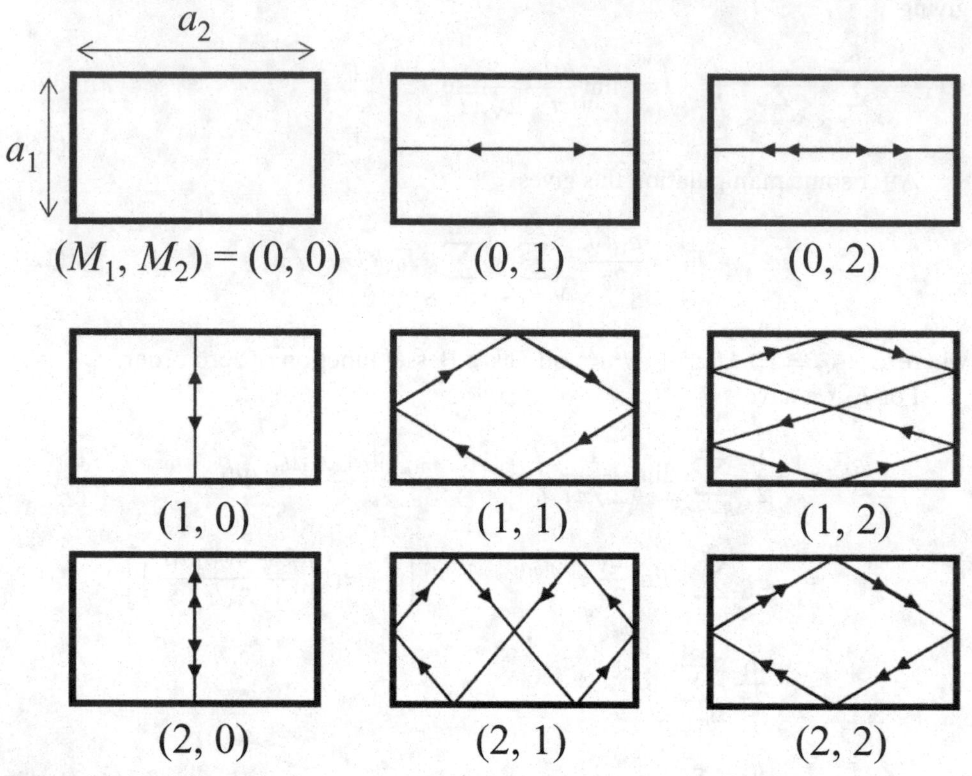

Figure 1.1. Periodic orbits for a rectangular enclosure.

leaving the remainder

$$\rho_{\mathrm{osc}}(k) = \frac{a_1 a_2 k}{2\pi} \sum_{M_1, M_2=-\infty}^{\infty}{}' \mathrm{J}_0(k L_{M_1,M_2}) \pm \sum_{i=1,2} \sum_{M=-\infty}^{\infty}{}' \frac{a_i}{2\pi} \cos(2k M a_i), \tag{1.13}$$

where the primes on the summations indicate that the terms in which all indices are zero are omitted. The smooth components can be integrated to give

$$\begin{aligned} \overline{N}(k) &= \frac{a_1 a_2}{4\pi} k^2 \mp \frac{a_1 + a_2}{2\pi} + \frac{1}{4} \\ &= \frac{A}{4\pi} k^2 \mp \frac{L}{4\pi} k + \frac{1}{4}, \end{aligned}$$

which is the well-known formula for the average number of modes in a rectangular enclosure with area A and perimeter L (see, e.g., Morse & Ingard 1968). The oscillating component can also be integrated to give

$$N_{\mathrm{osc}}(k) = \frac{a_1 a_2 k}{2\pi} \sum_{M_1, M_2=-\infty}^{\infty}{}' \frac{\mathrm{J}_1(k L_{M_1,M_2})}{L_{M_1,M_2}} \pm \sum_{i=1,2} \sum_{M_i=-\infty}^{\infty}{}' \frac{\sin(2k M_i a_i)}{4\pi M}, \tag{1.14}$$

where the second term can be recognized as the Fourier series representation of a sawtooth wave.

Partial sums of (1.14) plus $\overline{N}(k)$ are compared with the true modecount, calculated by evaluating (1.3) explicitly, in Figure 1.2.

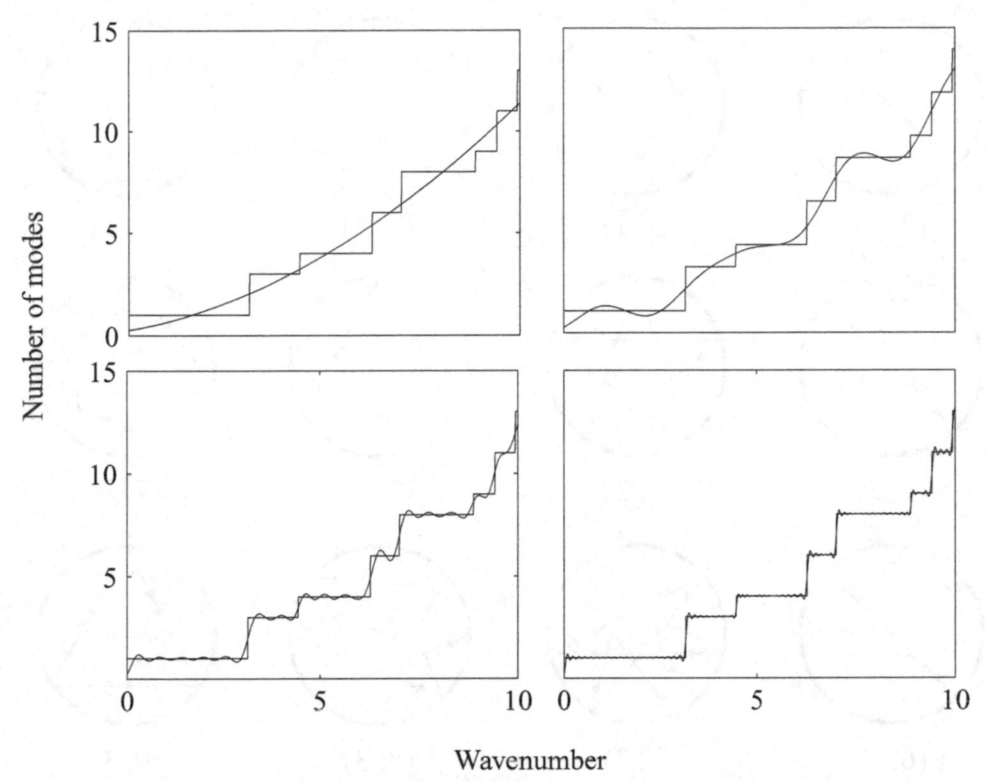

Figure 1.2. Partial sums of the semiclassical approximation to the modecount for a rectangular membrane with maximum values of M_i in all the summations of 0, 1, 4, and 20 respectively. After Wright (2001). Copyright 2001, the Acoustical Society of America.

1.2.2 The Length Spectrum of a Circle

Rather than try to derive a similar formula for the circle we will, for now, conjecture that such a formula exists and that it is of the form

$$\rho(k) \approx \sum_{\mathrm{PO}} A_{\mathrm{PO}}(k) \cos(k L_{\mathrm{PO}} + \phi_{\mathrm{PO}}), \tag{1.15}$$

where L_{PO} is the length of a periodic orbit and the sum is over all such orbits. Define the "length spectrum" $R(L)$ as the Fourier transform of $\rho(k)$. Then, if the conjecture is correct it ought to display peaks at $L = L_j$. The periodic orbits in the circle are shown in Figure 1.3, parameterized by v, the number of vertices, and w, the winding number about the center. The length of each orbit is given by

$$L_{vw} = 2vR \sin \frac{\pi w}{v}, \tag{1.16}$$

where R is the radius of the circle, taken to be unity henceforth.

Because the eigenwavenumbers of the circular membrane are zeros of Bessel functions, which can be found numerically, the length spectrum can be easily calculated as

$$R(L) = \int_{-\infty}^{\infty} \sum_{m,n} \delta(k - \mathrm{j}_{mn}) \mathrm{e}^{\mathrm{i}kL}\, \mathrm{d}k = \sum_{m,n} \mathrm{e}^{\mathrm{i}\,\mathrm{j}_{mn} L}. \tag{1.17}$$

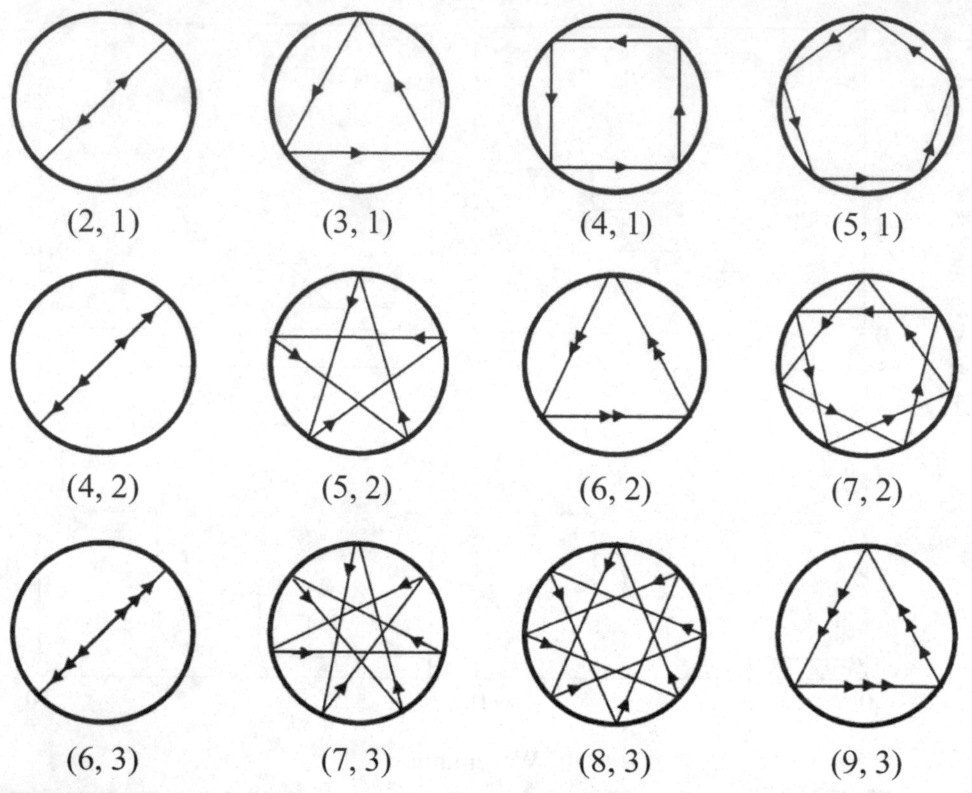

Figure 1.3. Periodic orbits for a circular domain. After Balian and Bloch (1972).

The absolute value of this is plotted in Figure 1.4. As expected from the preceding conjecture, it shows peaks at values of L satisfying Equation (1.16) for integer v and w, that is, 4, $3\sqrt{3}$, $4\sqrt{2}$, $10\sin\pi/5$, and so on.

With this evidence we are ready to sketch the derivation of a formula like Equation (1.15) for any shape of membrane or cavity. First, however, we will find it helpful to review the quantum theory that gave rise to this result, and the analogy between quantum billiards and acoustical systems.

1.3 The Quantum–Acoustic Analogy

A widely studied problem in quantum physics is that of a scalar particle in a potential field, which obeys Schrödinger's equation:

$$\frac{-\hbar^2}{2m}\nabla^2\psi_n + V(\mathbf{r})\psi_n = E_n\psi_n, \tag{1.18}$$

where $2\pi\hbar = 6.6 \times 10^{-34}$ Js is Planck's constant, m is the particle's mass, V is the potential at a point $\mathbf{r}$, and E_n is the nth discrete energy level. The complex wavefunction ψ_n can then be interpreted so that $|\psi_n(\mathbf{r})|^2\,\mathbf{dr}$ is the probability of finding a particle with energy E_n in the volume $\mathbf{dr}$ surrounding the point $\mathbf{r}$. If the potential takes the form of an infinite well, so that it is zero within a domain B and infinite outside it, then the boundary condition will be $\psi_n = 0$ on ∂B, and the wavefunctions will be normalized such that $\int_B |\psi_n(\mathbf{r})|^2\,\mathbf{dr} = 1$ because the particle must exist

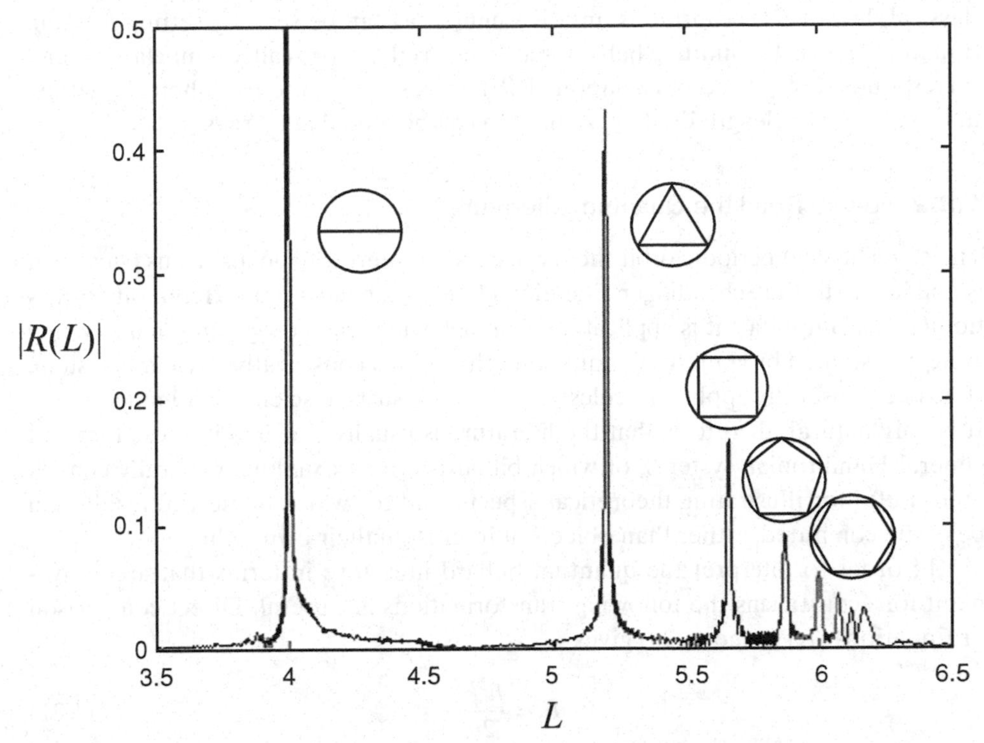

Figure 1.4. Normalized length spectrum of a circle. Periodic orbits have been sketched next to the peaks to which they correspond.

somewhere within B. Such a domain–particle system is known as a quantum billiard. The allusion is specifically to billiards (strictly carom billiards) rather than, say, snooker because in the closed systems studied here there are no pockets by which the particle can leave the domain. Open systems are considered in Chapter 6.

Because the potential is zero inside B we can rewrite Equation (1.18) as

$$\nabla^2 \psi_n + \frac{2m}{\hbar^2} E_n \psi_n = 0, \tag{1.19}$$

which is identical in form to the Helmholtz equation

$$\nabla^2 \psi_n + k_n^2 \psi_n = 0. \tag{1.20}$$

Therefore the problem of quantizing the energy levels of a two-dimensional billiard is the same as that of finding the modal frequencies of a membrane of the same shape.

1.3.1 The Semiclassical Limit

Physically, Planck's constant $\hbar$ determines the scale over which the energy levels are quantized. It is because it is very small in everyday units that quantum effects are not observed in everyday motions. Although it is a universal constant, it is convenient to allow it to vary by rescaling other quantities. As $\hbar$ becomes small, quantum effects become less noticeable, and the behavior of the system gets closer to that predicted by classical physics as $\hbar \to 0$. The behavior never becomes exactly classical, however, because this limit is singular, as can be seen by setting $\hbar = 0$ in Equation (1.19). The limiting behavior as $\hbar \to 0$ is therefore called semiclassical and corresponds to $E_n \to \infty$ in Equation (1.19) or $k_n \to \infty$ in the Helmholtz equation. In this short-wavelength limit, wave motion can be modeled by rays.

1.3.2 How to Read the Quantum Literature

The semiclassical periodic orbit theory may be of interest to acousticians because it is applicable to the Schrödinger Equation (1.18) and hence to the Helmholtz Equation (1.20). But in fact it is applicable to a much wider class of systems, namely, any system described by Hamilton's equations, that is, any conservative dynamic system. It has been usefully applied to celestial dynamics, surface science, and much more. It is only natural, therefore, that the literature is usually couched in these terms of general Hamiltonian systems, of which billiards are an example, generally considered useful for illustrating theoretical aspects, and for which numerical results can be easily computed, rather than objects of interest in their own right.

In order to interpret the quantum billiard literature in terms that are convenient for acousticians the following transformations are useful. Direct comparison of Equations (1.19) and (1.20) gives

$$E = \frac{\hbar^2 k^2}{2m} \tag{1.21}$$

in a zero potential. If $\tilde{\rho}(E)$ is the spectral density as a function of energy, then

$$\rho(k) = \frac{\hbar^2 k}{m} \tilde{\rho}(E). \tag{1.22}$$

Because of wave–particle duality the momentum p of the particle is $\hbar k$. Because the speed of the particle is the ratio of its momentum to its mass, the time period T of a path or orbit is given by

$$T = \frac{mL}{\hbar k}. \tag{1.23}$$

Finally, the action S of a path or orbit is

$$S = pL = \hbar k L. \tag{1.24}$$

Using these expressions it should be possible to evaluate the expressions appearing in the quantum billiard literature in such a way as to eliminate S, p, E, m and $\hbar$ and recover a formula that is applicable to the acoustics of membranes and cavities.

As has been pointed out by van Tiggelen (2005) the quantum–acoustic analogy, even between a membrane and a two-dimensional billiard, is not perfect. In the acoustic case the wave function ψ is the directly measurable quantity, whereas in the quantum case, only $|\psi|^2$ can be observed. In lossless acoustics, energy is conserved, whereas probability is conserved in the quantum case; that is, the eigenfunctions are normalized such that

$$\int_B |\psi|^2 \, \mathrm{d}^d \mathbf{x} = 1, \tag{1.25}$$

that is, the particle must be found somewhere in the d-dimensional billiard B.

The concept of *integrability* of a shape is important to the subject but hard to define without reference to the Hamiltonian of a point particle in a billiard of that shape, in which case it means that there are as many constants of the classical motion as there are degrees of freedom. In practice it is sufficient but not necessary for the wave equation to be separable in a particular shape of enclosure for the dynamics of a particle in it to be integrable. An example of a non-separable but integrable shape is the equilateral triangle. Other irregular polygons belong to the class known as *pseudo-integrable* (Richens & Berry 1981). Shapes that are neither integrable nor pseudo-integrable may exhibit chaotic ray dynamics, which is discussed in the next chapter.

1.4 The Semiclassical Trace Formula

The membrane has a Green's function G satisfying

$$\nabla^2 G + k^2 G = \delta(\mathbf{r} - \mathbf{r}_0) \quad \text{in } B, \tag{1.26}$$

$$G(\mathbf{r}, \mathbf{r}_0; k) = 0 \qquad \mathbf{r}, \mathbf{r}_0 \text{ on } \partial B. \tag{1.27}$$

The Green's function describes wave propagation from $\mathbf{r}$ to $\mathbf{r}_0$ and can be related to G_0, the Green's function for free space, by writing the following double-layer potential (Filippi et al. 1989):

$$G(\mathbf{r}, \mathbf{r}_0; k) = G_0(\mathbf{r}, \mathbf{r}_0; k) + \int_{\partial B} \frac{\partial G_0(\mathbf{r}, \alpha)}{\partial n_\alpha} f(\alpha, \mathbf{r}_0)\, \mathrm{d}\sigma_\alpha, \tag{1.28}$$

where f is to be determined. Balian and Bloch (1972)observed that the solution can be found by successive approximation as follows:

$$\begin{aligned} G(\mathbf{r}, \mathbf{r}_0; k) = {} & G_0(\mathbf{r}, \mathbf{r}_0; k) \\ & - 2 \int_{\partial B} \frac{\partial G_0(\mathbf{r}, \alpha)}{\partial n_\alpha} G_0(\alpha, \mathbf{r}_0)\, \mathrm{d}\alpha \\ & + 2^2 \iint_{\partial B \times \partial B} \frac{\partial G_0(\mathbf{r}, \alpha)}{\partial n_\alpha} \frac{\partial G_0(\alpha, \beta)}{\partial n_\beta} G_0(\beta, \mathbf{r}_0)\, \mathrm{d}\alpha\, \mathrm{d}\beta \\ & - 2^3 \iiint_{\partial B \times \partial B \times \partial B} \frac{\partial G_0(\mathbf{r}, \alpha)}{\partial n_\alpha} \frac{\partial G_0(\alpha, \beta)}{\partial n_\beta} \frac{\partial G_0(\beta, \gamma)}{\partial n_\gamma} G_0(\gamma, \mathbf{r}_0)\, \mathrm{d}\alpha\, \mathrm{d}\beta\, \mathrm{d}\gamma \\ & + \cdots, \end{aligned} \tag{1.29}$$

where each successive integral corresponds to another reflection of waves from the boundary; because the waves spread in all directions the integrations are around the entire boundary. In the semiclassical limit the Green's function will tend to its large argument asymptote; for example, in two dimensions

$$G_0(\mathbf{r}, \mathbf{r}_0; k) = \frac{1}{4\mathrm{i}} \mathrm{H}_0^{(1)}(k|\mathbf{r} - \mathbf{r}_0|) \sim -\frac{(1+\mathrm{i})}{4\sqrt{\pi k |\mathbf{r} - \mathbf{r}_0|}}\, \mathrm{e}^{\mathrm{i}k|\mathbf{r} - \mathbf{r}_0|}, \quad \text{as } k \to \infty, \tag{1.30}$$

where $\mathrm{H}_0^{(1)}$ is the zero-order Hankel function of the first kind. Whatever the number of dimensions the Green's function will behave like a complex exponential at large argument, and therefore the integrals in Equation (1.29) will all take the form

$$\int \cdots \int g(\mathbf{r}) \mathrm{e}^{\mathrm{i}k|\mathbf{r}-\mathbf{r}_0|} \, \mathrm{d}\alpha \cdots \mathrm{d}\omega,$$

in which case the method of stationary phase (Self 2005) can be used. Under this approximation, each integral will be dominated by the contribution from specularly reflecting paths and the semiclassical approximation to the Green's function will be

$$G(\mathbf{r}, \mathbf{r}_0) \approx \sum_j a_j(\mathbf{r}, \mathbf{r}_0) \mathrm{e}^{\mathrm{i}kL_j + \mathrm{i}\phi_j}, \tag{1.31}$$

where a_j is a geometrical prefactor, which can be obtained from the geometry of the orbit; L_j is the length of specularly reflecting ray paths; and ϕ_j, known as the Maslov phase, is related to phase changes undergone by a ray in traversing the path of length L_j.

Consider now the exact spectral density of the system, defined by

$$\rho(k) = \sum_{n=0}^{\infty} \delta(k - k_n). \tag{1.32}$$

This can be related to the Green's function of the system by the trace formula

$$\rho(k) = -\frac{1}{\pi} \lim_{\epsilon \to 0} \Im \left\{ \int_B G(\mathbf{r}, \mathbf{r}; k + \mathrm{i}\epsilon) \, \mathrm{d}\mathbf{r} \right\}, \tag{1.33}$$

where $G(\mathbf{r}, \mathbf{r}; k)$ is known as the trace Green's function, which is singular for $k = k_n$, necessitating the limiting process. The spectral density can be written as the sum of a smooth part and an oscillatory part:

$$\rho(k) = \overline{\rho}(k) + \rho_{\mathrm{osc}}(k), \tag{1.34}$$

and substituting the semiclassical Green's function of Equation (1.31) into Equation (1.33) gives the semiclassical trace formula

$$\rho(k) \approx \overline{\rho}(k) + \sum_j A_j(k) \mathrm{e}^{\mathrm{i}kL_j + \mathrm{i}\phi_j} + \cdots, \tag{1.35}$$

where L_j is the length of the jth periodic orbit, A_j gives the amplitude of its contribution (discussed later), and ϕ_j is the phase change accumulated over one period. The value of this expression is that it quantifies the effect of the geometry on the spectrum through A_j and L_j and those of the type of boundary conditions through ϕ_j. This is, in principle, true for a very wide class of shapes* and in particular does not depend on separability, though convergence of the series is not guaranteed in all cases for which such a formula can be obtained. Furthermore the fact that the geometry and boundary conditions enter the formula through different variables means that once the periodic orbits have been determined for a particular shape the approximate spectrum can be determined for any boundary condition type.

* When the shape is near to an integrable shape the relevant formula takes a different form (Tomsovic et al. 1995, Ullmo et al. 1996). For systems with mixed chaoticity the question is still open.

1.4.1 Smooth Spectral Density and Modecount

The smooth part of the spectral density, and hence the smooth modecount, has been extensively reviewed by Baltes and Hilf (1976). The relevant results can be briefly summarized: first for two dimensions,

$$\overline{N}(k) = \frac{|B|}{4\pi}k^2 \pm \frac{|\partial B|}{4\pi}k + \left[\frac{1}{12\pi}\int_B K(s)\,\mathrm{d}s + \frac{1}{24\pi}\sum_i\left(\frac{\pi}{\alpha_i} - \frac{\alpha_i}{\pi}\right)\right] + O(k^{-1}), \tag{1.36}$$

where $|B|$ and $|\partial B|$ are the area and boundary length of the domain, $K(s)$ is the radius of curvature as a function of distance s along the boundary, and α_i is the included angle of the ith corner. In three dimensions the corresponding expression is

$$\overline{N}(k) = \frac{|B|}{6\pi^2}k^3 \pm \frac{|\partial B|}{16\pi}k^2 + \left[\frac{1}{24\pi}\sum_i\left(\frac{\pi}{\alpha_i} - \frac{\alpha_i}{\pi}\right)L_i\right]k + O(k^0), \tag{1.37}$$

where $|B|$ and $|\partial B|$ are now the volume and surface area and L_i is the length of the ith edge with included angle α_i. The leading term of these expressions is called the Weyl law; the full series is often called the Weyl series.

One way to examine how closely the eigenvalues of a particular system adhere to the predictions of the Weyl series would be to calculate both N and $\overline{N}$ and plot one against the other. A similar effect, however, can be obtained with a discrete set by examining the values of $x_n = \overline{N}(k_n)$. This procedure is known as "unfolding the spectrum." The expected value of x_n is $n - (1/2)$, and a graph of the staircase obtained from x_n (in the same way that N is obtained from k_n) has the same form as a graph of N versus $\overline{N}$. In this way it is straightforward to compare the departures from the average of eigenvalues from different systems with one another and to examine phenomena such as spectral rigidity, as will be explored in the following chapter.

1.5 The Nature of the Approximation

It is only in very rare circumstances, such as for the rectangle, that an exact expression in terms of periodic orbits can be found. More usually when rules to find the amplitude and phase terms (discussed in Section 1.6) are followed, the result is an approximation. There are two reasons for this approximation: first that only the leading-order semiclassical approximation to the Green's function is used (though this will only be the case in two dimensions) and second that the method of steepest descent has been applied. This second source of approximation is the dominant one and means that only terms to leading order in k are obtained. Furthermore, the formula is only asymptotically true as $k \to \infty$.

It might be thought that this means that the trace formula is only useful at very high frequencies. However, the number of periodic orbits needed to resolve steps in the modecount function increases rapidly with wavenumber. This means, as noted by Fulling (2002), that the trace formula is more useful at low frequencies. This "proliferation of orbits" is even more marked in the chaotic shapes to be considered later than it is for the integrable shapes considered so far and in the next section. Furthermore, convergence of the series is only likely in the case of integrable or pseudo-integrable shapes.

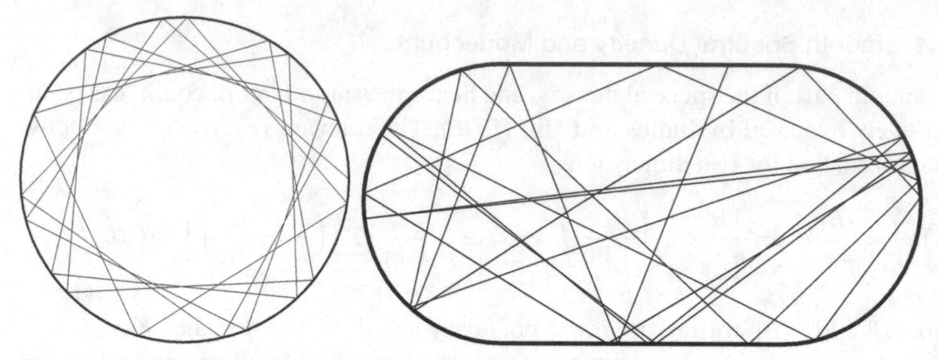

Figure 1.5. Two widely studied billiards. The ray dynamics in the circle are integrable, and all its periodic orbits are marginally stable and lie in continuous families. It was proven by Bunimovich (1974) that the ray dynamics in the stadium are entirely chaotic. All its periodic orbits are unstable, and all are isolated with the exception of the "bouncing ball" orbits that run perpendicularly between the two straight sides.

Boasman (1994) examined the accuracy of the semiclassical approximation within the context of the boundary integral method and found it to be good. Primack and Smilansky (1998) showed that the accuracy of the semiclassical trace formula depends only weakly, if at all, on dimensionality.

1.6 Derivation of Trace Formulas for Given Shapes

A number of methods have been developed to calculate expressions for the amplitude and phase terms in the trace formula. They are as follows:

(i) Gutzwiller's (1970) method for isolated orbits, which involves determining the stability matrix for each orbit;
(ii) Balian and Bloch's (1972) method, based on multiple reflections, which derives amplitude terms for a wide range of conditions in a three-dimensional billiard;
(iii) Berry and Tabor's (1976) method for integrable Hamiltonian systems, based on action-angle variables (these are developed in Chapter 11); and
(iv) Creagh and Littlejohn's (1991, 1992) generalization of Gutzwiller's method to include continuous families of orbits.

We will concentrate on the last of these because of its wide applicability and because it highlights the geometry of the orbits. Balian and Bloch (1972) gave a worked example of the application of the first method to a sphere, whereas worked examples for a circle were given using the second method by Richter et al. (1996), and using the third method by Creagh (1996).

1.6.1 Billiard Dynamics

The dynamics of billiards is a large subject that was first studied by Birkhoff (1927) and has since been treated in a number of surveys and monographs (Kozlov & Treshchëv 1980, Gutkin 2003, Tabachnikov 2005). Here only the elements necessary to the trace formula will be developed. Two widely studied billiards are shown in Figure 1.5; it is clear that the dynamics in the circle and the stadium are very different.

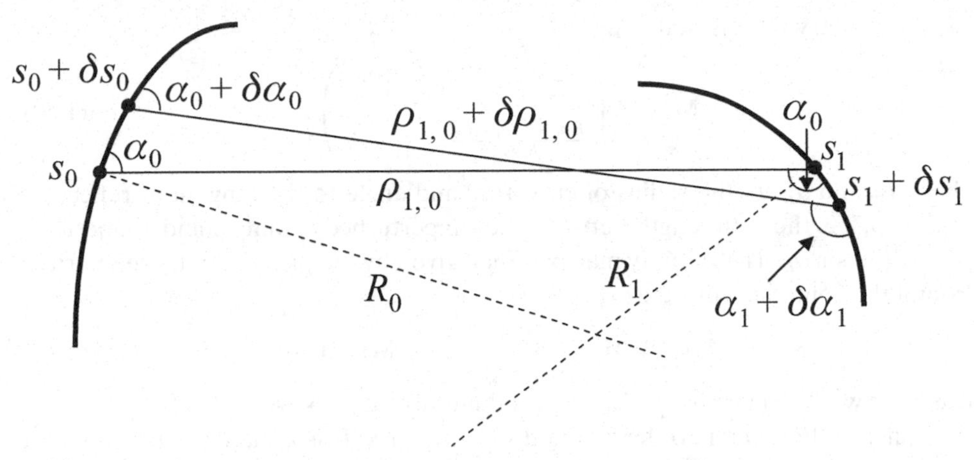

Figure 1.6. A segment of a periodic orbit between two bounces labeled 0 and 1.

The dynamics of any billiard can be characterized by taking its boundary as a Poincaré surface of section because between bounces the motion of the particle is completely determined. For some plane billiard B, let s denote the distance around its boundary from some chosen origin. Let s_n be the location of the nth bounce and α_n the angle between the path of the particle/ray and the tangent to the billiard wall at the point of impact. These two discrete variables completely define the particle/ray path, but we prefer to use $p_n = \cos\alpha_n$ for two reasons: first because it corresponds to a component of the momentum of a particle that is of physical interest in quantum mechanical problems, and it is convenient for us to follow the same notation; second because the nonlinear mapping M of the phase space $(s, p)^T$ induced by the billiard and defined by

$$\begin{pmatrix} s_{n+1} \\ p_{n+1} \end{pmatrix} = M \begin{pmatrix} s_n \\ p_n \end{pmatrix} \tag{1.38}$$

is area preserving. These coordinates define the phase space for the dynamics in the billiard. To study the dynamics we need to study the stability of rays to small perturbations, which is governed by the stability matrix.

1.6.2 Stability Matrix

The following procedure for evaluating the stability matrix directly from the orbits has been given by Berry (1981b) and in appendix C of Brack and Bhaduri (1997). Consider a part of an orbit that makes two successive bounces with the boundary, which we can number 0 and 1 without loss of generality (see Figure 1.6). The stability matrix depends on the stability of the orbit, which can be investigated by making a small change to point 0 and examining the resulting change in point 1. In the limit of small changes the transformation will be linear and can be written as

$$\begin{pmatrix} \delta s_1 \\ \delta\cos\alpha_1 \end{pmatrix} = \mathbf{M}_{1,0} \begin{pmatrix} \delta s_0 \\ \delta\cos\alpha_0 \end{pmatrix}. \tag{1.39}$$

Some geometry reveals that

$$\mathbf{M}_{1,0} = \begin{pmatrix} \frac{\rho_{1,0}}{R_0\eta_1} - \frac{\eta_0}{\eta_1} & -\frac{\rho_{1,0}}{\eta_0\eta_1} \\ \frac{R_0\eta_0+R_1\eta_1-\rho_{1,0}}{R_0 R_1} & \frac{\rho_{1,0}}{R_1\eta_0} - \frac{\eta_1}{\eta_0} \end{pmatrix}, \tag{1.40}$$

where R_i and α_i are the radius of curvature and angle to the tangent at reflection point i, $\rho_{1,0}$ is the path length between the (unperturbed) points 0 and 1, and $\eta_n = \sqrt{1-p_n^2} = \sin\alpha_n$. The stability matrix is then given by the product of these matrices around the complete periodic orbit:

$$\mathbf{M}_{\mathrm{PO}} = \mathbf{M}_{0,N-1}\mathbf{M}_{N-1,N-2}\cdots\mathbf{M}_{2,1}\mathbf{M}_{1,0}. \tag{1.41}$$

The orbit will be stable if $|\operatorname{tr}\mathbf{M}_{\mathrm{PO}}| < 2$ and unstable if $|\operatorname{tr}\mathbf{M}_{\mathrm{PO}}| > 2$.

Many authors refer to the "monodromy" matrix. It is a matrix consisting of a submatrix that is the stability matrix and another submatrix that is the 2×2 identity matrix.

1.6.3 Isolated Orbits

Consider first an isolated orbit such as those that occur in the stadium (see Figure 1.5) other than the "bouncing ball" orbits running between the two straight sides, which are not isolated. For a system consisting only of such orbits we have the Gutzwiller trace formula:

$$\rho_{\mathrm{osc}}(k) \approx \frac{1}{\pi}\sum_{\mathrm{PO}} \frac{L_{\mathrm{PPO}}}{\sqrt{|\det(\mathbf{M}_{\mathrm{PO}}-\mathbf{I})|}} \cos\left(kL_{\mathrm{PO}} - \sigma_{\mathrm{PO}}\frac{\pi}{2}\right), \tag{1.42}$$

where "PO" means periodic orbit and "PPO" means primitive periodic orbit, that is, the shortest orbits that are not repetitions of other orbits; L_{PPO} is the length of the primitive periodic orbit in question; $\mathbf{M}_{\mathrm{PO}}$ is its stability matrix; and σ_{PO} is the Maslov phase. The Maslov phase can be written as the sum of two contributions $\sigma = \mu + \nu$. For each reflection from a boundary with Dirichlet boundary conditions μ increases by 2; Neumann boundaries leave it unchanged. Crossing a focal point inside the domain increases μ by 1.

1.6.4 Families of Orbits

For many wave-bearing enclosures, particularly those whose shapes are integrable, orbits lie in continuous families in phase space. For example, any orbit in the rectangle can be perturbed by moving its point of origin along a side while keeping its initial angle fixed so as to form another, equally valid orbit. In the circle any orbit can be rotated through any angle without changing its length or shape. We will restrict ourselves here to one-parameter families of orbits in two-dimensional domains. For the full theory for multiple-parameter families in n-dimensional space, we refer the reader to Creagh and Littlejohn (1991). The contribution to the spectral density from one-parameter families is then

$$\rho_{\mathrm{osc}}(k) \approx \frac{1}{\pi}\sqrt{\frac{k}{2\pi}}\sum_{\mathrm{PO}}\left(\int |\mathcal{J}_{\mathrm{PO}}|^{-1/2}\,\mathrm{d}q\,\mathrm{d}r_\perp\right), \cos\left(kL_{\mathrm{PO}} - \sigma_{\mathrm{PO}}\frac{\pi}{2} - \frac{\pi}{4}\right), \tag{1.43}$$

where q and $r_\perp$ are coordinates that lie along and perpendicular to the periodic orbit, and the Jacobian representing the sensitivity of the periodic orbit's transverse final position to the angle at which it started is

$$\mathscr{J}_{\mathrm{PO}} = \left(\frac{\partial r_\perp}{\partial \alpha}\right)_{\mathrm{PO}}, \tag{1.44}$$

where α is the angle at which the orbit is launched. The integration is performed over the area of the domain through which this family of orbits passes. For example, with an area-enclosing orbit in the circle the integration would be over the annulus between the boundary and the caustic created by rotating the orbit about the center.

For two-dimensional domains with rotational symmetry the formula simplifies further to

$$\rho_{\mathrm{osc}}(k) \approx \sqrt{\frac{2kR}{\pi}} \sum_{\mathrm{PO}} \frac{L_{\mathrm{PO}} \cos^{1/2}\alpha}{a_{\mathrm{PO}} \left|(\partial\Theta/\partial\alpha)_{\mathrm{PO}}\right|^{1/2}} \cos\left(k L_{\mathrm{PO}} - \sigma_{\mathrm{PO}}\frac{\pi}{2} - \frac{\pi}{4}\right), \tag{1.45}$$

where R is the radius of the billiard, a_{PO} is the number of discrete rotational symmetries possessed by the orbit, and $\partial\Theta$ is the change in angle subtended at the boundary of the domain by an initially periodic orbit launched from the boundary whose launch angle is perturbed by $\partial\alpha$. This formula was used to obtain the trace formula for the concentric annulus by Wright and Ham (2007).

Most problems discussed in the literature are concerned with billiards with Dirichlet, or occasionally Neumann, boundary conditions. Sieber et al. (1995) examined the case with mixed boundary conditions that are a linear combination of the two, the contribution of each type being allowed to vary along the boundary.

1.6.5 Finding Periodic Orbits

In most of the shapes considered so far it has been possible to deduce all the families of periodic orbits by inspection and, by geometry, derive expressions for their lengths as a function of simple parameters. In less regular shapes, particularly the chaotic ones, to be discussed in the next section, the periodic orbits do not lie in parameterizable families; instead, each one has to be determined separately. Furthermore, in such shapes the number of orbits proliferates rapidly with the orbit length. Nonetheless, strategies have been devised for finding the periodic orbits.

One possibility is to form the length spectrum, as was done for the circle in Section 1.2.2. This shows peaks corresponding to the lengths of the periodic orbits, and the knowledge of these lengths may be enough to deduce (or help deduce) their location. This method suffers from the obvious disadvantage that it is necessary to know the eigenvalues to find the length spectrum. This is likely to be computationally expensive if many orbits are required, and of course once the eigenvalues have been found the spectral density and modecount can be directly computed without the need for the trace formula.

It might be thought that periodic orbits could be found by a shooting method, where orbits that fail to close have their reflection points moved until they do. This, however, only works for stable periodic orbits, and as shall be seen in the next section, many periodic orbits in cases of interest are unstable, so that, as the saying goes, a miss is as good as a mile.

A better strategy, described by Biham and Kvale (1992), uses symbolic dynamics. In its simplest form this can be considered a "rubber band" method. The boundary is divided into a number of elements whose symbols are $1, 2, \ldots, N$. It is then possible to enumerate all the members of a tree of strings of symbols $s_1 s_2 \cdots s_N s_1$, and these represent a closed trajectory whose nth bounce occurs somewhere in element s_n, though the reflections need not be specular. Depending on the shape under study the tree can then be "pruned" to remove all strings corresponding to impossible orbits (such as those where there is no line of sight between successive elements). For the remaining strings the reflection points can be optimized to minimize the length of the path (i.e., to relax the rubber band). Once these minima have been found the corresponding orbit can be examined: if its reflections are all specular, it is a periodic orbit; if not, it can be discarded. This procedure can still be consuming for long orbits. Improvements to this method have been proposed (Hansen 1993a–c, 1995, Hansen & Cvitanović 1995).

1.7 Early Work in Sound and Vibration

Acoustical research in this area can be loosely classified into a number of areas:

(i) reports of acoustical phenomena that can be explained in terms of the trace formula or in terms of quantum chaos;
(ii) explanations of these phenomena in terms of the trace formula and the like; and
(iii) experiments on acoustical systems performed in order to confirm theoretical predictions in semiclassical physics.

These categories are not necessarily exclusive or exhaustive. Kuhl et al. (2005) reviewed experimental work in the field of chaotic scattering, including, but not limited to, acoustic and elastic experiments.

1.7.1 Acoustics

As pointed out by Baltes and Hilf (1976) "[t]he need for the study of averaged mode numbers was first met in the acoustics of rooms." Early investigations were made by Bolt (1947) and Bolt and Roop (1950) who examined the spacing of natural frequencies in an idealised rectangular room and observed that "irregularly" shaped rooms have more evenly distributed resonances. Schroeder and Kuttruff (1962) examined average modal frequency spacing in relation to reverberation time measurements.

Balian and Bloch, in their seminal 1972 paper, were aware that the theory they were developing was as applicable to acoustics as it was to quantum systems and made a number of related observations. They pointed out that

> [t]he role of standing waves arising after multiple reflection on the boundary surface is known (and intuitively obvious) in a very different context. In the problem of designing concert halls, it is essential to minimize the oscillations in the density of eigenmodes,

> since such oscillations tend to emphasize some frequencies with respect to others. An elementary rule to follow for this purpose is to avoid shapes favoring the existence of standing waves, along closed paths involving a number of reflections on the walls of the hall. We shall precisely show here that oscillations in the density of eigenvalues are associated with the closed optical paths which may be constructed with any number of mirror reflections on the boundary surface.

They also made a more tenuous claim in a footnote discussing the dependence of amplitude of oscillations on shape:

> The general question of determining the domains for which the oscillation amplitude is the largest possible has not been investigated. The use of shapes approximating that of Fig. 10 [theirs], for improving the regularity of distribution of harmonics and hence the quality of sound, was apparently discovered in the XVI-th century by makers of lutes, both for the shape of the belly and for the cross-section.

The shape in question is the intersection of two concentric ellipses with major axes at right angles. It is hard to recognize this shape in a lute; furthermore the acoustic modes of the cavity at frequencies at which the effects described would be noticeable are unlikely to be important, compared with structural modes. In any case the comment is surprising because they calculated that this shape, which has a strongly degenerate short orbit, will have a particularly large amplitude of contributions ρ_{osc}, implying *uneven* distribution of resonances.

Berry (1981a, Appendix I) discussed the need to suppress clustering of resonances in an auditorium, noting that nonisolated periodic orbits tend to increase this clustering, and predicted that a fractal auditorium, having only diffractive periodic orbits, would have the most smoothly distributed resonances. He returned to the theme in Berry (1983), saying

> [a] physical problem for which the closed path sum is a promising approach is the acoustics of auditoriums, where absorption at the walls causes individual eigenfrequencies to be broadened into resonances much wider than their separation, thus effecting a natural smoothing.

Weaver (1989a) reviewed models for the transmission functions of reverberation rooms and concluded that the effect of spectral rigidity was significant. In the 1990s Legrand and coworkers (Legrand & Sornette 1990, Mortessagne et al. 1993a, Legrand & Mortessagne 1996, Mortessagne & Legrand 1996) examined Sabine's formula for reverberation time in the context of a stadium-shaped room with absorbing walls. The relevance of ergodicity and mixing in auditoria had previously been proposed by Joyce (1975) in the context of the geometrical theory of acoustics. These ideas are considered further in Chapters 2 and 4. Delande and Sornette (1997) discussed the radiation of sound from a stadium-shaped membrane in an infinite baffle and showed that so-called scarred modes were responsible for directional radiation. Berry's semiclassical theory of spectral rigidity has been used to derive expressions for the deviations of the true modecount from its Weyl average in rectangular (Wright 2001), circular (Wright et al. 2003b), and annular ducts (Wright & Ham 2007).

1.7.2 Plates

Work on plates by physicists has been motivated by two factors. First, experiments on plates in vacuo, although delicate, are nonetheless feasible and yield very high Q values, allowing many individual resonances to be distinguished and statistical comparison with the predictions of random matrix theory (RMT) to be made. Second, plates, at least under certain conditions, obey simplified governing equations such as the biharmonic equation for thin plates, allowing progress to be made in obtaining semiclassical descriptions, which for the full elastic problem, or even for shells, are formidably difficult.

Several groups have reported experimental measurements of resonances in plates and the comparison with the predictions of RMT (Bertelsen 1997, Schaadt 1997, Schaadt & Kudrolli 1999, Bertelsen et al. 2000, Andersen et al. 2001, Neicu et al. 2001, Schaadt et al. 2001, Schaadt et al. 2003). Brodier et al. (2001) also used numerical estimates to investigate the statistics of a "clover"-shaped plate. Neicu and Kudrolli (2002) were able to observe the presence of periodic orbits in the vibration of a plate. Pato et al. (2005) compared data from four different experiments, including plates and solids, wherein each had a parameter that could be varied causing the eigenvalues to vary. The authors showed how the curvature of the variation of eigenvalues with parameter could be described by a universal distribution after suitable normalization.

The theoretical study of semiclassical plates was undertaken by Bogomolny and Hugues (1998). An approach to the evaluation of the semiclassical trace formula for thin plates based on their work follows later. In order to evaluate modecount staircases in this way it is necessary to know the lower-order terms of the Weyl series, which will depend on boundary conditions. These were first calculated by Vasil'ev (1987).

The extension of the semiclassical approach from systems such as those already considered to thin elastic plates, whose transverse vibrations are governed by the biharmonic equation

$$\left(\nabla^4 - k^4\right) w = 0 \tag{1.46}$$

can be justified as follows: The biharmonic equation can be factorized as

$$\left(\nabla^2 + k^2\right)\left(\nabla^2 - k^2\right) w = 0, \tag{1.47}$$

in which the first factor describes propagating dispersive waves, whereas the second describes evanescent waves. Clearly any solution of the Helmholtz equation will also satisfy the biharmonic equation. The periodic orbits for a rectangular plate are therefore the same as those for a rectangular membrane, and most of the periodic orbit derivation can be repeated in this case with minor changes. The boundary conditions, however, must be treated carefully. The Maslov phase will depend on the type of boundary condition satisfied by that edge. For a simply supported edge it will be zero, which is to be expected because the rectangular simply supported plate is exactly analogous to the rectangular membrane. For clamped boundary conditions, however, the phase ϕ depends on the angle θ that the incoming ray makes with the

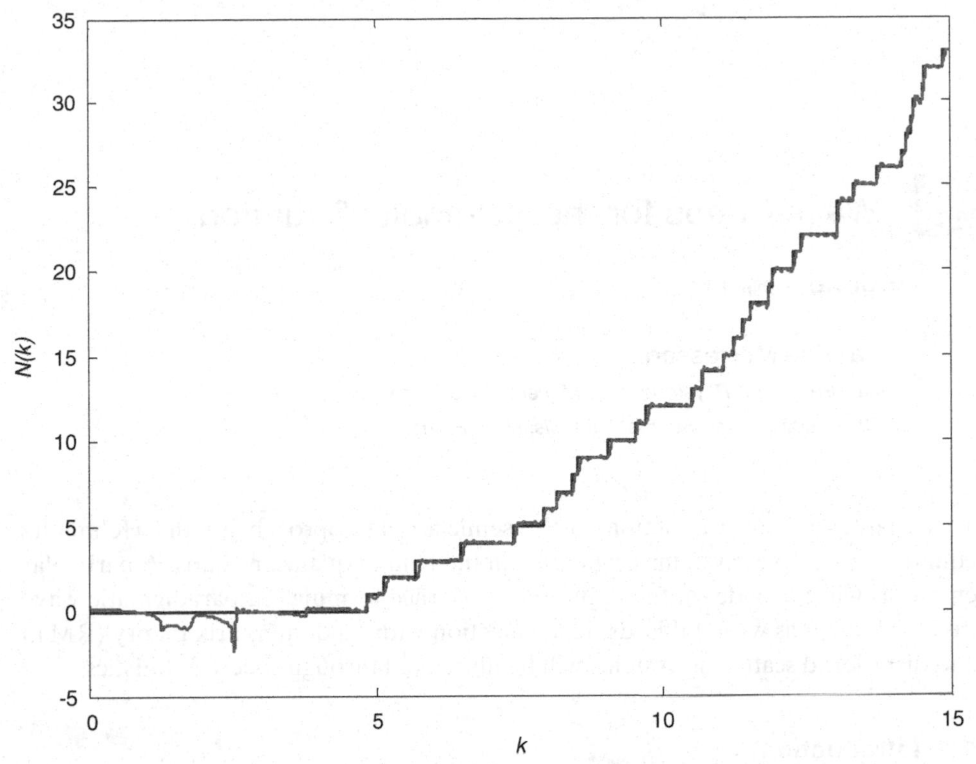

Figure 1.7. Semiclassical and exact modecounts for a rectangular plate measuring 1 m × 2.5 m and 5-mm thick clamped at the long edges and simply supported at the other two. Young's modulus is 70 GPa, density 2,700 kg m^{-3} and Poisson's ratio 0.3. From Wright et al. (2003a).

edge as (Bogomolny & Hugues 1998)

$$\phi_c(\theta) = -2\tan^{-1}\left(\frac{\sin\theta}{\sqrt{1+\cos^2\theta}}\right), \tag{1.48}$$

whereas for a free edge the formula is

$$\phi_f(\theta) = -2\tan^{-1}\left\{\frac{\sin\theta}{\sqrt{1+\cos^2\theta}}\left[\frac{1-(1-\nu)\cos^2\theta}{1+(1-\nu)\cos^2\theta}\right]\right\}, \tag{1.49}$$

where ν is the material's Poisson's ratio and θ is the angle made by the periodic orbits reflecting from the edges and the tangent to that edge at the reflection point. The total phase accumulated during the orbit must be summed and added to the arguments of the oscillatory functions in Equation (1.14). The resulting trace formula can then be used to calculate modecount functions for rectangular plates in the same way as was done for the rectangular membrane. Results are shown in Figure 1.7 for a rectangular plate with mixed boundary conditions, where the modecount is compared with the exact modecount for this structure (Wright et al. 2003a). The agreement for a similar plate with two adjacent clamped edges and two adjacent simply supported edges is considerably worse. Elastic formulations are explored in more detail in Chapter 7.

2 Wave Chaos for the Helmholtz Equation

Olivier Legrand

Fabrice Mortessagne

Laboratoire de Physique de la Matière Condensée,
Université de Nice Sophia-Antipolis, Nice, France

This chapter is an introduction to the semiclassical approach for the Helmholtz equation in complex systems originating in the field of quantum chaos. A particular emphasis will be made on the applications of trace formulae in paradigmatic wave cavities known as wave billiards. Its connection with random matrix theory (RMT) and disordered scattering systems will be illustrated through spectral statistics.

2.1 Introduction

The study of wave propagation in complicated structures can be achieved in the high-frequency (or small-wavelength) limit by considering the dynamics of rays. The complexity of wave media can be due either to the presence of inhomogeneities (scattering centers) of the wave velocity or to the geometry of boundaries enclosing a homogeneous medium. It is the latter case that was originally addressed by the field of *quantum chaos* to describe solutions of the Schrödinger equation when the classical limit displays chaos. The Helmholtz equation is the strict formal analog of the Schrödinger equation for electromagnetic or acoustic waves, the geometrical limit of rays being equivalent to the classical limit of particle motion. To qualify this context, the new expression *wave chaos* has naturally emerged. Accordingly, *billiards* have become geometrical paradigms of wave cavities.

In this chapter we will particularly discuss how the global knowledge about ray dynamics in a chaotic billiard may be used to explain universal statistical features of the corresponding wave cavity, concerning spatial wave patterns of modes, as well as frequency spectra. These features are for instance embodied in notions such as the *spatial ergodicity* of modes and the *spectral rigidity*, which are indicators of particular spatial and spectral correlations. The spectral study can be done through the so-called trace formula based on the *periodic orbits* (POs) of chaotic billiards. From the latter we will derive universal spatial and spectral features in agreement with the predictions of *random matrix theory*. We will also see that deviations from a universal behavior can be found, which carry information about the specific geometry of the cavity. Finally, as a first step toward disordered scattering systems, a description of spectra of cavities dressed with a point scatterer, will be given in terms of *diffractive orbits*.

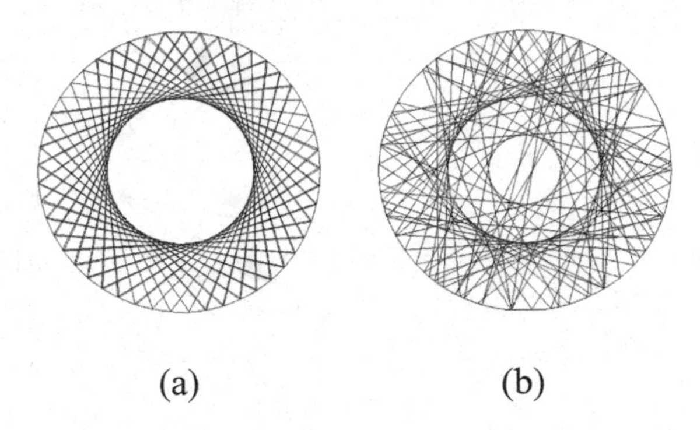

Figure 2.1. Examples of a single ray trajectory after a propagation of 150 in units of the radius R: (a) inside a circular billiard, where a caustic is clearly observed; (b) inside a circular billiard cut by a small straight segment of length $10^{-2}R$. The caustic is destroyed because of the chaotic motion.

2.2 Ray Chaos and Trace Formula for the Helmholtz Equation

2.2.1 Ray Chaos in Cavities – Periodic Orbits

The Helmholtz equation describes a variety of stationary wave phenomena studied in electromagnetism, acoustics, seismology, and quantum mechanics; its general form for a (complex) scalar wavefunction ψ (ψ being the pressure variation in a fluid, a component of the electromagnetic field in a cavity, the elevation of a membrane, etc.) reads

$$(\nabla^2 + k^2(\mathbf{r}))\psi(\mathbf{r}) = 0. \tag{2.1}$$

The wavenumber k depends explicitly on space according to the dispersion relation $k^2(\mathbf{r}) = \omega^2/c^2(\mathbf{r})$, where $c(\mathbf{r})$ is the spatially varying wave velocity. By writing the wavefunction in the form $\psi(\mathbf{r}) = A(\mathbf{r})\exp[\mathrm{i}S(\mathbf{r})]$, and by neglecting terms of the form $\nabla^2 A/A$, (2.1) leads to the so-called eikonal equation for the phase:

$$(\nabla S)^2 = k^2(\mathbf{r}). \tag{2.2}$$

In the geometrical limit, the rays may be viewed as the characteristic curves $[\mathbf{r}(t), \mathbf{k}(t)]$ of a fictitious particle of variable mass $(2c^2)^{-1}$ and pseudo-momentum $\mathbf{k}$ whose dynamics is controlled by the Hamiltonian (Tanner & Søndergaard 2007)

$$H(\mathbf{r}, \mathbf{k}) = c^2(\mathbf{r})\mathbf{k}^2 = \omega^2. \tag{2.3}$$

The ray trajectories are given by the Hamilton's equations of motion:

$$\frac{\mathrm{d}}{\mathrm{d}t}\mathbf{r} = \frac{c^2}{\omega}\mathbf{k}, \qquad \frac{\mathrm{d}}{\mathrm{d}t}\mathbf{k} = -\frac{c}{\omega}k^2\nabla c = -\frac{\omega}{c}\nabla c. \tag{2.4}$$

Acoustic enclosures or rooms are common examples of complex wave cavities where the dynamics of rays may generically display chaos. Using the formal analogy between the geometrical limit of rays and classical mechanics, the simplest paradigms of such enclosures are billiards, which are closed homogeneous domains containing a particle specularly bouncing on the walls. According to the shape of the billiard, the motion may be regular or chaotic. Without going into technical details, we wish now to illustrate the particular dynamics of chaotic billiards. Let us first recall the regular motion of rays in the billiard with the shape of a circle. Figure 2.1(a) shows a typical trajectory within such a billiard after a propagation length of 150 in units of the radius R. One can clearly observe the presence of a caustic. The latter encloses a region of space that this trajectory never visits (whatever the number of

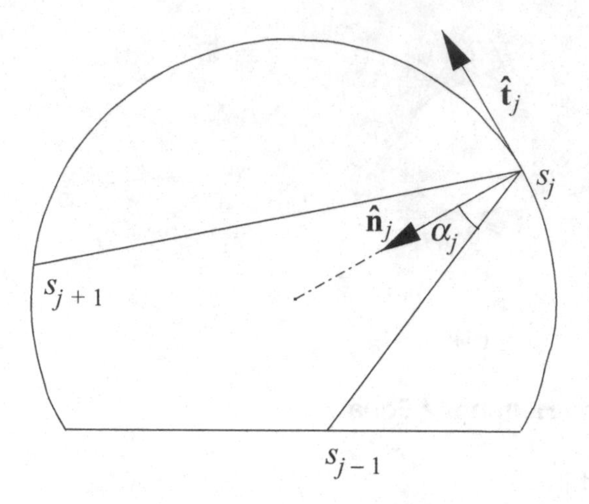

Figure 2.2. Representation of the dynamics in a billiard through the coordinates associated, at each rebound, to the curvilinear abscissa s along the boundary, and the sine of the angle of reflection α with respect to the inward boundary normal.

reflections). This kind of structure is destroyed in chaotic billiards. This is exemplified by considering the following modification of the previous billiard. A new shape is obtained by cutting a small straight segment of length $10^{-2}R$. Whereas the change of boundary is not visible on Figure 2.1(b), its effect on the dynamics is dramatic: for the same initial conditions (position and direction) the formerly forbidden region is invaded after a finite number of reflections. In the theory of Hamiltonian chaos, it is shown that this effect stems from the *extreme sensitivity to initial conditions*, which appears for any nonvanishing size of the cut (except for a cut of length $2R$, which corresponds to the semicircle billiard).

The qualification of chaos is more conveniently studied through a phase space representation. A common representation in billiards consists restricting the dynamics to the knowledge, at each impact, of the curvilinear abscissa s and of the sine of the angle of reflection α with respect to the inward boundary normal (see Figure 2.2). Thus, at jth reflection, defining $\hat{\mathbf{t}}_j$, the unit vector tangent to the oriented boundary at abscissa s_j, and $\hat{\mathbf{n}}_j$, the inward normal unit vector, the pseudo-momentum reads $\mathbf{k} = \sin\alpha_j\hat{\mathbf{t}}_j + \cos\alpha_j\hat{\mathbf{n}}_j$. The same trajectories as in Figure 2.1 are shown in the phase space $(s, \sin\alpha)$ in Figure 2.3 for a finite number of bounces: the regular motion is associated with the conservation of α in the circular billiard (Figure 2.3 (a)), while in the truncated billiard, which is chaotic, the whole phase space is eventually uniformly covered by almost any trajectory (Figure 2.3 (b)). It should be mentioned here that there exist particular trajectories that do not fit into this scheme, namely, POs. These orbits are trajectories that close upon themselves in phase space (hence also in real space). For a chaotic system they must, of course, be unstable in the sense that any small initial deviation from it must diverge exponentially with time.

It is important to stress here that if one considers rays as trajectories of point-like particles carrying wave energy, then the assumption of uniform and isotropic distribution of wave energy, commonly used in reverberant enclosures, is fulfilled in chaotic billiards. This is the basis for all exponential reverberation laws used in room acoustics since the pioneering works of Sabine (see, for instance, Mortessagne et al. (1993a, b) and references therein).

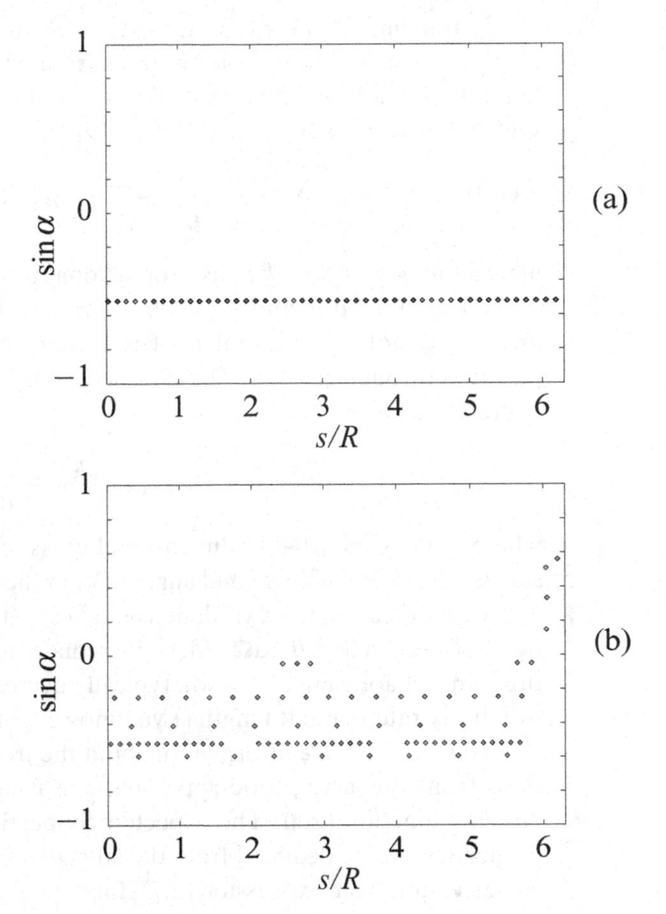

Figure 2.3. Same trajectories as in Figure 2.1 using the phase space coordinates $(s, \sin\alpha)$ introduced in Figure 2.2. (a) The regular motion in the circular billiard is associated with the conservation of α, (b) whereas in the chaotic billiard, the whole phase space is asymptotically uniformly covered by almost any ray trajectory.

2.2.2 The Semiclassical Approach for Chaotic Systems – The Trace Formula

The aim of a semiclassical (high-frequency) analysis is to obtain approximate solutions for the response of the stationary wave equation, only by means of classical trajectories and the wavelength λ when the latter can be assumed to be small enough compared with all sizes of the billiard. In the following, we will be interested in homogeneous media where the Helmholtz scalar wave equation holds:

$$(\nabla^2 + k^2)\psi(\mathbf{r}) = 0 \quad \text{inside the enclosure} \quad \mathscr{D}, \tag{2.5}$$

where $k = 2\pi/\lambda$ is the wavenumber that can only take on discrete values, because of the Dirichlet conditions $\psi = 0$ prescribed at the boundary $\partial\mathscr{D}$ of the enclosure. Other boundary conditions would also produce a discrete spectrum of values for k as long as they correspond to lossless conditions.

The eigenfunctions ϕ_n of this boundary-value problem form a complete set enabling the following expansion for the Green's function of Equation (2.5):

$$G(\mathbf{r}_B, \mathbf{r}_A; k) = \sum_n \frac{\phi_n^*(\mathbf{r}_B)\phi_n(\mathbf{r}_A)}{k^2 - k_n^2}. \tag{2.6}$$

In the limit $k|\mathbf{r}_B - \mathbf{r}_A| \gg 1$, the Green's function can be approximated as a sum over all geometrical ray contributions from multiply reflected trajectories connecting points A and B (the so-called semiclassical Green's function; see, for instance, Gutzwiller 1990 or Berry and Mount 1972):

$$G_{\mathrm{sc}}(\mathbf{r}_B, \mathbf{r}_A; k) = \sum_{q:\, A \to B} \frac{1}{\mathrm{i}(2\pi \mathrm{i})^{(d-1)/2}} \left|\Delta_{AB,q}\right|^{1/2} \exp\left[\mathrm{i}S_q(\mathbf{r}_B, \mathbf{r}_A; k) - \mathrm{i}\frac{\pi}{2} m_q\right], \quad (2.7)$$

where the *action* $S_q = kL_q$ is proportional to the path length $L_q(\mathbf{r}_B, \mathbf{r}_A)$ along the trajectory labeled q, and $m_q = 2n_{r,q} + n_{c,q}$ is the so-called Maslov index, with $n_{r,q}$ and $n_{c,q}$ denoting the number of reflections at the boundary and the number of passages through caustics. The *divergence* factor Δ_{AB} has the following geometrical expression:

$$\Delta_{AB} = \frac{k^{(d-3)}}{4} \frac{\mathrm{d}\Omega_A}{\mathrm{d}A_B}, \quad (2.8)$$

where $\mathrm{d}A_B$ is the $(d-1)$-dimensional cross section at point B of the tube of rays starting from A within a solid angle $\mathrm{d}\Omega_A$ in the vicinity of a given trajectory.

To fix ideas, in the two-dimensional case, for a direct trajectory without reflection between A and B, $|\mathrm{d}\Omega_A/\mathrm{d}A_B|^{1/2}$ reduces to $1/\sqrt{L_q}$, whereas, after many reflections, in a chaotic billiard, it will typically decrease like $\exp(-hL_q/2)$, where h is the instability rate per unit length (Lyapunov exponent).

To retrieve some information about the frequency spectrum and the eigenfunctions from this asymptotic approach, one needs a more global quantity than the Green's function itself. The expected connection between ray and modal spectral properties can be deduced from the so-called trace formula. Let us briefly describe its derivation from expression (2.7). First, one needs to handle the pole singularities in (2.6) by using

$$\frac{1}{k^2 - k_n^2} \to \lim_{\varepsilon \to 0^+} \frac{1}{(k^2 + \mathrm{i}\varepsilon) - k_n^2} = \mathscr{P}\left(\frac{1}{k^2 - k_n^2}\right) - \mathrm{i}\pi\delta(k^2 - k_n^2), \quad (2.9)$$

where $\mathscr{P}(\cdot)$ denotes the Cauchy principal value. Relation (2.9) implies that the imaginary part of the Green's function, once integrated over the domain (often referred to as a *trace* in quantum physics), yields the density of eigenvalues:

$$\frac{1}{\pi}\Im\left\{\int_{\mathscr{D}} G(\mathbf{r}, \mathbf{r}; k)\, \mathrm{d}\mathbf{r}\right\} = \sum_n \delta(k^2 - k_n^2). \quad (2.10)$$

This quantity represents the eigenvalue-density function $D(k^2)$ in terms of the variable k^2 (also known as the density of states in quantum physics), which is related to the modal density $\rho(k) = \sum_n \delta(k - k_n)$ through $\rho(k) = 2kD(k^2)$.

In the high-frequency limit, if one uses the semiclassical approximation (2.7) in the preceding equation, one notices the rapidly oscillating factor $\exp(\mathrm{i}kL_q)$. One can therefore evaluate the spatial integral by stationary phase. Indeed, the equations of motion for the ray trajectory from A to B imply that the partial derivative of $L_q(\mathbf{r}_B, \mathbf{r}_A)$ with respect to $\mathbf{r}_B$ is the unit vector $\hat{n}_B$ tangent to the trajectory at B, whereas its derivative with respect to $\mathbf{r}_A$ is minus the unit vector $\hat{n}_A$ tangent to the trajectory at A. Thus, the stationary-phase condition imposed at $\mathbf{r} = \mathbf{r}_B = \mathbf{r}_A$ selects closed trajectories in phase space, that is, POs. All this finally yields the celebrated semiclassical trace formula over POs, also known as the Gutzwiller trace formula (Gutzwiller 1990):

$$\rho(k) = \overline{\rho}(k) + \frac{1}{\pi} \sum_{\text{PPO}q} L_q \sum_{\text{repetitions } r} \frac{\cos\left[r(kL_q - m_q\pi/2)\right]}{|\det\left(\mathbf{I} - \mathbf{M}_q^r\right)|^{1/2}}, \tag{2.11}$$

where the summation is performed over all primitive periodic orbits (PPOs) and their repetitions and $\mathbf{M}_q$ is the monodromy matrix of the PPO describing the linear change of the transverse coordinates (in phase space) after one period.

In formula (2.11), $\overline{\rho}(k)$ is the contribution of the *zero-length* trajectories. For such trajectories, the semiclassical approximation is not valid, and the space integral must be performed by using the free-space Green's function:

$$\overline{\rho}(k) = \frac{2k}{\pi} \Im \left\{ \int_D \lim_{\mathbf{r}_B \to \mathbf{r}_A} G_0(\mathbf{r}_B, \mathbf{r}_A; k)\, \mathrm{d}\mathbf{r}_A \right\}. \tag{2.12}$$

This corresponds to the average behavior of the modal density, varying smoothly with k, and given, in billiards, by an asymptotic series in powers of k^{-1}, whose leading term is known as the Weyl's law (also known as the Thomas–Fermi approximation in nuclear physics), and reads

$$\rho_{\text{Weyl}}(k) = \frac{V_d \Omega_d}{(2\pi)^d} k^{d-1}, \tag{2.13}$$

where V_d is the d-dimensional volume of the billiard and Ω_d is the surface of the d-dimensional sphere of unit radius. The sum over periodic orbits describes the oscillatory behavior, denoted ρ^{osc}, of this density around its smooth component. Longer periodic orbits contribute to finer oscillations, and the largest length required to describe the modal density at the scale of the mean spacing $1/\overline{\rho}(k)$ is called the Heisenberg length $L_H(k) \equiv 2\pi\overline{\rho}(k)$. For periodic orbits longer than L_H, as noted by Bogomolny and Keating (1996), their contributions should not provide any significant information about the modal density other than to account for the discreteness of the spectrum: this implies subtle compensations between terms, indicating in fact that most of the information contained in long orbits can be retrieved from shorter ones. We will come back to this point later in the section devoted to the spectral correlations. Note that formula (2.11) is only valid at the lowest order in the small parameter $(kL)^{-1}$, where L is a typical size of the billiard, and also that it should be restricted to isolated orbits, which is typically the case in truly chaotic billiards. In the case of continuous families of orbits, especially for integrable billiards, a sum over periodic orbits can be analytically performed through the use of Poisson sum rules (see, e.g., Stöckmann 1999, Chapter 7 Section 1; see also Wright and Ham 2007 and Chapter 1 of the present book). Another related problem is the apparent lack of convergence of the trace formula in chaotic billiards. Indeed, long orbits have amplitudes behaving like $L_q \exp(-hL_q/2)$, whereas it is known that the number of periodic orbits with length less that L proliferates like $(hL)^{-1}\exp(hL)$ (Berry 1991). This should imply that the sum does not converge unless, as already mentioned earlier, the fact that the POs are not independent and that the information they contain is structured conspire in such a way as to ensure convergence.

2.2.3 Speckle-Like and Scarred Wavefunctions

The chaotic exploration of phase space by rays illustrated in Section 2.2.1 should be expected to govern the statistical spatial distribution of eigenmodes. Such an ergodic behavior was, indeed, rigorously demonstrated in the late 1970s by Voros (1979) and was popularized by Berry (1977) through an analogy with laser speckle patterns. Berry (1983) provided evidence that the wavefunction of a typical mode may be viewed as a Gaussian random function resulting from a random superposition of plane waves. The ergodicity of eigenmodes emerges through the normalized local density of states defined as follows:

$$\tilde{\rho}(\mathbf{r}, \omega^2) = \frac{\int \delta\left(\omega^2 - H(\mathbf{r}, \mathbf{k}')\right) \mathrm{d}\mathbf{k}'}{\int \delta\left(\omega^2 - H(\mathbf{r}', \mathbf{k}')\right) \mathrm{d}\mathbf{r}' \, \mathrm{d}\mathbf{k}'}, \tag{2.14}$$

where $H(\mathbf{r}, \mathbf{k})$ is the Hamiltonian given in (2.3). In the semiclassical limit $kL \to \infty$, Equation (2.14) turns into

$$\tilde{\rho}(\mathbf{r}, \omega^2) \simeq \lim_{kL \to \infty} \left\langle |\psi(\mathbf{r})|^2 \right\rangle_{\omega^2} = \frac{1}{N} \sum_n |\phi_n(\mathbf{r})|^2, \tag{2.15}$$

where the sum runs over N eigenmodes centered on "energy" ω^2 (on wavenumber $k(\mathbf{r}) = \sqrt{\omega^2/c^2(\mathbf{r})}$, equivalently) in an interval small enough for the density of states to remain constant and large enough to ensure a large value of N. In practice, fewer and fewer modes are required, as the central mode is far in the semiclassical domain, and eventually, the ergodic behavior is obtained for individual modes. As an illustration, Equation (2.5) has been numerically solved for a domain $\mathcal{D}$ with the shape of the chaotic billiard shown in Figure 2.2. Recall that in the billiard case the velocity is uniform inside the domain: $c(\mathbf{r}) \equiv c$ and, consequently, $\omega^2 = c^2k^2$. In Figure 2.4, the square amplitude of mode number $n = 1513$ ($k_{1513} = 87.89R^{-1}$) is depicted. The ergodic nature of the mode is clearly shown: apart from the axial symmetry, such an eigenmode can be viewed, locally, as a superposition of planes waves with fixed k and random phases and directions. As a consequence, the field autocorrelation function defined as

$$C_\psi(\mathbf{r}, \mathbf{r}_0; k) = \left\langle \psi^*\left(\mathbf{r} - \frac{1}{2}\mathbf{r}_0\right) \psi\left(\mathbf{r} + \frac{1}{2}\mathbf{r}_0\right) \right\rangle_k \tag{2.16}$$

reads (Berry 1977, Srednicki and Stiernelof 1996)

$$C_\psi(\mathbf{r}, \mathbf{r}_0; k) = \frac{\int \exp\left[\mathrm{i}\mathbf{k}' \cdot \mathbf{r}_0\right] \delta\left(c^2k^2 - H(\mathbf{r}, \mathbf{k}')\right) \mathrm{d}\mathbf{k}'}{\int \delta\left(c^2k^2 - H(\mathbf{r}', \mathbf{k}')\right) \mathrm{d}\mathbf{r}' \, \mathrm{d}\mathbf{k}'}. \tag{2.17}$$

In a two-dimensional billiard the Hamiltonian is uniform, and in Equation (2.17) the Dirac δ function only fixes the norm of $\mathbf{k}$, giving rise to the important result (Berry 1977)

$$C_\psi(\mathbf{r}, \mathbf{r}_0; k) = \mathrm{J}_0(kr_0), \tag{2.18}$$

Figure 2.4. A typical *ergodic* eigenmode (squared amplitude), solution of Equation (2.5) with Dirichlet boundary conditions, in the truncated chaotic billiard with $k \times R = 87.89$. Apart from the obvious symmetry, such an eigenmode can be viewed as a superposition of plane waves at a given k with random phases and directions.

where J_0 is the zeroth-order Bessel function and r_0 is the norm of $\mathbf{r}_0$. Using an ergodic hypothesis, the average in (2.16) can be replaced by a spatial average over the midpoint $\mathbf{r}$, which, in practice, should be evaluated over a domain encompassing a sufficiently large number of oscillations (Srednicki & Stiernelof 1996).

In the asymptotic limit, a random superposition of plane waves with random uncorrelated phases is expected to yield a Gaussian random field. In the case of real eigenmodes, this implies that the probability $P(\psi)\,\mathrm{d}\psi$ that the eigenfunction has a value between ψ and $\psi + \mathrm{d}\psi$ is given by

$$P(\psi) = \frac{1}{\sqrt{2\pi\,\langle\psi^2\rangle}} \exp\left(-\frac{\psi^2}{2\langle\psi^2\rangle}\right), \tag{2.19}$$

where $\langle\cdots\rangle$ denotes a spatial average on the domain $\mathscr{D}$. One should note that a Gaussian distribution does not imply the stronger requirement (2.17). Result (2.19) is also recovered by RMT for the Gaussian orthogonal ensemble of real symmetric $N \times N$ matrices in the limit $N \to \infty$ (Haake 2001 see also Chapter 3 of the present book). Indeed, RMT leads to the so-called Porter–Thomas distribution for the squared eigenvectors components. The latter distribution is obtained from (2.19) for the intensity $I = \psi^2$ and reads

$$P(I) = \frac{1}{\sqrt{2\pi I/\langle I\rangle}} \exp\left(-\frac{I}{2\langle I\rangle}\right). \tag{2.20}$$

To check this behavior, we first numerically solve the Helmholtz equation (2.5) with Dirichlet boundary conditions using a plane wave decomposition method. This method (Heller 1991) has allowed the calculation of the first 2,000 eigenmodes of the truncated circle billiard (Doya et al. 2002). Because of Dirichlet boundary conditions, the eigenmodes are real.

Using these calculated modes, we have evaluated the *radial* field autocorrelation function $C_\psi(r_0; k)$,

$$C_\psi(r_0; k) = \frac{1}{2\pi} \int_0^{2\pi} C_\psi(\mathbf{r}_0; k)\,\mathrm{d}\theta \tag{2.21}$$

with θ the polar angle and where the field autocorrelation function $C_\psi(\mathbf{r}_0; k)$ is equivalent to (2.16) with a spatial average over $\mathbf{r}$,

$$C_\psi(\mathbf{r}_0; k) = \left\langle \psi^*\left(\mathbf{r} - \frac{1}{2}\mathbf{r}_0\right)\psi\left(\mathbf{r} + \frac{1}{2}\mathbf{r}_0\right)\right\rangle_{\mathbf{r}}, \tag{2.22}$$

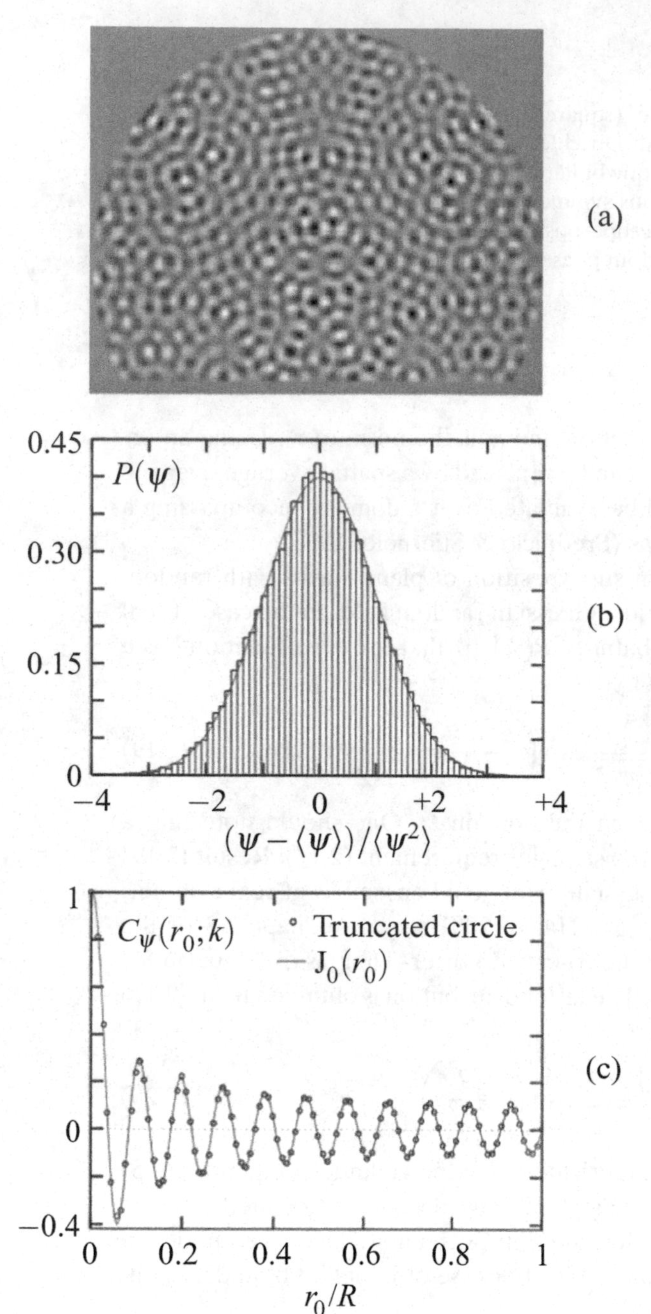

Figure 2.5. (a) A high-energy eigenmode (amplitude) with $k \times R = 87.89$ in the truncated chaotic billiard, (b) its associated probability distribution $P(\psi)$, compared with a Gaussian distribution (continuous line), and (c) the radial field autocorrelation function $C_\psi(r_0; k)$.

where the average $\langle\cdots\rangle_{\mathbf{r}}$ reads $\iint_{\mathscr{D}} \cdots d\mathbf{r} / \iint_{\mathscr{D}} |\psi(\mathbf{r})|^2 \, d\mathbf{r}$, with $\mathscr{D}$ the domain of integration.

In Figure 2.5, we have represented one typical high-energy eigenmode (amplitude) of the truncated circle billiard for, a value of k equal to 87.89 in units of inverse radius R, its probability distribution, and the corresponding radial field autocorrelation function following Equations (2.21) and (2.22). The assumption of a random superposition of plane waves is confirmed by the good agreement between the probability distribution $P(\psi)$ and the Gaussian distribution as can be seen in

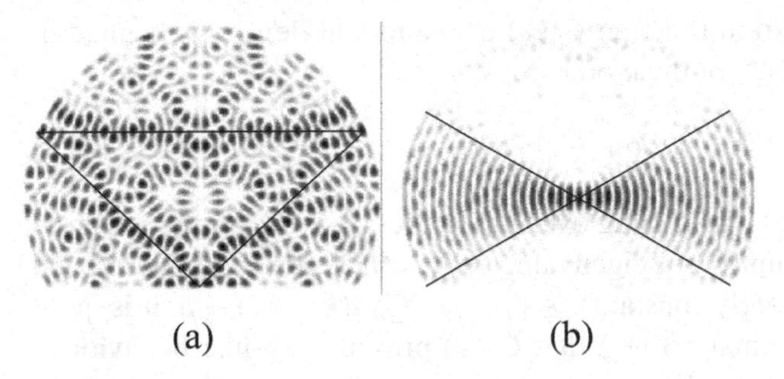

Figure 2.6. Examples of eigenmodes displaying an intensity enhancement in the vicinity of (a) an unstable periodic orbit (superimposed as a solid line), (b) the continuous family of diameters (boundaries shown as solid lines).

Figure 2.5(b). The radial autocorrelation function $C_\psi(r_0; k)$ is compared with the expected zero-order Bessel function $J_0(kr_0)$ for $k \times R = 87.89$. Note that Prediction (2.18) is perfectly verified.

As any prediction concerning average behaviors, the results just presented suffer rare but important exceptions. Indeed, inspecting Figure 2.6(a), a clear deviation from ergodicity is seen, which is in fact associated with a particular periodic orbit (superimposed as a solid line). This intensity enhancement in the vicinity of a single periodic orbit is coined *scarring* (Bogomolny 1988, Heller 1991). This unexpected behavior has led the quantum chaos community to reconsider the semiclassical limit (2.14). They have established that the semiclassical skeleton of eigenmodes is built on all the periodic orbits of the system. Thus the one-to-one relationship shown in Figure 2.6, between an eigenmode and a periodic orbit, has to be considered as an exception because, as the number of POs proliferates exponentially with their lengths, eigenmodes must build upon many of them. A thorough description of the subtle scarring phenomenon is given by E. Vergini in the present book (Chapter 5).

2.3 Two-Point Spectral Correlations and Form Factor (Space-Averaged Time Response)

2.3.1 Spectral Rigidity *à la* Berry (Diagonal Approximation)

2.3.1.1 Form Factor

In a complex wave system, either a chaotic cavity or a disordered medium, it is not possible to give a detailed description of the frequency spectrum by providing a determined sequence of numbers. Hence, the frequency spectrum is too complicated to be explained level by level but may nevertheless be studied through a statistical approach, in a way quite analogous to the statistical approach of a gas of interacting particles. In the study of spectral properties, a key role is played by the spectral *correlations*, and their description in terms of adequate quantities will be our concern in this section. With these quantities we will be in a position to establish how certain universal features predicted by RMT can be recovered from a global knowledge of chaotic dynamics but also in what respect some nonuniversal behavior can be related to the shortest POs of the system.

The spectral autocorrelation function $C_k(\kappa)$ of the modal density is defined in terms of the oscillating part $\rho^{\rm osc}$ of the modal density as

$$C_k(\kappa) = \langle \rho^{\rm osc}(k+\frac{\kappa}{2})\rho^{\rm osc}(k-\frac{\kappa}{2})\rangle_k, \tag{2.23}$$

where the brackets denote local averaging over k (in practice, over an interval large enough to include a large number of eigenvalues but sufficiently small as to keep the modal density approximately constant). As $\rho^{\rm osc} = \sum_n \delta(k-k_n) - \overline{\rho}$, it is quite easy to show (see, e.g., Stöckmann 1999) that $C_\rho(\kappa)$ presents a δ-like behavior at the origin, which can be accounted for by defining an ancillary quantity called the two-level cluster function $Y_2(\kappa)$ through

$$C_k(\kappa) = \overline{\rho}^2\left[\delta\left(\kappa L_H\right) - Y_2\left(\kappa L_H\right)\right]. \tag{2.24}$$

For the following calculations, it is convenient to rewrite the trace formula for $\rho^{\rm osc}$ as

$$\rho^{\rm osc}(k) = \sum_{{\rm PO}j} A_j {\rm e}^{{\rm i}kL_j}, \tag{2.25}$$

where the sum runs over all POs including repetitions and complex conjugate terms, stability and phase factors being integrated into the amplitudes A_j. The form factor is defined through the Fourier transform of the two-point autocorrelation function $C_k(k)$ as

$$\begin{aligned} K(L) &= \frac{1}{\overline{\rho}(k)}\int C_k(\kappa)\exp\left({\rm i}\kappa L\right){\rm d}\kappa \\ &= 1 - \frac{1}{2\pi}\int Y_2(s)\exp\left({\rm i}s\frac{L}{L_H}\right){\rm d}s \equiv 1 - b\left(\frac{L}{L_H}\right), \end{aligned} \tag{2.26}$$

which, using (2.25), becomes

$$K_{\rm sc}(L) = \frac{2\pi}{\overline{\rho}(k)}\sum_{j,l} A_j^* A_l {\rm e}^{{\rm i}k(L_l-L_j)}\delta\left[L - \left(\frac{L_j+L_l}{2}\right)\right]. \tag{2.27}$$

For very short lengths (L of the order of the length of the shortest PO), $K(L)$ displays a series of sharp peaks. At longer lengths, because of the proliferation of POs, the peaks tend to overlap, and one can try to evaluate the smooth behavior of the form factor by means of a classical sum rule. For large k, the terms of expression (2.27) with $L_j \neq L_l$ rapidly oscillate and cancel out in the process of averaging over a k-interval as mentioned above. This leads one to ignore all the off-diagonal terms of the sum, and in this *diagonal* approximation, one is left with the following expression:

$$K_{\rm sc}^{\rm diag}(L) = \frac{2\pi}{\overline{\rho}(k)}\sum_j |A_j|^2\delta\left(L - L_j\right). \tag{2.28}$$

For the PPOs, the squared modulus of the amplitude behaves like $|A_j|^2 \simeq (\frac{L_j}{2\pi})^2\exp(-hL_j)$, whereas their density (number of PPOs with period between L and $L+{\rm d}L$ over ${\rm d}L$) increases as $\exp(hL)/L$. Thus, ignoring the contributions of

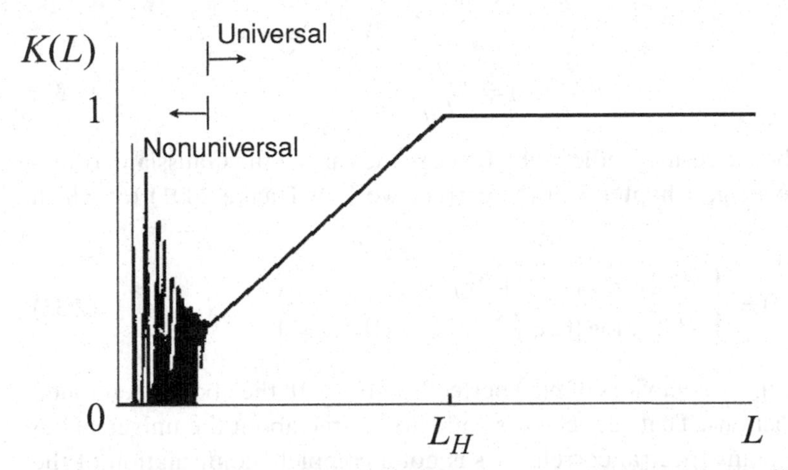

Figure 2.7. Schematic representation of the semiclassical form factor, illustrating how the diagonal approximation predicts a nonuniversal part showing peaks due to short POs and a universal part due to long POs but for lengths shorter than L_H. The unit value for $L \gg L_H$ is predicted by semiclassical arguments beyond the diagonal approximation.

the repetitions of smaller orbits at a given lengths, one finally obtains

$$K_{\rm sc}^{\rm diag}(L) = \frac{L}{L_H}. \tag{2.29}$$

Of course, such an expression becomes unphysical for arbitrarily long lengths and cannot be valid for $L \gtrsim L_H$. Indeed, for such long periods, a complete cancellation of non-diagonal terms cannot be achieved because of the increasing number of POs whose lengths nearly coincide. Delande (2001) has proposed another semiclassical *sum rule*, which shows that

$$K_{\rm sc}(L) \to 1 \quad \text{when} \quad L \gg L_H. \tag{2.30}$$

The argument goes as follows: if the integral defining $K(L)$ is performed over an interval Δk containing $N = \overline{\rho}(k)\Delta k$ eigenvalues, it approximately reads (using the Wiener–Khinchin theorem)

$$\begin{aligned} K(L) &= \frac{1}{N}\left|\int_{k-\Delta k/2}^{k+\Delta k/2} [\rho(k') - \overline{\rho}(k')]{\rm e}^{-{\rm i}k'L}\,{\rm d}k'\right|^2 \\ &\approx \frac{1}{N}\left|\sum_n \exp(-{\rm i}k_n L)\right|^2. \end{aligned} \tag{2.31}$$

At long lengths $L \gg L_H$, the oscillating terms in the preceding expression add incoherently, therefore yielding the limit (2.30). All the above results concerning the form factor are summarized in Figure 2.7.

For time-reversal invariant systems, which is generally the case for classical wave cavities, one has to correct the preceding argument because, for each PO, there is a time-reverse PO with an identical amplitude and phase. Therefore, the corresponding off-diagonal terms in the double sum contribute the same amount as the diagonal terms, yielding a result that is doubled. At short times, expression (2.29) thus becomes

$$K_{sc}^{diag}(L) = 2\frac{L}{L_H}. \tag{2.32}$$

This turns out to be the result predicted by RMT in the case of the Gaussian orthogonal ensemble (see, e.g., Chapter 3 of the current work or Haake 2001) for which

$$b(x) = \begin{cases} 1 - 2x + x\log(1+2x) & 0 < x < 1 \\ -1 + x\log[(2x+1)/(2x-1)] & x > 1. \end{cases}, \tag{2.33}$$

The small-x and large-x behaviors of $b(x)$ perfectly agree with the above-mentioned semiclassical predictions. That semiclassics and RMT agree about the universal behavior of the two-point spectral correlations is not a complete confirmation of the Bohigas–Giannoni–Schmit conjecture (Bohigas et al. 1984), which states that the spectral fluctuations of classically chaotic systems should be described by the relevant ensembles of random matrices. For example, the nearest-neighbor spacing distributions derived from RMT have never been totally justified in semiclassical arguments. As has been shown with different classes of pseudo-integrable systems (see the billiards with point scatterers given later), non-chaotic systems can very well mimic level repulsion or short-range spectral rigidity. Thus a clear-cut separation between classically integrable or chaotic systems on the basis of their spectral properties is no longer an issue.

2.3.1.2 Length Spectrum

In the nonuniversal regime corresponding to the lengths of the shortest POs, expression (2.28) can also be rewritten as the squared modulus of the Fourier transform of expression (2.25) for ρ_{osc}. When restricting the integral to the neighborhood of a given k, $K_{sc}(L)$ can be written as

$$K_{sc}(L) = \frac{1}{N}\left|2\pi \sum_j A_j \delta\left(L - L_j\right)\right|^2. \tag{2.34}$$

From the last expression or from expression (2.28), the so-called length spectrum is revealed from the short length behavior of a given billiard's form factor. This behavior is clearly nonuniversal and can be used to identify the shortest POs with the largest contribution to the spectral density. This can be a genuine practical tool to gain some extra knowledge about the specific geometry of a cavity when one only knows part of its spectral response deduced from scattering experiments.

As an illustration, a length spectrum computed from the first 2,000 eigenvalues of the truncated circle (see Figures 2.4 and 2.6) with Dirichlet boundary conditions is shown in Figure 2.8. As shown in Section 2.2.1, this billiard is known to be chaotic in the strongest sense. A few POs among the shortest are indicated by arrows and are displayed in Figure 2.9. We remark that the vast majority of periodic orbits contribute to the generic ergodic behavior described in Section 2.2.3. Nonetheless, a special family of POs, namely, the continuous family of diameters, which survived the truncation and constitute marginally unstable periodic orbits, are responsible for some of the high peaks at short length. They are thus also responsible

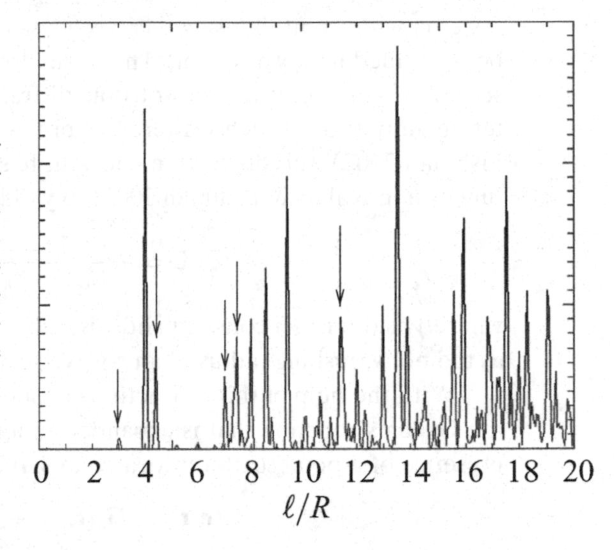

Figure 2.8. The *length spectrum* or Fourier transform of the density of states $n(\kappa)$, for the eigenvalue problem (2.5) in the truncated chaotic billiard shown in Figures 2.4 and 2.6. The *trace formula* permits to show that the length spectrum should have peaks at the period lengths of the periodic orbits. The arrows indicate lengths corresponding to the periodic orbits shown in Figure 2.9.

for nonuniversal features of the long range spectral correlations as well as for the non-Gaussian statistics of the eigenmode (b) shown in Figure 2.6.

2.3.2 From Ballistic to Scattering Systems: Point-Like Scatterers and Diffractive Orbits

In a billiard where one or more point-like scatterers are added, an approach similar to what has been described in the previous sections is feasible. In diffractive

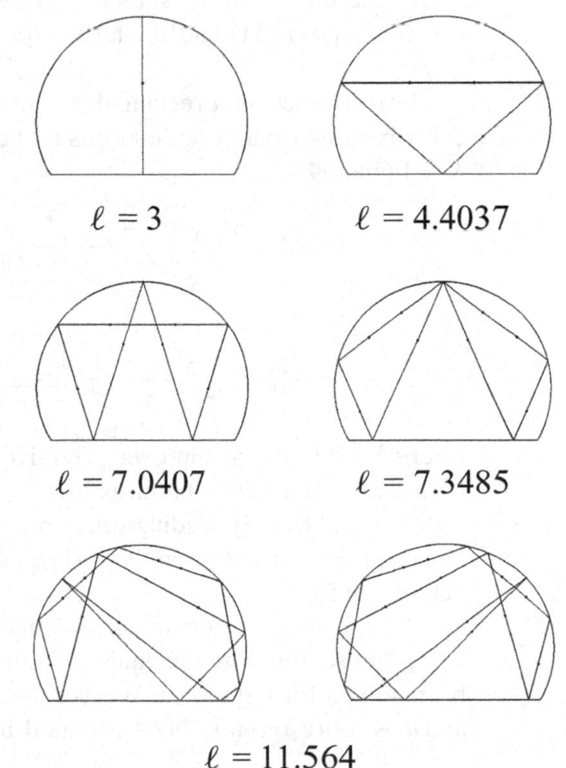

Figure 2.9. A few periodic orbits whose periods (in units of R) correspond to peaks of the length spectrum shown in Figure 2.8.

systems with point-like singularities, classical trajectories that hit those points can be continued in any direction. These can nonetheless be treated within the wave description by introducing an isotropic diffraction coefficient D, which fixes the scattering amplitude at each scatterer. In previous works (Exner & Šeba 1996, Rahav & Fishman 2002), this diffraction constant has been calculated (with the free Green's function in Rahav & Fishman 2002) to yield

$$\mathsf{D} = \frac{2\pi}{-\ln(ka/2) - \gamma + \mathrm{i}\pi/2}, \tag{2.35}$$

where γ is the Euler constant and a is a characteristic length that may be interpreted as the nonvanishing radius of an s-wave scattering disk (Exner & Šeba 1996).

With the help of this diffraction constant, a Dyson's formula can be written for the Green's function that is expanded in a multiple-scattering series, which, in the presence of a point scatterer located at $\mathbf{s}$, reads

$$G(\mathbf{r}, \mathbf{r}') = G_0(\mathbf{r}, \mathbf{r}') + G_0(\mathbf{r}, \mathbf{s})\mathsf{D}G(\mathbf{s}, \mathbf{r}'), \tag{2.36}$$

where G_0 is the unperturbed Green's function of the bare billiard. If two consecutive scattering events are more distant than a wavelength, a semiclassical approximation for the unperturbed Green's function can be used, so that a semiclassical trace formula can be obtained for the perturbed billiard, which is based on POs and diffractive periodic orbits (DOs) as well. These DOs are composed of closed orbits all starting and ending at the same scatterer. Hence the oscillating part of the modal density now includes DOs, yielding

$$\rho^{\mathrm{osc}}(k) = \rho^{\mathrm{osc}}_{\mathrm{PO}}(k) + \rho^{\mathrm{osc}}_{\mathrm{DO}}(k), \tag{2.37}$$

where the term with the subscript PO is the corresponding part of the Gutzwiller trace formula (2.11) and the term with the subscript DO is a similar sum over all DOs.

For instance, in a rectangular domain of area $\mathscr{A}$ with a single point scatterer, contributions from periodic orbits and diffractive orbits respectively read (Pavloff & Schmit 1995)

$$\rho^{\mathrm{osc}}_{\mathrm{PO}} = \frac{\mathscr{A}}{\pi} {\sum_{\mathrm{PO}}}' \sum_{r=1}^{\infty} \frac{k}{\sqrt{2\pi k r L_{\mathrm{PO}}}} \cos(kr L_{\mathrm{PO}} - r n_{\mathrm{PO}}\pi - \pi/4), \tag{2.38}$$

and

$$\rho^{\mathrm{osc}}_{\mathrm{DO}} = {\sum_{\mathrm{DO}}}' \frac{L_{\mathrm{DO}}}{\pi} \frac{\mathsf{D}}{\sqrt{8\pi k L_{\mathrm{DO}}}} \cos(k L_{\mathrm{DO}} - n_{\mathrm{DO}}\pi - 3\pi/4), \tag{2.39}$$

where $\sum'$ denotes a sum over primitive periodic (diffractive) orbits of length L_{PO} (L_{DO}) and number of bounces n_{PO} (n_{DO}) and r is the number of repetitions. In formula (2.39), only leading-order one-scattering events are included, repetitions or concatenations of primitive orbits of order ν being of order $k^{-\nu/2}$ (Pavloff & Schmit 1995).

The problem of numerically calculating the eigenwavenumbers in the presence of a point scatterer in a rectangular billiard with Dirichlet boundary conditions has been solved for instance in Weaver and Sornette (1995) and Legrand et al. (1997) and has more recently been revisited in Laurent et al. (2006) where the authors

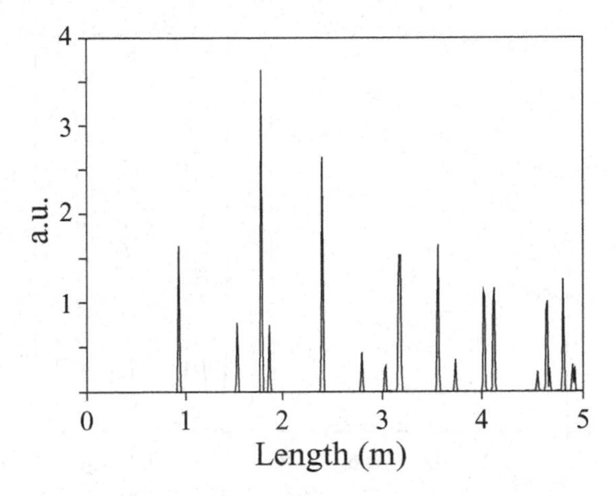

Figure 2.10. Length spectrum computed in a rectangular cavity with a single point scatterer. Approximately 12,000 resonances have been used.

provided a comparison with experimental results in a two-dimensional microwave cavity.

An example of a length spectrum corresponding to a rectangular cavity with a single point scatterer is given in Figure 2.10. Here, the dimensions of the cavity in which we have performed our calculations are those of an actual microwave cavity used in our experiments, with perimeter $\mathscr{L} = 2.446$ m and area $\mathscr{A} = 0.3528\,\mathrm{m}^2$. A large number of modes (approximately 12,000) have been used so that the length resolution is excellent. To illustrate that such a length spectrum is still dominated by the POs of the empty cavity, the amplitude scale chosen in Figure 2.10 is such that the contributions of DOs are much too small to be seen.

POs are easily identified on the length spectrum shown in Figure 2.10. At first sight, it could even seem that no other contribution can be seen as if DOs were absent from it. Somehow, it could even be expected because no long-range correlations are observed in the frequency spectrum, thereby indicating that if DOs should contribute, especially at short lengths, they should do so only in a negligible way. This is what can be observed by closely inspecting a typical length spectrum for lengths smaller or of the order of the size of the cavity in the presence of a single point scatterer. In Figure 2.11, the contributions of DOs are displayed on the length range from 0 to 1.6 m, using an enlarged scale for the amplitude of the peaks. Lines with different styles indicate the lengths of DOs (dotted lines) and POs (dashed lines) within this range.

In Legrand et al. (1997), R. L. Weaver and the present authors proposed a heuristic way of semiclassically understanding the short- and mid-range spectral correlations of rectangular billiards with point scatterers. In particular, a semiclassical prediction for the form factor in the universal regime was proposed. It corresponded to the typical two-point spectral correlations observed in such pseudo-integrable billiards and furthermore produced the linear behavior of the spacing distribution at small spacings (known as the *level repulsion*, which is typical of Gaussian Orthogonal Ensemble spectra). At large spacings, however, the spacing distribution of spectra calculated in such rectangular billiards with point scatterers decreases exponentially, thus behaving like the so-called Poisson distribution generically observed for uncorrelated spectra associated with integrable billiards (such as the rectangle; see Chapter 3).

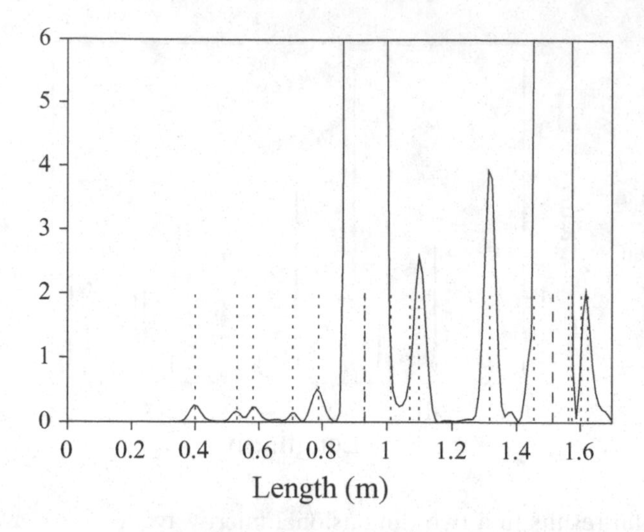

Figure 2.11. Zoom ($\times 10^4$) of the length spectrum shown in Figure 2.10 on the length range from 0 to 1.6 m, using an enlarged scale for the amplitude of the peaks: dashed lines, POs; dotted lines, DOs.

The argument proposed in Legrand et al. (1997) is now briefly summarized. First, one has to acknowledge that the wave problem associates a finite size of the order of the cross section σ (here a length) to the point scatterer. From expression (2.35) of the diffraction constant, one can show (Laurent et al. 2006) that apart from a logarithmic correction, the scattering cross section essentially scales as the wavelength,

$$\sigma = |\mathrm{D}|^2/4k, \tag{2.40}$$

thus making the scatterer practically equally efficient at all frequencies in a given band. Thus, in a coarse-grained way obtained, for instance, by averaging the modal density over a small k-interval, the orbits that are effectively scattered at lengths much larger than the shortest DOs are those that meet a small disk of diameter σ centered on the scatterer. The rate at which a typical ray hits this disk is given by a Sabine-like expression:

$$\Gamma = \frac{\pi\sigma}{\pi\mathscr{A}}. \tag{2.41}$$

One then evaluates the form factor by considering that it results from two contributions: one from the original POs of the bare rectangle that have not met the disk at length L, and another one from *new* POs that have met the disk. For the first *regular* part of the form factor, the fraction of orbits that have survived decays like $\exp(-\Gamma L)$, thus yielding

$$K^{\mathrm{reg}}(L) \approx K^{\mathrm{rect}}(L)\exp(-\Gamma L) = \exp(-\Gamma L) \tag{2.42}$$

(where $K^{\mathrm{rect}}(L) = 1$ yields *Poisson* statistics for uncorrelated spectra; see, for instance, Berry & Tabor 1977), whereas, for the *scattered* part, the probability of having met the disk at length L behaves like $1 - \exp(-\Gamma L)$, thus yielding

$$\begin{aligned} K^{\mathrm{scat}}(L) &\approx 2\frac{L}{L_H}[1 - \exp(-\Gamma L)] \quad \text{if} \quad L \ll L_H, \\ &\approx [1 - \exp(-\Gamma L)] \qquad\quad \text{if} \quad L \gg L_H, \end{aligned} \tag{2.43}$$

where the proliferation of scattered periodic orbits was assumed to follow the same sum rule as the POs of a genuinely chaotic billiard (following an argument introduced in Argaman et al. 1993). In a two-dimensional billiard with area $\mathscr{A}$, $L_H = \mathscr{A}k$ such that $\Gamma L_H = |\mathsf{D}|^2/4$ is practically a constant in a given frequency band, therefore leading to a long-length (i.e., $L \gg \Gamma^{-1}$) behavior of the form factor very similar to the one of chaotic systems displaying level repulsion. At shorter lengths (i.e., $L \ll \Gamma^{-1}$), $K(L) = K^{\text{reg}}(L) + K^{\text{scat}}(L)$ is practically unity like $K^{\text{rect}}(L)$, therefore leading to an absence of long-range spectral correlations. This was analytically demonstrated by Bogomolny et al. (2001) who used the fact that the eigenvalues of such singular billiards can be considered as the zeros of random meromorphic functions with a large number of poles, when these poles are independent random variables.

If the billiard in which a point scatterer is added is already chaotic, Bogomolny et al. (2000) have shown, by using both a random matrix argument and a semiclassical approach, that the spectral correlations are essentially not modified with respect to those of the bare chaotic billiard. In a certain way, diffractive orbits cannot randomize more a system that is already fully chaotic.

2.4 Conclusion

In the present chapter we have tried to provide a self-contained introduction to the semiclassical approach for the Helmholtz equation in complex systems, known as chaotic billiards. These systems are paradigms of wave cavities where the chaotic ray motion of the geometrical limit has implications on the spectral response and on the distribution of wavefunctions. We have introduced the trace formula, yielding the modal density in terms of a sum over the unstable POs of the billiard, and placed a particular emphasis on its applications in spectral and spatial correlations. Its connection with RMT has been discussed. To bridge the gap between chaotic billiards and scattering systems, we have also presented how the spectral correlations of billiards in the presence of a point scatterer can be influenced by DOs, leading to level repulsion even though the unperturbed cavity is not chaotic.

The Unreasonable Effectiveness of Random Matrix Theory for the Vibrations and Acoustics of Complex Structures

Richard Weaver

Department of Physics, University of Illinois at Urbana-Champaign, Urbana, IL, USA

3.1 Introduction

The applicability of random matrix theory (RMT) to acoustic systems has been demonstrated in recent years by a number of acoustical researchers, e.g., Weaver, Ellegaard, Langley, and Soize. Nevertheless, the field remains obscure to acousticians, and newcomers often find themselves perplexed. Why ought we believe this contention that a simple ensemble of random matrices describes an ensemble of structural or acoustic systems? Or more pointedly, why ought a single sample from such an ensemble represent one such system? What are the chief predictions of RMT; of what utility are they, and of what generality? The contention is, at first thought, absurd. And yet there is much empirical evidence of the relevance of RMT. There are furthermore several examples of its utility and correctness; it makes useful and accurate predictions.

There has been much attention paid in the acoustics community to uncertain systems. Although the broad features of an elastic or acoustic structure may be described with some confidence, details can vary, owing to randomness in manufacturing or to accumulated wear, or merely to irregularities that exceed the ability or patience of a numerical model. Such observations are widely invoked to justify the applicability of statistics. Random matrices represent a limit of uncertain systems. It is perhaps unsurprising that in the limit of maximum uncertainty, universal (i.e., pertaining to a class of systems) nontrivial statistics may be derived. These include statistics for the eigenfrequencies and eigenfunctions and for the responses. It is perhaps surprising that the same statistics can arise even when the broad features of a system are well known. It is perhaps even more striking that these statistics can apply even to a single system that a casual observer would not have considered random. In many a structure, regardless of whether or not its properties are known exactly or appear random, RMT seems to do a good job.

RMT found its first applications among nuclear theorists of the 1950s. At that time, nuclear physics was intensely engaged in measuring nuclear reaction cross sections and nuclear energy levels. Among the many observations was that the energy levels of excited heavy nuclei – which appeared at first sight to be highly random – tended to repel each other; degeneracies were rare. Although it might have been hypothesized that this feature was characteristic of the nuclear forces, it was established that it was instead a generic feature of the eigenvalues of sufficiently complex systems.

In a series of contributions (most notably Wigner 1951, 1955, 1957, 1958, 1959, Rosenzweig & Porter 1962, Porter 1965) it was shown that certain ensembles of random matrices had eigenvalue fluctuations away from their mean spectral density, identical to those of the energy levels of excited nuclei. Together with the quantum mechanician's understanding that nuclear energy levels are the eigenvalues of a large matrix describing nuclear dynamics, it was surmised that the observed statistics merely indicate that the dynamics is complicated. They say little about the forces themselves.

A series of later contributions (Mehta 1960, Mehta & Gaudin 1960, Gaudin 1961, Dyson 1962, Balian 1968) put RMT on a firm mathematical and statistical footing and derived various now-well-known features of the statistics of the eigenvalues and eigenvectors of random matrices. Many of these are reviewed and illustrated next.

All of this, it has been discovered, has implications for the modes and vibrations of closed and nearly closed acoustic systems. Manfred Schroeder, as early as 1954 (Schroeder 1954a, 1954b), noted that eigenvalue statistics are relevant for predictions of response statistics and proceeded to measure spectral fluctuations in scaled reverberation rooms using microwaves. He observed repulsion between the eigenvalues (i.e., small probability of near degeneracy) but attributed it to the inadequate resolving power of his measurements. He went on to develop a theory (Schroeder 1959, 1962, 1969a, 1969b, Schroeder & Kuttruff 1962) for the statistics of responses in a high-damping regime, in which modes are not resolvable. This regime is simpler; there is little dependence on eigenvalue statistics. Lyon (1969) and Davy (1981, 1986, 1987, 1990) noted that the need for response statistics is not confined to Schroeder's high-modal-overlap regime and explicitly included nontrivial (but we would now say naïve) eigenvalue correlations, noting further that developments in nuclear physics suggested, at least there, that these statistics are not trivial. In 1989, Weaver (1989b) followed by Ellegaard et al. (Bertelsen et al. 2000, Andersen et al. 2001, Ellegaard et al. 1995a, 1995b, 2001, Schaadt et al. 2001, 2003) measured the eigenfrequency statistics of solid bodies in the ultrasonic range and showed, with high accuracy, their correspondence to the predictions of RMT. Many laboratories have measured RMT eigenstatistics for the electromagnetic modes of certain microwave cavities (Stöckmann & Stein 1990, Richter 1998), most of which are directly analogous to the acoustics of membranes; see the review by Stöckmann (1999) for details. Lyon (1969), Davy (1981, 1986, 1990), Weaver (Weaver 1989a, Lobkis et al. 2000), and Langley (Langley & Brown 2004a, 2004b, Langley & Cotoni 2005) all showed the relevance of eigenstatistics for the statistics of power transmission in structures. Schroeder (1965) and Lobkis et al. (2003) showed how an understanding of eigenvector statistics in dissipative systems can explain and predict nonexponential decays.

This chapter is intended to introduce RMT, and in particular the Gaussian orthogonal ensemble (GOE), to a wider acoustics community, to present its chief results, and to argue why it is relevant for acoustics. More general discussions of RMT, in the context of nuclear physics or microwave physics or the quantum mechanics of classically chaotic systems, may be found elsewhere. The reader interested in a more extensive treatment is directed to the monographs by Stöckmann (1999), Mehta (1991), and Haake (2001) or to the review articles by Guhr et al. (1998), Brody et al. (1981), and Bohigas (1991).

Some work in acoustics has assumed that the higher eigenfrequencies of complex systems are distributed independent of each other, their statistics characterized solely by a local modal density $n(\omega)$. Such an assumption leads to the simple conclusion that there are no correlations between neighboring eigenfrequencies. In particular it implies that the probability density of nearest-neighbor spacings is exponential: $p(\Delta\omega)\,\mathrm{d}\Delta\omega = n\exp(-n\Delta\omega)\,\mathrm{d}\Delta\omega$ and that near degeneracies are not rare. Such eigenstatistics are called Poisson. The eigenvalue fluctuations of uniform rectangular membranes with incommensurate sides are described, at high frequency, by such statistics. RMT, however, suggests that eigencorrelations in more generic structures are much more interesting.

3.2 The Gaussian Orthogonal Ensemble

There are many different ensembles of random matrices. Perhaps the simplest and the most relevant is the GOE, the matrices of which are real and symmetric with otherwise-uncorrelated elements, all centered Gaussian random numbers. Diagonal elements have variance $1/2\nu$; off-diagonal elements have variance $1/4\nu$, where the parameter $1/\sqrt{\nu}$ serves to specify an eigenvalue scale.

As a simple and analytically very tractable example, one considers the 2×2 Gaussian Orthogonal Ensemble matrix (Wigner 1951, 1955, 1957, 1958)

$$\mathbf{H} = \begin{pmatrix} H_{11} & H_{12} \\ H_{12} & H_{22} \end{pmatrix} \tag{3.1}$$

with probability

$$\mathrm{d}P = p(\mathbf{H})\mathscr{D}\mathbf{H} = \exp(-\nu H_{11}^2 - \nu H_{22}^2 - 2\nu H_{12}^2)\,\mathrm{d}H_{11}\,\mathrm{d}H_{22}\,\mathrm{d}H_{12}. \tag{3.2}$$

Further, $\mathbf{H}$ has eigenvalues

$$\lambda_{1,2} = \left(\mathrm{tr} \pm \sqrt{\mathrm{tr}^2 - 4\det}\right)/2 \tag{3.3}$$

whose difference is

$$\lambda_2 - \lambda_1 = \sqrt{\mathrm{tr}^2 - 4\det} = \sqrt{(H_{11} - H_{22})^2 + 4H_{12}^2}. \tag{3.4}$$

The vanishing of the eigenvalue difference requires the vanishing of *two* independent random numbers; the probability density for that is therefore zero. It is not difficult to show,† using the assumed statistics (3.2) on the elements of $\mathbf{H}$, that the

† It is done by constructing $p(s) = \int p(\mathbf{H})\delta\left(s - \sqrt{(H_{11} - H_{22})^2 + 4H_{12}^2}\right)\mathscr{D}\mathbf{H}$ and performing the triple integral.

distribution of normalized eigenvalue difference $s = |\lambda_1 - \lambda_2| / \langle |\lambda_1 - \lambda_2| \rangle$ (normalized by its mean) is precisely

$$p(s) = \frac{\pi}{2} s \exp(-\pi s^2/4). \tag{3.5}$$

Thus the 2×2 case explicitly exhibits eigenvalue repulsion, with a vanishing probability for degeneracy: $\lambda_1 = \lambda_2$. It was Wigner's surmise that Equation (3.5), so different from the Poisson prediction $\exp(-s)$, also describes the distribution of normalized nearest-neighbor spacings for $N \times N$ GOE matrices.

A tendency for eigenvalue repulsion in $N \times N$ systems may be seen by the following simple argument: Consider a set of eigenvalues λ_r and eigenvectors $\mathbf{u}^r$ of a matrix $\mathbf{H}$,

$$\mathbf{H}\mathbf{u}^r = \lambda_r \mathbf{u}^r, \tag{3.6}$$

and consider a perturbation to $\mathbf{H}$, $\mathbf{H} \mapsto \mathbf{H} + \boldsymbol{\Delta}$. Then, by perturbation theory to order $\boldsymbol{\Delta}^2$ (see Pierre, 1988 or any text on quantum mechanics and time-independent perturbation theory), the rth eigenvalue becomes

$$\lambda_r \mapsto \lambda_r + (\mathbf{u}^r)^T \boldsymbol{\Delta} \mathbf{u}^r + \sum_{s \neq r} \frac{\left((\mathbf{u}^r)^T \boldsymbol{\Delta} \mathbf{u}^s\right)^2}{\lambda_r - \lambda_s} + \cdots. \tag{3.7}$$

If $\boldsymbol{\Delta}$ is a centered random perturbation, the first correction has zero mean, but the second correction tends to repel the eigenvalues. The expected change in an eigenvalue is *away* from its near neighbor.

The original GOE was derived on the basis of two hypotheses, that the ensemble ought to be invariant under orthogonal transformations (i.e., that two matrices $\mathbf{H}$ and $\mathbf{H}'$ related by an orthogonal transformation $\mathbf{H}' = \mathbf{O}^T \mathbf{H} \mathbf{O}$, occur with equal probability density) and that the elements be statistically independent. The first of these requirements was seen as natural; the ensemble should have no preferred basis. The second seemed artificial and was later rendered unnecessary by a maximum information entropy argument by Balian (1968) and by the introduction of a related ensemble by Dyson, the circular orthogonal ensemble (COE) (Dyson 1962).

The argument from maximum information entropy (Balian 1968) is particularly attractive – and we note that the approach is at the foundation of the methods of Soize (Chapter 13). We denote the volume measure in the space of all real symmetric $N \times N$ matrices by $\mathscr{D}\mathbf{H} = \prod_{i \geq j} \mathrm{d}H_{ij}$. (This measure follows from a metric invariant under orthogonal transformations, $\delta l^2 = \mathrm{tr}(\delta\mathbf{H}\delta\mathbf{H})$). We define the probability of a matrix to be in that volume by $\mathrm{d}P = p(\mathbf{H})\, \mathscr{D}\mathbf{H}$ and construct the information entropy

$$S = -\int p(\mathbf{H}) \log_2(p(\mathbf{H}))\, \mathscr{D}\mathbf{H}, \tag{3.8}$$

which is to be maximized subject to the constraints

$$\int p(\mathbf{H})\, \mathscr{D}\mathbf{H} = 1, \qquad \int p(\mathbf{H}) \|\mathbf{H}\|^2\, \mathscr{D}\mathbf{H} = \langle \|\mathbf{H}\| \rangle^2, \tag{3.9}$$

that the probability density $p(\mathbf{H})$ is normalized to unit total probability and that $\mathbf{H}$ has a specified mean square norm, that is, an eigenvalue scale.

Introducing Lagrange multipliers α and ν associated with the two constraints, one finds that the functional

$$\Lambda = \int p(\mathbf{H})[\ln(p(\mathbf{H})) + \alpha + \nu\|\mathbf{H}\|^2]\, \mathscr{D}\mathbf{H} \tag{3.10}$$

must be stationary, which then implies that the probability density p must be

$$p(\mathbf{H}) = \exp(-\alpha - 1 - \nu\|\mathbf{H}\|^2). \tag{3.11}$$

We choose a norm for $\mathbf{H}$, which is invariant under orthogonal rotations, and find that the ensemble weights are also invariant under such transformations. In particular if we choose the usual L_2 norm

$$\|\mathbf{H}\|^2 = \mathrm{tr}(\mathbf{H}^2) = \sum_{i,j} H_{ij}^2 = \sum_i H_{ii}^2 + 2\sum_{i>j} H_{ij}^2 \tag{3.12}$$

we recover the GOE in which the elements of $\mathbf{H}$ are independent centered Gaussian random numbers (subject to the condition that $\mathbf{H}$ is symmetric) for which the diagonal elements have variance $1/2\nu$ and the off-diagonal elements have variance $1/4\nu$. Equations (3.9) serve to determine α and ν.

Thus the GOE follows from minimal conditions on the ensemble of Hamiltonians; it is the ensemble carrying the least information.

3.2.1 Other Ensembles

Closely related to the GOE is the COE of Dyson (1962) formulated with no a priori eigenvalue scale. It nevertheless has spectral fluctuations identical to those of the GOE. Dyson posited an ensemble of symmetric unitary matrices U related uniquely and smoothly to symmetric matrices $\mathbf{H}$, for example, but not necessarily by $U = \exp(\mathrm{i}H)$. The eigenvalues $\lambda_j = \exp(\mathrm{i}\theta_j)$ of U are assumed to be distributed uniformly on the unit circle. On positing a measure of volume in that ensemble that is invariant under unitary transformations $U \to W^T U W$ (where W is any unitary matrix), and a hypothesis of uniform probability in that space, he derived the COE and showed that the corresponding $\mathbf{H}$ had spectral fluctuations identical to those of the GOE. The COE is mathematically and conceptually more attractive but is analytically more challenging.

For systems with imperfectly broken symmetries we expect the dynamical matrix to be dominated by a (block) diagonal, with small coupling between the blocks. In the limit of zero coupling we have exact block diagonality and two independent GOEs. If coupling is sufficiently strong (mixing rates are fast compared with Heisenberg times) we have a single GOE. Despite their practical importance, the generic spectral features of systems with moderate coupling are not yet well studied.

Another well-studied ensemble is the Gaussian unitary ensemble (GUE), consisting of complex Hermitian matrices $\mathbf{H}^T = \mathbf{H}^*$, with Gaussian random elements and with weights that are invariant under unitary transformations, that is, such that there is no preferred basis. The matrices occur with differential probability

$$\mathscr{D}P = p(\mathbf{H})\mathscr{D}\mathbf{H} \propto \exp(-\nu\,\mathrm{tr}\,\mathbf{H}^2) \prod_{i>j} \mathrm{d}\Im\{H_{ij}\}\mathrm{d}\Re\{H_{ij}\} \prod_i \mathrm{d}H_{ii}. \tag{3.13}$$

Such matrices describe time-reversal non-invariant systems (often called gyroscopic in the structural acoustics literature), as are found in mechanical structures with Coriolus forces, or acoustics in the presence of steady flow, or electrons in a magnetic field.

Recent years have supplemented the theoretical work of Wigner and Dyson and Mehta calculating eigenvalue statistics for the GOE and GUE. More recent years have seen development of powerful and technically complicated methods for performing averages over these matrix ensembles. The interested reader is directed to tutorials on the subject (Efetov 1983, Verbaarschot et al. 1985, Fyodorov 1995). These techniques, termed replica trick and supersymmetry, have been used to construct estimates for the statistics of responses (Green's functions G) in dissipative systems. In the absence of dissipation, or in the presence of uniform dissipation, G may be expressed in terms of the modes and eigenfrequencies that are described by the usual GOE or GUE. Response statistics are then derived from these. In the presence of dissipation, however, a modal perspective can be problematic. Nevertheless, responses can be expressed directly in terms of $\mathbf{H}$ and the dissipation operators $\mathbf{\Gamma}$. Various moments of $\mathbf{G} = (\mathbf{H} + \mathrm{i}\omega\mathbf{\Gamma} - (\omega - \mathrm{i}\varepsilon)^2\mathbf{I})^{-1}$ can then be evaluated by taking averages over the ensemble using the methods of Verbaarschot et al. (1985), Efetov (1983), and Fyodorov (1995).

3.2.2 Eigenvalue Statistics

Derivations of the eigenstatistics of the $N \times N$ GOE are outside the scope of this chapter but may be found elsewhere (Mehta 1960, 1991, Mehta & Gaudin 1960, Dyson 1962, Brody et al. 1981, Bohigas 1991, Guhr et al. 1998, Stöckmann 1999, Haake 2001). They are often elegant and well worth the attention of the interested student. They are particularly interesting in the $N \to \infty$ limit (and with ν scaling with N; $\nu = 2\gamma^{-2}$ at fixed γ.) Among the most salient are the mean eigenvalue density

$$\rho(\lambda) = \frac{\nu}{\pi}\sqrt{\frac{2N}{\nu} - \lambda^2} = \frac{2N}{\gamma\pi}\sqrt{1 - (\lambda/\gamma)^2}. \tag{3.14}$$

This is the Wigner semicircle law. At large N, there is a weak tail for $|\lambda| > \sqrt{2N/\nu}$.

The joint density of two distinct eigenvalues (close on the scale of the variations in ρ but possibly far apart on the scale of ρ itself) is

$$\rho(\lambda, \lambda') = \rho(\lambda)^2\,[1 - Y_2(\rho(\lambda)|\lambda - \lambda'|)], \tag{3.15}$$

where Y_2 is a dimensionless function of a dimensionless argument and is given by

$$Y_2(s) = \mathrm{j}_0(\pi s)^2 - J(s)D(s), \tag{3.16}$$

$$D(s) = \frac{\mathrm{d}}{\mathrm{d}s}\,\mathrm{j}_0(\pi s), \quad I(s) = \int_0^s \mathrm{j}_0(\pi s')\,\mathscr{D}s', \quad J(s) = I(s) - \frac{1}{2}\,\mathrm{sign}(s), \tag{3.17}$$

j_0 being the usual spherical Bessel function: $\mathrm{j}_0(x) = \sin(x)/x$. Further, Y_2 is plotted in Figure 3.1.

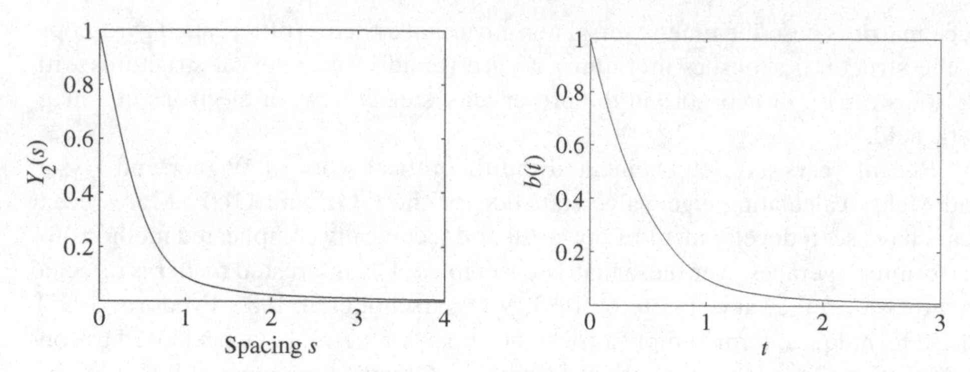

Figure 3.1. The function $Y_2(s)$ for the GOE and its Fourier transform, the form factor $b(t)$.

Its Fourier transform is of note,

$$b(t) \equiv \int_{-\infty}^{\infty} Y_2(\xi)\exp(2i\pi\, t\xi)\mathrm{d}\xi$$
$$= \begin{cases} 1 - 2|t| + |t|\log(1+2|t|), & |t| < 1, \\ -1 + |t|\log\dfrac{2|t|+1}{2|t|-1}, & |t| > 1. \end{cases} \tag{3.18}$$

Its argument, scaling as it does inversely with eigenfrequency spacing ξ, may be interpreted as a dimensionless time.

We note in particular that $b(0)$, the integral of $Y_2(s)$ over all s, is unity. There is a spectral hole that does not relax even at long range. The expected number of additional eigenvalues within a long range L around a specified eigenvalue is $L-1$. This may be contrasted with the expectation (L) of the number of eigenvalues in an *arbitrary* span of length L. It may also be contrasted with the Poisson case of uncorrelated eigenvalues in which the expected number of additional eigenvalues within this span around a specified eigenvalue is L itself and is unaffected by the condition that there is one eigenvalue known to lie in the middle of the range.

The nearest neighbour spacing distribution $p(s)$ differs, for the $N \times N$ GOE, from the Wigner surmise (3.5), but for most purposes the difference is negligible. It is plotted in Figure 3.2.

The so-called number variance $\Sigma^2(L)$ characterizes fluctuations in eigenvalue density. An arbitrary range of normalized length L has an expected number of eigenvalues L. The variance away from the mean is denoted by Σ^2. For the RMT it is given in terms of an integral of Y_2 and is well approximated for the GOE by its asymptotic form

$$\Sigma^2(L) \sim 2(\log L)/\pi^2 + 0.44, \tag{3.19}$$

which is much less than its Poisson value, L. Small values for number variance correspond to *spectral rigidity*; indeed a periodic spectrum would have number variance essentially zero; Σ^2 is plotted in Figure 3.3.

Another statistic that indicates rigidity is the Delta-3 statistic, $\Delta_3(L)$, which describes the mean square deviation of a staircase function (see the below-given

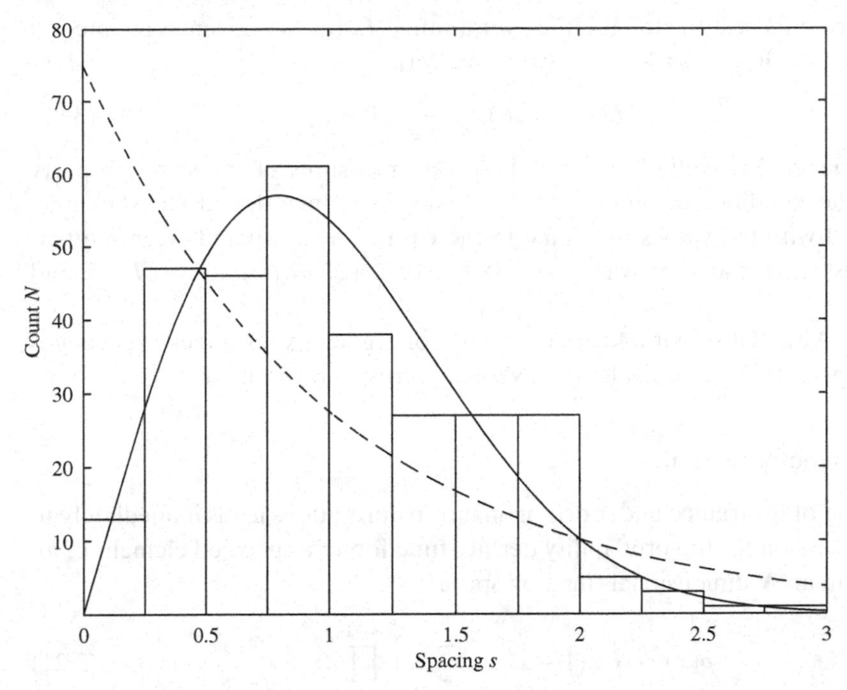

Figure 3.2. The nearest-neighbor spacing distribution for the 300 eigenvalues λ_i of a sample from the 300×300 GOE with $\nu = 1/4$. The Wigner surmise, Equation (3.5) (solid line), and Poisson prediction $\exp(-s)$ (dashed line) are superimposed.

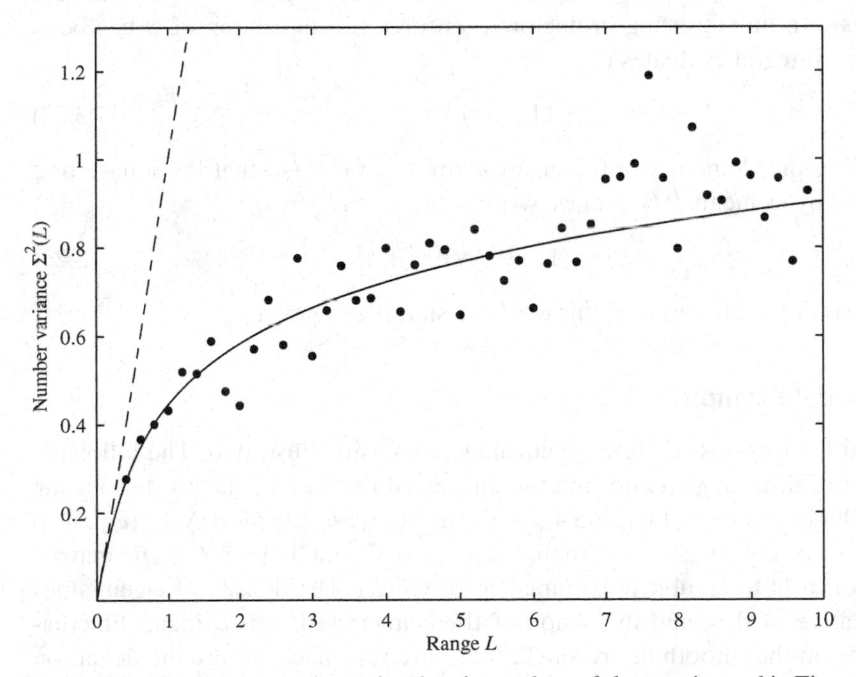

Figure 3.3. The number variance for the eigenvalues of the matrix used in Figure 3.2 is compared with the prediction of the GOE (whose asymptotic form is given by Equation (3.19)) (solid line) and Poisson ($\Sigma^2 = L$) (dashed line).

definition) from its best-fit straight line over a range L. It too is given as an integral of Y_2 and for the GOE has a simple asymptotic form

$$\Delta_3(L) \sim \log(L)/\pi^2 - 0.007. \tag{3.20}$$

Number variance Σ^2, Delta-3 Δ_3, and Y_2 are all measures of the same property related to the conditional number density. The above-mentioned statistics may be contrasted with the values they take in the case of uncorrelated eigenfrequencies, the Poissonian statistics, where $Y_2 = 0$, $b = 0$, $\Sigma^2(L) = L$, $\Delta_3(L) = L/15$, and $p(s) = \exp(-s)$.

At large N the GOE has a further property of ergodicity; averages over ranges of spectral variable λ are equivalent to averages across the ensemble.

3.2.3 Eigenvector Statistics

The condition of invariance under orthogonal transformations leads immediately to a simple expression for the probability density function of a specified element of an eigenvector in an N-dimensional function space:

$$p(x) \sim \int \delta(1 - x^2 - \sum x_j^2) \prod_{i=1}^{N-1} \mathrm{d}x_i. \tag{3.21}$$

The specified element is x; x_j are the other elements, and the delta function enforces normalization of the eigenvector. Enforcement of orthogonality is not required unless one is interested in joint statistics between different eigenvectors of different eigenvectors. An uninteresting prefactor to enforce normalization of p has been dropped. The integral evaluates to

$$p(x) \sim (1 - x^2)^{N/2-3/2}. \tag{3.22}$$

Renormalizing the element x of the eigenvector $\xi = x\sqrt{N}$ (so that the square of ξ is normalized by its mean, $\langle \xi^2 \rangle = 1$) we write

$$p(\xi) \sim (1 - \xi^2/N)^{N/2-3/2}. \tag{3.23}$$

In the important limit of large N, this is a Gaussian $p(\xi) \sim \exp(-\xi^2/2)$.

3.3 Numerical Examples

The spectral fluctuations of GOE eigenvalues are easily illustrated. The following plots are not difficult to generate, and the interested reader can quickly do so using readily available software. Figure 3.4 shows the staircase function $N(\lambda)$ (equal to the number of eigenvalues λ_j less than or equal to λ) of a 300×300 GOE matrix. Superposed is a fit to a cubic polynomial $N^{\text{smoothed}}(\lambda)$. The density of eigenvalues is the derivative of this, and the shape of the semicircle law is evident. Fluctuations away from the smooth fit are small; there are few places where the deviation is greater than one. The spectrum may be unfolded by use of the known density of eigenvalues, Equation (3.14). But in cases in which eigenvalues are taken from laboratory measurements, the density is not known with sufficient accuracy for the unfolding, and the experimentalist will take the smoothed density from a low-order

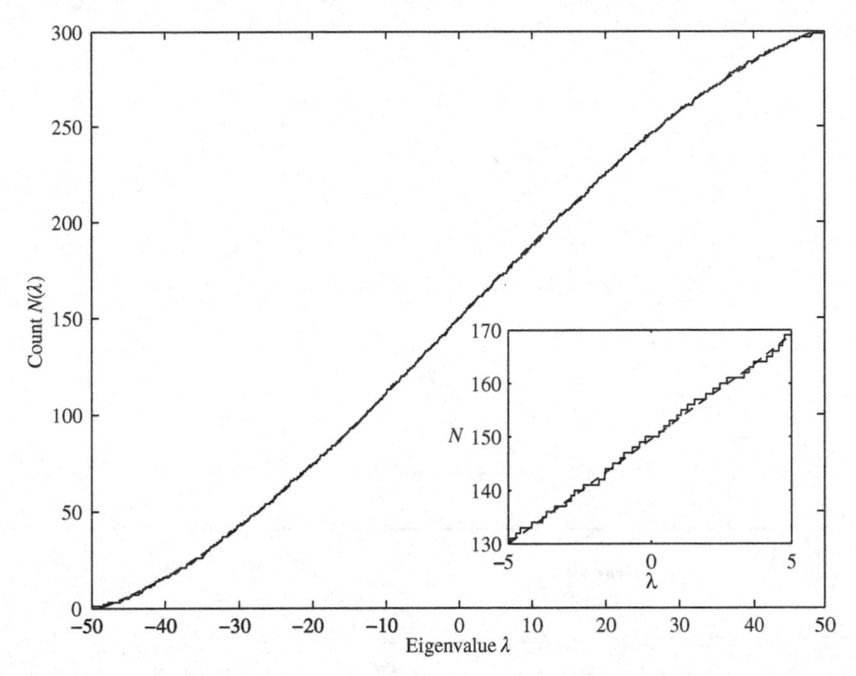

Figure 3.4. The staircase function $N(\lambda)$ for the 300 eigenvalues of the matrix used in the previous two figures. The dashed curve is a fit to a cubic polynomial. The inset shows a close-up of the same data.

fit to the staircase function. The spectrum of Figure 3.4 is unfolded by constructing the normalized eigenvalues $z_i = N^{\text{smoothed}}(\lambda_i)$. A histogram of the normalized spacings $s_i = z_{i+1} - z_i$ is given in Figure 3.2.

It is noteworthy that any smooth function of λ (smooth over any range much greater than the mean spacing), for example, $\omega_i = f(\lambda_i)$, will, after unfolding, give essentially the same normalized eigenvalues z. Thus if the eigenvalues λ of an acoustic stiffness matrix had the fluctuations of the GOE, so would the corresponding frequencies $\omega = \sqrt{\lambda}$.

It is also noteworthy that the fluctuations of the GOE are reproduced with other ensembles of symmetric matrices. Figure 3.5 shows the same numerical process as applied to a sparse 300×300 symmetric matrix with entries being random numbers taken from a distribution: 0 with probability 94%, −1 with probability 3%, and +1 with probability 3%.

Finally, for comparison, Figure 3.6 shows the staircase function and its cubic polynomial fit and the corresponding deviations and histogram for 300 uncorrelated numbers ("eigenvalues") taken from a uniform distribution [0, 300].

3.4 Relevance for Acoustics

It is thus apparent that the GOE corresponds in some sense to a maximal entropy ensemble of symmetric matrices and that the eigenvalue fluctuations of numerical samples from that ensemble correspond well with established analytical theory. Why, though, is this relevant to acoustics? Why ought we appeal to maximal ignorance and impose so few conditions on the matrices that describe closed

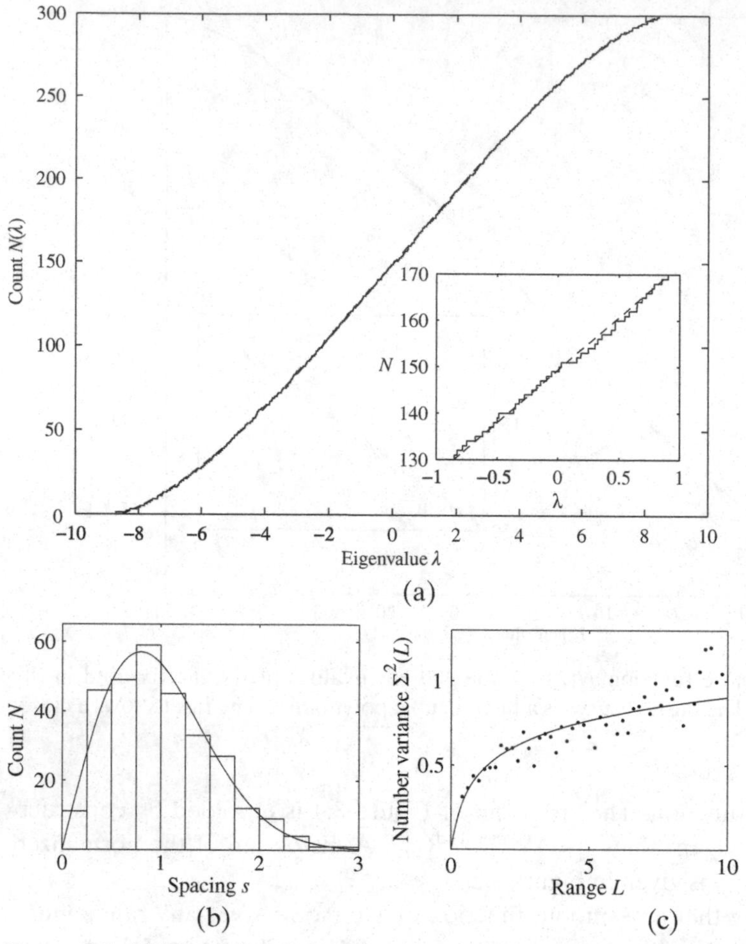

Figure 3.5. (a) The staircase function from a 300×300 random symmetric matrix from a different ensemble shows the same behaviors as those of Figure 3.4. The inset shows a close-up of the same data. (b) Spacing histogram for the non-GOE matrix. The solid line is the Wigner surmise. (c) Number variance for the non-GOE matrix. The solid line is the prediction of the GOE.

acoustic and structural acoustic systems? Indeed, as Soize (Chapter 13) has emphasized, stiffness matrices in acoustics are positive definite; so the GOE is unacceptable for some purposes. Furthermore, stiffness matrices, even of the most complicated acoustic systems, are highly sparse when expressed in terms of basis vectors with finite spatial support (as with for example a finite element or a finite difference description.) Real systems have particular features, like the sparseness in the non-GOE ensemble discussed earlier, which one might suppose ought to be incorporated in the ensemble. Furthermore, acoustic systems are usually damped, and modes are resolvable only at low frequency. Nevertheless, there is substantial empirical evidence that RMT (and indeed the GOE) does a good job of describing universal fluctuations of the eigenfrequencies (or of their real parts if damped) of generic acoustic systems away from a system-specific nonuniversal local mean eigenfrequency density.

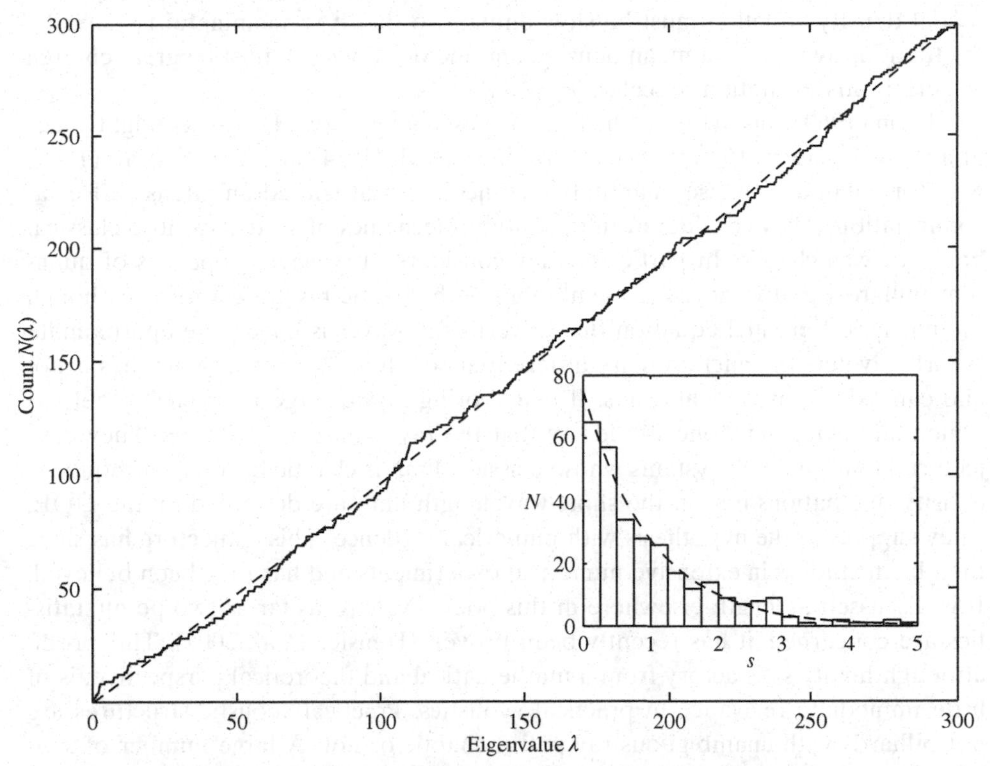

Figure 3.6. The staircase function for 300 independent uniform random numbers λ_j taken in the interval from 0 to 300. The dashed line is a fit to a cubic. Deviations from smoothness are stronger than they are for GOE eigenvalues. The inset shows the spacing histogram for the random numbers used to construct it, with Poisson prediction superimposed (dashed line).

This evidence comes from numerous laboratory measurements. Weaver (1989b) and, extensively, Ellegaard et al. (Ellegaard et al. 1995a, 1995b, 2001, Bertelsen et al. 2000, Andersen et al. 2001, Schaadt et al. 2001, 2003) investigated eigenstatistics for ultrasonic vibrations in solids. Davy (1990) did so in reverberation rooms and Langley et al. (Langley & Brown 2004a, 2004b, Langley & Cotoni 2005) in structural acoustics. In all of these cases, the smoothed density of eigenfrequencies was taken from a fit (theoretical estimates being insufficiently precise), and the residual fluctuations successfully compared with those of the GOE. It has also been confirmed in numerous numerical studies of the eigenfrequencies of membranes of certain shapes (McDonald & Kaufman 1979, Heller & Tomsovic 1993).

The GOE, to the extent it is applicable in acoustics, describes local spectral features. One does not expect the semicircle law $\rho(\lambda)$ (Equation (3.14)) of the GOE to describe the density of eigenfrequencies $n(\omega)$ (per unit angular frequency ω) of an acoustic system any more than it describes the mean density of nuclear energy levels. It is the spectral fluctuations, when scaled by the local density of eigenfrequencies, that are conjectured to correspond to those of the GOE; the parameter ν must be chosen to match the local modal density. It may of course vary with frequency but must do so slowly if GOE fluctuations are to apply over any significant range. The modal density variation must be slow enough so that it is meaningful to describe fluctuations away from a mean density, and the density $n(\omega)$ must be great enough to permit this separation of scales. $n^2 \gg \mathrm{d}n/\mathrm{d}\omega$.

Lyon (1969) appears to be the first to guess that nontrivial statistics might occur in acoustic systems. However it was Bohigas et al. (1984) and Casati et al. (1980) who formulated a precise quantitative conjecture that ignited an intense effort at confirmation. They considered the quantum mechanics of systems whose classical behavior was chaotic. In particular they considered the wave properties of quantum billiards, that is, acoustic membranes with chaotic ray trajectories. Although the partial differential equation that governs the waves is linear, the approximate (short-wavelength limit) ordinary differential equation that governs the rays is not and can exhibit dynamical chaos. If neighboring rays diverge from each other exponentially with time, one would say that the ray dynamics is chaotic. They conjectured that quantum systems whose classical limit is chaotic have eigenfrequency density fluctuations that in the short-wavelength limit are described by the GOE. They supported the hypothesis with numerical evidence. This conjecture has since then been studied in extensive numerical experiments and has long been believed. It is discussed at length elsewhere in this book. At least as far as two-point statistics are concerned, it has recently been proven (Heusler et al. 2007). This proof, although highly satisfactory from a mathematical and theoretical perspective, is of little immediate relevance in practical acoustics. Practical acoustic structures are not billiards with unambiguous ray paths, chaotic or not. A large number of simple systems are of mixed character, with parts of phase space being chaotic and other parts being regular. Many systems may not be chaotic but nevertheless mix quickly. Many structures have no clear relevant short-wavelength limit (having heterogeneities on many scales) and so lack classical counterparts in the sense of quantum mechanics. Furthermore, practical interest is rarely confined to the short-wavelength limit. Nevertheless, RMT still appears to apply. The correspondence applies not just to ray-chaotic systems, even though that is where the conjecture has been pursued the most. Indeed the first acoustic demonstration of GOE spectral statistics was for a three-dimensional elastic body at frequencies far below the short-wavelength limit (Weaver 1989b). Similar measurements on microwave systems are also at finite wavelength (Stöckmann & Stein 1990, Richter 1998). At shorter wavelengths, modes tend to be unresolvable; so experiments in this limit are not possible. Even in the short-wavelength limit, the body studied in Weaver (1989b) was not ray chaotic (Bohigas et al. 1991). There is good reason to believe that RMT applies, for practical purposes, to structures far more general than the ray-chaotic billiards that have attracted so much theoretical and experimental attention, and also to systems at finite frequency. Precise understanding of the limits of applicability of GOE to such systems remains so far unclear, but that has not prevented its application.

It appears from numerical and laboratory experiments, and from theoretical arguments, that the GOE describes spectral fluctuations out to a finite range $\Delta\omega$ equal to (2π times) the inverse of the time t_{mix} required for the diffuse field to mix over available phase space. We recognize that the so-called Heisenberg time T_{H} is equal to the time required to resolve typical nearest-neighbor eigenfrequencies, $T_{\text{H}} = 2\pi n(\omega)$, and conclude that GOE fluctuations apply if the mixing time is much less than the Heisenberg time. We further note that the normalized range L over which one expects these universal fluctuations (i.e., the number of neighboring eigenfrequencies that are correlated) is the number of mixing times in a Heisenberg time.

Eigenvalues can be independent only if they correspond to modes in regions of phase space that do not overlap. For this reason chaotic systems, whose rays mix well, have well-correlated eigenfrequencies. However, diffuse fields with different frequencies do not occupy the same region in a properly generalized phase space because they are on different "shells" in frequency space. Resolution (by an amount of frequency Δf) of a diffuse field to a shell of specified frequency f is only possible after sufficient time $1/\Delta f$.

A picture develops of waves emanating from a transient source and producing diffuse fields that are slowly, or perhaps rapidly, filling phase space volumes. These volumes are simultaneously spreading in space $\mathbf{r}$ and momentum $\mathbf{k}$ (among any additional descriptors of position in phase space, for example, an azimuthal number in a nominally axisymmetric structure) and shrinking in the frequency dimension. It is only if fields spread fast enough to fill phase space laterally before shrinking too much to overlap in the frequency dimension that the corresponding eigenvalues can correlate. A wave disturbance spreads from an initial condition to span a region of phase space; at the same time it resolves itself in frequency. Rapid mixing leads to high correlations over long ranges in frequency.

This picture is well illustrated by numerical simulations of the Šeba billiard (Šeba 1990, Albeverio & Šeba 1991, Weaver & Sornette 1995, Legrand et al. 1997). This is a rectangular membrane (with incommensurate sides and symmetry breaking boundary conditions, e.g., Dirichlet on the bottom and left sides and Neumann on the top and right) with an isotropic point scatterer having a cross section of the order of a wavelength (see also Chapter 2). In the short-wavelength limit the dynamics is regular; rays conserve their values of $|k_x|$ and $|k_y|$ until they hit the infinitesimal point – which on average requires an infinite time. At any finite frequency, however, lifetime against scattering is of the order of the Heisenberg time – the inverse of the mean eigenfrequency spacing. If several (N) point scatterers are placed in the rectangle, the lifetime against scattering into new values of $|k_x|$ and $|k_y|$ is of the order of T_H/N. Although there are technical challenges to solving this system, they are not insurmountable, and the spectral fluctuations are clear: GOE correlations fail for ranges L greater than the number of scatterers N.

Thus we are led to conjecture that – in structural acoustic systems in which we expect energy to diffuse in a time t_{mix} over a substructure both spatially and in phase space (e.g., in angular momentum or reflection classes) – structural acoustic modes will be correlated and in particular will exhibit eigenfrequency repulsion and spectral rigidity, over a frequency range $1/t_{mix}$. Modes separated by more than $1/t_{mix}$ will have no direct correlation.

Sparseness in a configuration-space representation such as an acoustician may employ when using finite elements is thus not relevant. Sparseness in a basis that spans a finite-frequency band would be more relevant and would presumably correspond to deviations from RMT over that frequency range.

As an immediate and unsurprising corollary to this picture, we see that if a structure has a good symmetry, for example, a reflection symmetry or axisymmetry, mixing across all of phase space never occurs, and one does not expect to find GOE fluctuations at all. It has long been recognized that RMT applies only to individual symmetry classes, that the matrix **H** will be block diagonal in a basis that respects that symmetry. Therefore the modes within any one class may be GOE-like, but the modes from different classes are uncorrelated. The net consequence is that the experimentalist must either separate the modes into their respective classes or look for spectral statistics corresponding to the superposition of uncorrelated spectra. Another consequence is that spectral correlations corresponding to superposed GOE spectra imply that there is some symmetry that the experimentalist may not have recognized a priori.

Another corollary to this picture arises if one considers two weakly coupled substructures (perhaps two reverberation rooms or two symmetry classes) between which the typical time for a diffuse field to mix is greater than the Heisenberg time. Under these conditions, there are no correlations between the eigenfrequencies corresponding to modes in the separate rooms, and the modes in each room are localized (Weaver & Lobkis 2000a). Although a diffuse field or simple SEA picture may suggest that the fields eventually equilibrate between the two rooms after a transient source acts in one room, in fact they do not. If mixing time is merely comparable to Heisenberg time one expects something subtler. This is perhaps inevitable at low frequencies in structural acoustics where T_{H} tends to be short. One then posits merely short-range eigencorrelations. This regime has not been well explored.

3.4.1 Applications

Applications of RMT arise in predicting power density fluctuations (Weaver 1989a, Legrand et al. 1995, Lobkis et al. 2000, Langley & Brown 2004a, 2004b, Langley & Cotoni 2005) crucial for estimating confidence with which averages of power density can be measured or predicted. Fluctuations as a function of both receiver position and frequency are of interest. RMT has also been shown to predict the sensitivity of a diffuse waveform to small changes in a structure (Gorin et al. 2006a) and in particular how the correspondence diminishes with the length of time the acoustics has been probing the structure.

Chief among applications is the modal statistics information it brings to those who in the tradition of Schroeder (Schroeder 1959, 1962, 1965, 1969a, 1969b, Schroeder & Kuttruff 1962, Lyon 1969, Davy 1981, 1986, 1987, Kubota & Dowell 1992, Legrand et al. 1995) wish to describe the statistics of structural responses by means of modal expansions (Weaver 1989a, Weaver & Burkhardt 1994, de Rosny et al. 2000, Weaver & Lobkis 2000b, Langley & Brown 2004a, 2004b, Langley & Cotoni 2005). We present one such here, with a remarkable prediction.

Consider a modal expansion for the time-domain response of an undamped structure, at position $\mathbf{x}$, to a concentrated transient force $f(t)$, at position $\mathbf{y}$ (with Fourier transform $F(\omega)$ centered on a frequency ω_c and having bandwidth $\Delta\omega \gg 1/n$; $\Delta\omega \ll \omega_c$):

$$\psi(\mathbf{x}, \mathbf{y}; t) = \Im\left\{\sum_r \phi_r(\mathbf{x})\phi_r(\mathbf{y}) \exp(\mathrm{i}\omega_r t) F(\omega_r)/\omega_r\right\}. \tag{3.24}$$

The sum is over all modes $\phi_r(\mathbf{x})$ with natural frequency ω_r, $r = 1, 2, \ldots, \infty$. The expansion (3.24) is valid at all times after the force $f(t)$ ceases.

On squaring and averaging over short times of the order of $1/\Delta\omega$ one finds

$$\psi^2(\mathbf{x}, \mathbf{y}; t) = \frac{1}{2}\sum_r\sum_s \frac{\phi_r(\mathbf{x})\phi_r(\mathbf{y})\phi_s(\mathbf{x})\phi_s(\mathbf{y})\exp(\mathrm{i}\omega_r t - \mathrm{i}\omega_s t)F(\omega_r)F^*(\omega_s)}{\omega_r\omega_s}. \quad (3.25)$$

We first consider the case $\mathbf{x} \neq \mathbf{y}$. According to the large-N limit of RMT, the modes $r \neq s$ are uncorrelated even at the same position $\mathbf{x}$; so their contributions vanish. The contribution from the $r = s$ terms is

$$\sum_r \frac{[\phi_r(\mathbf{x})]^2[\phi_r(\mathbf{y})]^2|F(\omega_r)|^2}{2\omega_r^2}. \quad (3.26)$$

This quantity has an expectation that is time independent and may be written, if $\mathbf{x} \neq \mathbf{y}$, as

$$\begin{aligned}\langle\psi^2(\mathbf{x}, \mathbf{y}; t)\rangle &\approx \langle\phi^2\rangle^2 \int \frac{n(\omega)|F(\omega)|^2}{4\omega^2}\,\mathrm{d}\omega \\ &\approx \langle\phi^2\rangle^2 \frac{n(\omega_c)\Delta\omega F(\omega_c)|^2}{2\omega_c^2},\end{aligned} \quad (3.27)$$

where $\langle\phi^2\rangle$ is the expected value of the square of a mode at the generic uncorrelated points $\mathbf{x}$ or $\mathbf{y}$. By normalization $\int \phi^2\,\mathrm{d}\mathbf{x} = 1$, we may write $\langle\phi^2\rangle = 1/\text{volume}$ (the above-given expression may be modified if points $\mathbf{x}$ or $\mathbf{y}$ are special, e.g., on a free surface.)

At case $\mathbf{x} = \mathbf{y}$, the total expression is

$$\begin{aligned}\langle\psi^2(\mathbf{x}, \mathbf{y}; t)\rangle &= \sum_r \frac{[\phi_r(\mathbf{x})]^4|F(\omega_r)|^2}{2\omega_r^2} \\ &+ \sum_r\sum_s{}' \frac{[\phi_r(\mathbf{x})]^2[\phi_s(\mathbf{x})]^2\exp\{\mathrm{i}\omega_r t - \mathrm{i}\omega_s t\}F^*(\omega_s)F(\omega_r)}{2\omega_r\omega_s},\end{aligned}$$

where the prime on the double sum indicates that we exclude the $r = s$ term. The first term is time independent and has an expectation

$$\langle\phi^4\rangle \int \frac{n(\omega_r)|F(\omega_r)|^2}{2\omega_r^2}\mathrm{d}\omega_r \approx \langle\phi^4\rangle \frac{n(\omega_c)\Delta\omega|F(\omega_c)|^2}{2\omega_c^2}, \quad (3.28)$$

which exceeds (3.27) by a factor $\langle\phi^4\rangle / \langle\phi^2\rangle^2$. This quantity arises routinely in estimates for acoustic power variances in reverberant structures. It can be calculated exactly for the highly nongeneric case of a rectangular room. According to RMT, modes have Gaussian statistics and this quantity is three, at least in the limit of large N. For finite N it is somewhat less. Langley has indicated that his laboratory measurements support a value less than three. This is consistent with Equation (3.23); it would be interesting to correlate that with independent estimates of the effective value of N.

The $r \neq s$ terms have expectation

$$\langle\phi^2\rangle^2 \int \frac{n(\omega)n(\omega'|\omega)|F(\omega)|^2}{2\omega^2}\,\mathrm{d}\omega\,\mathrm{d}\omega, \quad (3.29)$$

where we have replaced $F^*(\omega')/\omega'$ with $F^*(\omega)/\omega$ under the presumption that these quantities vary little over the bandwidth $\Delta\omega$. The conditional number density $n(\omega'|\omega)$ is given by (3.15), and so the sum of the $r \neq s$ terms has expectation (we have changed variables, $\omega' = \omega + \delta$)

$$\langle\phi^2\rangle^2 \int \frac{n(\omega)^2[1 - Y_2(n\delta)]|F(\omega)|^2 \exp(i\delta t)}{2\omega^2} d\omega\, d\delta$$
$$= -b(t/2\pi n)\langle\phi^2\rangle^2 \frac{n(\omega_c)\Delta\omega|F(\omega_c)|^2}{2\omega_c^2} \quad (3.30)$$

(the delta function at $t = 0$ has been neglected as unimportant at all times $t \neq 0$).

We conclude that the mean square ψ at points $\mathbf{x} = \mathbf{y}$ exceeds that at points $\mathbf{x} \neq \mathbf{y}$ by a factor $3 - b(t/2\pi n) = 3 - b(t/T_H)$. Immediately after a transient source has stopped, a diffuse field has twice the energy density at the source than it does elsewhere. This *enhanced backscatter* ratio becomes three after times of the order of the Heisenberg time. It has been measured in laboratory and numerical experiments (Weaver & Burkhardt 1994, de Rosny et al. 2000, Weaver & Lobkis 2000b), and the prediction that it evolves from two at short times to three at late time has been confirmed.

3.5 Summary

The GOE is well understood, and its importance for spectral fluctuations in reverberant acoustic systems is clear. No statistical description of generic systems should neglect it. Uncertainties remain, however, in its applications to systems with non-negligible dissipation or systems in which the dynamics only slowly mixes different parts of phase space. In its application to practical structures, and in particular to dissipative (Lobkis et al. 2000, 2003, Lobkis & Weaver 2000, Barthélemy et al. 2005) or slowly mixing systems or systems with transport, it has yet to be fully mastered. What indeed are the universal features to be expected in practical large and complex reverberant acoustics systems? When, and with what detail, can we expect a given system to be of this simple universality class?

4 Gaussian Random Wavefields and the Ergodic Mode Hypothesis

Mark R. Dennis

H. H. Wills Physics Laboratory, University of Bristol, Bristol, UK

The modeling of field distributions by Gaussian random wavefields is reviewed. Basic properties of Gaussian random functions are discussed and applied to the specific examples of randomly scattered fields (speckle patterns) and random eigenfunctions in chaotic enclosures, according to Berry's ergodic mode hypothesis. Sabine's law for reverberation time is derived for an ergodic random mode.

4.1 Introduction

The fact that deterministic wavefields in complex geometries can be represented and analyzed by statistical methods is an important assumption in many physical systems. In acoustics, such random-seeming deterministic fields can arise in scattering from a rough surface or in the spatial structure of the eigenfunctions of a perfectly resonant ergodic cavity. The random fields in the former case originate from approximating the surface roughness with randomly placed secondary sources, whose superposition gives rise to a random "speckle pattern," typical for a wide range of wave scattering systems, especially laser light (Goodman 1985, 2007), and long studied in acoustics (Morse & Bolt 1944, Ebeling 1984). More surprising is Berry's hypothesis (Berry 1977), originally proposed in the context of quantum chaos, that a typical mode of an irregular cavity itself strongly resembles a typical Gaussian random function for high frequencies (i.e., in the semiclassical limit); again, this notion has existed in some sense for a long time in acoustics (Morse & Bolt 1944). This statement may be heuristically thought of as a consequence of the correspondence principle between geometric ray behavior and wave behavior for high frequencies, as applied to ergodic enclosures: the modeling of fields as superpositions of plane waves of fixed frequency, whose directions and phases are random, echoes chaotic ray dynamics where the straight line segments appear random in position and direction. Of course, there might be some overlap between the effects of random scattering and cavity mode ergodicity in experimental situations; I will nevertheless consider them separately here in order to distinguish the different physical assumptions behind each.

Random wavefields in general are represented by the sum of independent, identically distributed random functions, and consequently they may be modeled by Gaussian random fields. The Gaussian random wave model for ergodic modes

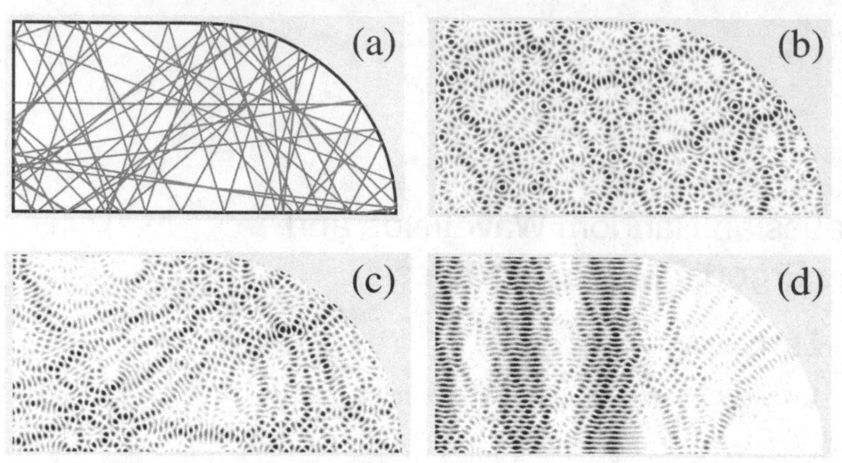

Figure 4.1. Rays and modes in an ergodic cavity (the desymmetrized stadium cavity, i.e., the join of a quarter disk, radius R, with a square of side R). (a) The trajectory of a typical ray reflected 60 times specularly from the walls, displaying irregular (ergodic) ray dynamics. (b)–(d) Three modes of the cavity with Neumann boundary conditions (i.e., even–even modes of the symmetric stadium), with consecutive dimensionless frequencies $kR =$ (b) 119.5321, (c) 119.5486, (d) 119.5857. The darker densities represent a higher modulus squared of the real mode φ. From the Weyl rule, these frequencies are around the 2080th level for this cavity. The spatial distribution of the field in (b) appears the most isotropic of the three; (d) is a "bouncing ball" mode, as discussed in Section 4.4.

is, in a sense, the counterpart for eigenfunctions of the random matrix theory hypothesis for eigenvalues (as discussed in Chapter 3). The nature of these statistical assumptions, and some basic techniques for calculating probabilities and averages of Gaussian random fields, will be the subject of this chapter. Particular emphasis will be placed on the ergodic mode hypothesis, which will be illustrated, as is standard in the field of quantum chaos, by the eigenfunctions of the chaotic two-dimensional stadium billiard. For the desymmetrized stadium, a sample ray trajectory and three consecutive cavity modes are shown in Figure 4.1 (computed using the method described in Heller 1991). The appearance of these functions is very different for the rays and field in a regular cavity, for which the ray dynamics is integrable and the Laplace operator is separable. An example in the rectangle is shown in Figure 4.2.

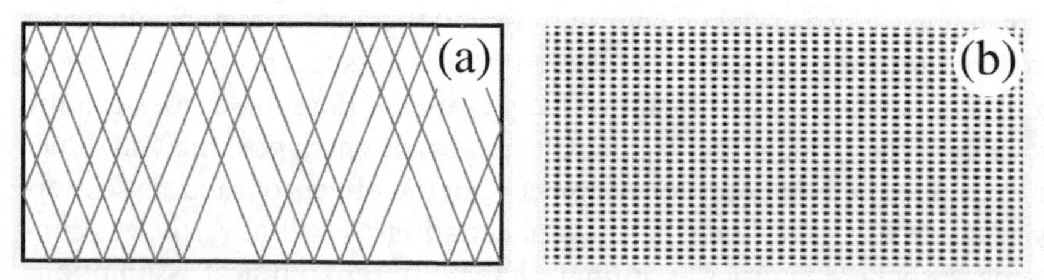

Figure 4.2. Rays and a mode in a regular cavity (the rectangle with dimensions $2R \times R$). (a) The trajectory of a typical ray reflected 31 times specularly from the walls, displaying regular ray dynamics. (b) Modes of the cavity with Neumann boundary conditions, with dimensionless frequency $kR = \sqrt{21^2 + 32^2}\pi = 120.2455$. As the Helmholtz equation is separable in Cartesian coordinates, all modes in this cavity factorize into a product of sinusoids in x and y.

Gaussian random fields are a universal tool in the physical sciences, used to represent spatially random functions more widely than acoustics and quantum chaos. Other examples from physical sciences include the distribution of heights of water waves on the sea (Longuet-Higgins 1957a, 1957b), randomly scattered optical fields (Goodman 2007), cosmology (Liddle & Lyth 2000) (modeling the distribution of matter in the universe and the direction-dependent temperature fluctuations of the cosmic microwave background radiation), and distribution of activity in the human brain (Cao & Worsley 2001). The universality of the Gaussian distribution arises from the central limit theorem of statistics (e.g., Papoulis 1991; also discussed later), according to which the sum of infinitely many independent, identically distributed random variables has a Gaussian (normal) probability distribution. When we say a particular field is Gaussian random, we mean it is an explicit function, which may be smooth and may satisfy a differential wave equation, and the distribution of its values over spatial points has a Gaussian distribution. More abstractly (and more correct probabilistically), there is an ensemble of such functions, and the value of the field at each point, over the ensemble, is determined by a Gaussian distribution. Not only is the Gaussian distribution universal, but it is extremely tractable mathematically, and it is possible to calculate many statistical properties of Gaussian random functions analytically.

In acoustics, the field in question is usually the complex field ψ, which represents the velocity potential, that is, velocity $\mathbf{u} = -\nabla\psi$, and pressure $p = \rho\partial_t\psi$. When the field is that of a pure tone (as will be assumed throughout this chapter) with angular frequency ω and wave speed c, then ψ satisfies the time-independent wave equation (Helmholtz equation) with wavenumber $k = \omega/c$,

$$\nabla^2\psi + k^2\psi = 0, \tag{4.1}$$

and the dependence $\psi \propto \exp(-i\omega t)$, and time dependence generally, will be ignored. Modes of closed cavities satisfying ideal boundary conditions, with no absorption, have a discrete spectrum of spatial frequencies k and are time-reversal symmetric. This symmetry implies that the field can be written as a real function φ; general complex fields will be represented by ψ. For perfectly hard walls, Neumann conditions are satisfied: $\partial_n\psi = 0$ (∂_n is the inward normal derivative). In studies of wave ergodicity inspired by quantum mechanics, ψ represents the complex wavefunction, and the cavity represents a potential well satisfying $\psi = 0$ on the boundary. Such Dirichlet conditions are more frequently studied in the literature. In realistic acoustic situations, there is absorption at the walls, breaking time-reversal symmetry; such cases will not be discussed in detail here, apart from in Section 4.5. Furthermore, the use of random waves in elastic media (Tanner & Søndergaard 2007), for instance, to model vibrating plates and inhomogeneous cavities, will not be discussed here.

The structure of this chapter is as follows: In the next section, the basic properties of Gaussian random fields will be defined and discussed, followed by descriptions of two important appearances in acoustics, namely, randomly scattered speckle fields (Section 4.3) and the Gaussian random wave model for ergodic modes (Section 4.4). The ergodic mode hypothesis is used to derive Sabine's law for reverberation time (Section 4.5).

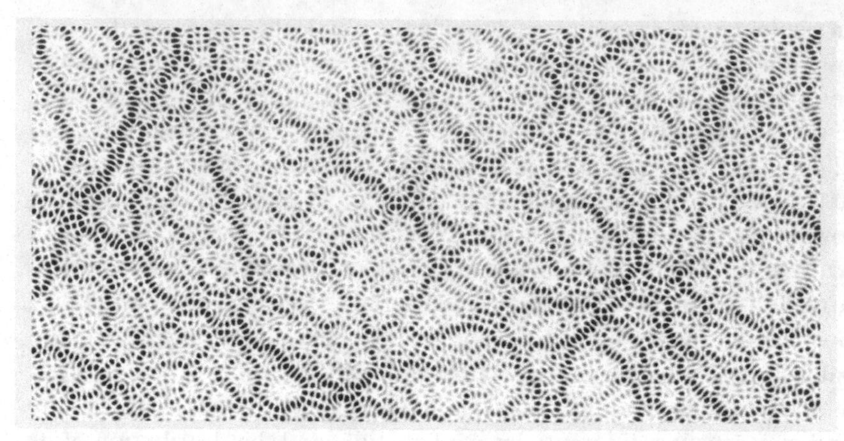

Figure 4.3. Isotropic random wave, satisfying the two-dimensional Helmholtz Equation (4.1). This wave is formed by a superposition of the form (4.3), with uniformly distributed phases and wavevector directions, with $J = 320$. The random field is plotted in an area $2R \times 4R$, with $kR = 120$. There is no significance of the plotting boundary, as the random wave does not satisfy boundary conditions of any natural cavity.

4.2 Definition and Properties of Gaussian Random Functions

The fields usually considered in statistical wave theory (acoustic, optical, quantum, etc.) are superpositions of many plane waves ($J \gg 1$),

$$\psi(\mathbf{r}) = \frac{1}{\sqrt{J}} \sum_{j=1}^{J} a_j \exp(\mathrm{i}\mathbf{k}_j \cdot \mathbf{r} + \chi_j), \tag{4.2}$$

when complex, and, for real functions,

$$\varphi(\mathbf{r}) = \sqrt{\frac{2}{J}} \sum_{j=1}^{J} a_j \cos(\mathrm{i}\mathbf{k}_j \cdot \mathbf{r} + \chi_j), \tag{4.3}$$

where the phases χ_j are independently and uniformly distributed between 0 and 2π. The simplest random functions considered here (the "isotropic random wave model") have wavevectors $\mathbf{k}_j$ with independent, uniformly random direction and fixed magnitude and amplitudes a_j with $\langle a_j^2 \rangle$ constant for all j, Rayleigh distributed for ψ and Gaussian distributed for φ. An example of a real random function φ so defined is shown in Figure 4.3.

As discussed next, in the limit of infinite superpositions $J \to \infty$, the probability distribution of these functions is given by a Gaussian distribution,

$$P(\psi)\,\mathrm{d}^2\psi = \frac{\mathrm{d}^2\psi}{2\pi\,\langle|\psi|^2\rangle} \exp\left(-|\psi|^2/2\langle|\psi|^2\rangle\right), \tag{4.4}$$

$$P(\varphi)\,\mathrm{d}\varphi = \frac{\mathrm{d}\varphi}{\sqrt{2\pi\,\langle\varphi^2\rangle}} \exp\left(-\varphi^2/2\langle\varphi^2\rangle\right). \tag{4.5}$$

Further, $\mathrm{d}^2\psi$ appears in (4.4) as the probability distribution of ψ, consisting of the real and imaginary parts (i.e., the area element in the Argand plane). Averages (expectation values) will be denoted $\langle\bullet\rangle$; ensemble averaging will be assumed unless stated otherwise.

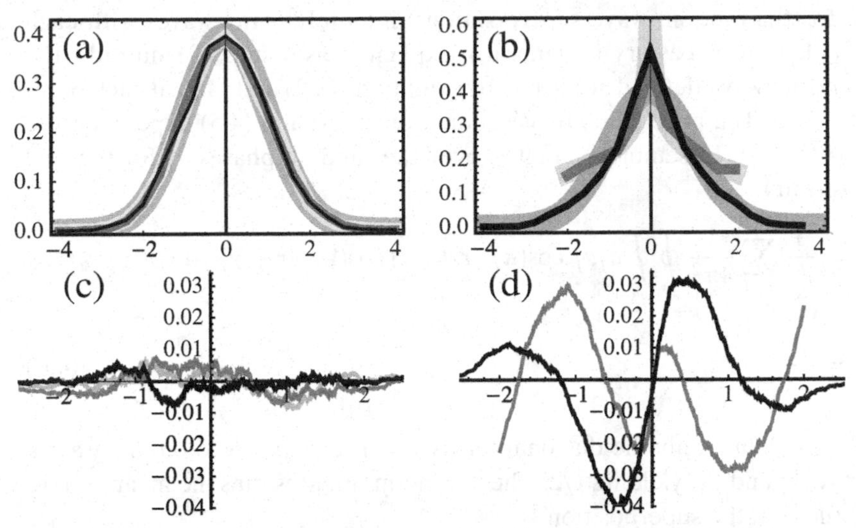

Figure 4.4. Distributions of the values of 10, 000 randomly sampled points in the fields of Figures 4.1–4.3, normalized according to standard deviations. (a) Gaussian curve (thick gray curve), compared against histograms of wavefields (piecewise linear curves): the random plane wave superposition of Figure 4.3 (light gray) and the stadium modes of Figures 4.1(b) (darker gray) and 4.1(c) (black). The fit of these fields with the normal Gaussian distribution is good. (b) Representation similar to previous part, with histograms of the rectangle mode of Figure 4.2 (dark gray) and the bouncing ball mode of Figure 4.1(d) (black) compared against a Gaussian curve (thick gray) and elliptic integral distribution (light gray). The rectangle mode fits the elliptic integral, and the bouncing ball distribution appears to follow a distribution intermediate between the Gaussian and elliptic integral. (c) Discrepancies of the cumulative distributions $\mathscr{P}$ with an error function for the three fields shown in (a), with the same color key. (d) The same as (c) but for the two fields shown in (b). The non-Gaussian nature of these distributions is very clear from this plot.

The probability distributions of the values of fields shown in Figures 4.1–4.3 are shown in Figures 4.4(a) and (b), sampled over equivalent sets of 10,000 points. The distributions for the random wave of Figure 4.3 and the two irregular modes of Figure 4.1 fit a Gaussian distribution very well; the bouncing ball mode and the rectangle mode do not. (In fact, it is straightforward to show that any rectangle mode of sufficiently high frequency has density function $P(\varphi) = K(1 - \varphi^2/4)/\pi^2$, with normalized variance, K being an elliptic integral of the first kind.) In practice, it is easier to check the agreement with theoretical probability distributions using the cumulative probability distribution $\mathscr{P}(X) \equiv \int^X P(x)\,\mathrm{d}x$, as this can be checked directly against the sorted list of sampled data more accurately than with a histogram (e.g., Berry & Robnik 1986, Li & Robnik 1994). For a Gaussian distribution, the cumulative distribution $\mathscr{P}$ is obviously an error function. The discrepancy in the cumulative distribution from the samples of the five fields from the ideal theoretical error function is shown in Figure 4.4(c).

The Gaussian probability density function follows from the central limit theorem, which gives a Gaussian probability density function for any random variable defined as the limiting infinite sum of independent identically distributed random variables; in the present case, the random variables, at a fixed point, are the random waves. (See Papoulis 1991 for a general proof and Stöckmann 1999 for an explicit justification for the random wave superpositions; infinitely many waves with each wavenumber $|\mathbf{k}_j|$ are necessary in general.) Explicitly, as a sum of sinusoids, the mean of a random wave defined according to Equations (4.2) and (4.3) at each point has $\langle\psi\rangle = \langle\varphi\rangle = 0$. The probability density functions (4.4) and (4.5) depend on the variances $\langle|\psi|^2\rangle$, $\langle\varphi^2\rangle$. Ensemble averaging over the random phases χ_j for the real field φ at position $\mathbf{r}$,

$$\begin{aligned}\langle\varphi^2\rangle &= \frac{1}{J^2}\sum_{j,j'}\frac{1}{4\pi^2}\oint\oint a_j a_{j'}\cos(\mathbf{k}_j\cdot\mathbf{r}+\chi_j)\cos(\mathbf{k}_{j'}\cdot\mathbf{r}+\chi_{j'})\,\mathrm{d}\chi_{j'}\,\mathrm{d}\chi_j \\ &= \frac{1}{2J}\sum_{j=1}^{J}a_j^2, \end{aligned} \tag{4.6}$$

where, in the final line, only the diagonal terms $j = j'$ give nonzero on integrating with respect to χ_j and χ_j', yielding 1/2. The second moment is thus the mean square of the amplitudes in the superposition.

Fields defined as derivatives of Gaussian random fields are also Gaussian random by another application of the central limit theorem, with variances given by an appropriate generalization of Equation (4.6). In general, covariances such as $\langle\varphi\partial_x\varphi\rangle$, where the fields are possibly evaluated at different points, do not vanish and can be calculated explicitly by analogy with Equation (4.6). The multivariate joint probability density for $\mathbf{v} = \{\varphi_1, \varphi_2, ..., \varphi_N\}$, is

$$P(\mathbf{v}) = \frac{1}{(2\pi)^{N/2}\sqrt{\det\mathbf{\Sigma}}}\exp\left(\frac{1}{2}\mathbf{v}\cdot\mathbf{\Sigma}^{-1}\cdot\mathbf{v}\right) \tag{4.7}$$

with correlation matrix $\mathbf{\Sigma}$, with entries $\Sigma_{ij} = \langle\varphi_i\varphi_j\rangle$. Averages for complex random functions can be taken with respect to the real and imaginary parts of Equation (4.2), which are independent. Zero-mean Gaussian random functions are particularly simple in that simultaneous averages of more than two random functions are completely determined by the two-point covariances, by the Gaussian moment theorem (finite-dimensional Wick's theorem),

$$\langle\varphi_1\varphi_2\cdots\varphi_N\rangle = \sum_{\text{all pairings }p}\prod_{(j_1,j_2)\in p}\langle\varphi_{j_1}\varphi_{j_2}\rangle \tag{4.8}$$

(so all such moments vanish when N is odd). This differs from random fields with other probability distributions, whose multipoint correlations $\langle\varphi_1\varphi_2\cdots\varphi_n\rangle$ cannot be determined from pairwise averages $\langle\varphi_i\varphi_j\rangle$, even though these may vanish because of symmetries.

The isotropic random wave model ensemble of Gaussian random functions defined by (4.2) and (4.3) has the following properties (Adler 1981, Papoulis 1991):

- *Stationarity*. All averages are independent of global position $\mathbf{r}$ (i.e., the probability measure of the ensemble is invariant to translation).
- *Isotropy*. Averages are independent of global direction (i.e., the probability is invariant to rotation).
- *Ergodicity*. Spatial averages (integrating over spatial position $\mathbf{r}$) and ensemble averages (integrating over the uniformly distributed directions $\mathbf{k}_j$, phases χ_j, and amplitudes a_j) are equivalent.

The spatial structure of any random field, Gaussian or otherwise, is determined by its two-point *correlation function* (also known as the autocorrelation function, covariance function, or covariance kernel). It is defined as the normalized average of the product of the field value at points $\mathbf{r}_1, \mathbf{r}_2$,

$$C(\mathbf{r}_1, \mathbf{r}_2) \equiv \frac{\langle \psi^*(\mathbf{r}_1)\psi(\mathbf{r}_2)\rangle}{\sqrt{\langle|\psi(\mathbf{r}_1)|^2\rangle\langle|\psi(\mathbf{r}_2)|^2\rangle}} \tag{4.9}$$

(with ψ substituted by φ for real fields). By definition, the correlation function is unity when $\mathbf{r}_2 = \mathbf{r}_1$. When ψ is statistically stationary, the correlation function becomes $C(\mathbf{r}_2 - \mathbf{r}_1)$, dependent only on the difference $\mathbf{r}_2 - \mathbf{r}_1$; when it is isotropic also, it depends only on the magnitude r of this difference.

In general, the operation of taking derivatives commutes with that of taking averages: if the random field satisfies the Helmholtz Equation (4.1), then so does the correlation function $C(\mathbf{r}_1, \mathbf{r}_2)$, with derivatives taken with respect either to $\mathbf{r}_1$ or $\mathbf{r}_2$. Thus the correlation function $C(r)$ for a stationary and isotropic field satisfies the radial part of the Helmholtz equation

$$C''(r) + \frac{d-1}{r}C'(r) + k^2 C(r) = 0, \tag{4.10}$$

in dimension d. In the cases $d = 2$, $d = 3$ relevant here, $C(r)$ is given by a zero-order planar or spherical Bessel function,

$$C(r) = \mathrm{J}_0(kr) \quad \text{in two dimensions,} \qquad C(r) = \frac{\sin(kr)}{kr} \quad \text{in three dimensions.} \tag{4.11}$$

Waves with different distributions of wavenumbers, or which are not isotropic, have different correlation functions.

Of course, when spatially averaging real functions (for instance, to test isotropy), finite sampling errors affect the result. Figure 4.5 shows an example of $C(\mathbf{r}_2 - \mathbf{r}_1)$ computed over equivalent areas for the fields shown in Figures 4.1–4.3, as well as the ideal isotropic Bessel function correlation function $\mathrm{J}_0(k|\mathbf{r}_2 - \mathbf{r}_1|)$. (These were computed by superposing the fields in disks around 500 sample points, weighted by the field at each sample point, similar to the approach of Li & Robnik 1994). Spatial inhomogeneities of the random superposition of Figure 4.3 mean that the numerically determined correlation function is not completely isotropic; in fact, the field of the mode in Figure 4.1(b) is more isotropic at smaller distances. Clearly, the spatial anisotropy of rectangle mode of Figure 4.2 and the bouncing ball mode of Figure 4.1(d) is demonstrated in the correlation function. Since all of the waves in question are solutions of Equation (4.1), the average of the correlation function over the direction of $\mathbf{r}_2 - \mathbf{r}_1$ is equal to $\mathrm{J}_0(k|\mathbf{r}_2 - \mathbf{r}_1|)$ in all cases.

The isotropic random wave model is the simplest random field to consider, as beyond the fixed wavenumber k (which simply sets the lengthscale) and dimension d, it has no other parameters and is as symmetric as possible. It may be generalized in many different ways: the wave may have a distribution of different wavenumbers k; the real and imaginary parts of ψ may have different distributions, representing phase rigidity (Brouwer 2003); and the field may be neither statistically isotropic nor stationary.

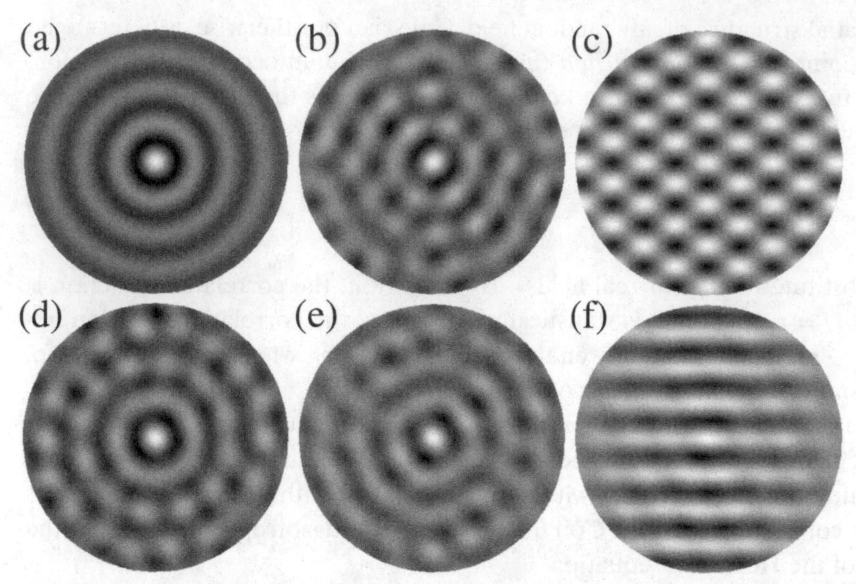

Figure 4.5. Correlation functions $C(\mathbf{r}_2 - \mathbf{r}_1)$ numerically computed from the wavefields of previous figures over equivalent spatial windows. (a) Ideal isotropic Bessel function $J_0(kr)$; (b) isotropic random superposition of Figure 4.3; (c) regular rectangular mode of Figure 4.2; (d)–(f) correspond to the stadium modes of Figures 4.1(b)–(d) respectively.

Important examples of the last case are the boundary-adapted ensembles considered by Berry (2002). These are random fields assumed to satisfy a homogeneous boundary condition along an infinite straight line (when $d = 2$) and plane surface (when $d = 3$); this boundary condition, chosen along the locus $x = 0$, will be considered to be Neumann (labeled $+$) or Dirichlet (labeled $-$). Random functions $\psi_\pm$ in each boundary-adapted random wave ensemble can be expressed in terms of isotropic random functions by $\psi_\pm(\mathbf{r}) = 1/\sqrt{2}\,(\psi(\mathbf{r}) \pm \psi(\bar{\mathbf{r}}))$, with $\bar{\mathbf{r}}$ representing the position vector $\mathbf{r}$ with $x \to -x$, reflected in the line/plane $x = 0$. Explicitly, they have representation

$$\psi_+(\mathbf{r}) = \frac{\sqrt{2}}{\sqrt{J}} \sum_{j=1}^{J} a_j \exp(\mathrm{i}\mathbf{k}_{\perp j} \cdot \mathbf{r} + \chi_j) \cos(k_{\| j} x), \tag{4.12}$$

$$\psi_-(\mathbf{r}) = \frac{\sqrt{2}}{\sqrt{J}} \sum_{j=1}^{J} a_j \exp(\mathrm{i}\mathbf{k}_{\perp j} \cdot \mathbf{r} + \chi_j) \sin(k_{\| j} x), \tag{4.13}$$

with $\mathbf{k}_{\perp j}$ the projection of the jth wavevector perpendicular to $x = 0$ and $k_{\| j}$ its projection parallel to the x-axis. The boundary-adapted real waves $\varphi_\pm$ are defined in a similar way, with the exponential functions replaced by cosine, as with Equations (4.2) and (4.3).

The x-direction of these boundary-adapted monochromatic random functions is distinguished, and so they are not isotropic and neither stationary nor ergodic with respect to x (although they are with respect to directions perpendicular to the wall). The intensity of the field $\langle \psi_\pm^2 \rangle / \langle \psi \rangle$, normalized with respect to the non-adapted mean intensity, is a function $B_{\pm}(x)$ of the distance x from the wall:

$$B_{\pm}(x) = 1 \pm \mathrm{J}_0(kx) \quad \text{in two dimensions,}$$

$$B_{\pm}(x) = 1 \pm \frac{\sin(2kx)}{2kx} \quad \text{in three dimensions.} \tag{4.14}$$

This directly follows from the definitions of the boundary-adapted fields before (4.12), $B_{\pm}(x) = C(0) \pm C(2x)$. As the distance from the wall increases, the boundary-adapted fields become increasingly isotropic, and their average values in this bulk limit are the same as the original isotropic ensemble. Clearly, for Dirichlet conditions, $B_{-}(0) = 0$, reflecting the fact that the field vanishes on the walls. For Neumann conditions, the average value of the field intensity on the wall is twice the bulk value. More complicated boundary-adapted random wave models have been discussed, including soft walls (Bies & Heller 2002), uniformly curved hard walls (Wheeler 2005), and Robin (mixed) boundary conditions (Berry & Ishio 2002).

Although lack of space forbids description of further calculations here, many physical and geometric properties of Gaussian random fields can be calculated analytically, including distributions of the intensities and flux densities of complex waves (Ebeling 1984, Saichev et al. 2002), various gradients of intensity and phase gradients (Goodman 2007), and numerous properties of nodes and extrema (Dennis 2007). Furthermore, large-scale numerical calculations are also possible with superpositions of random waves, allowing further properties of Gaussian random waves to be studied, such as the scaling of nodal domains in two-dimensional real waves (Bogomolny & Schmit 2002) and the topological linking of nodal line loops in three-dimensional complex waves (O'Holleran et al. 2009).

In the following two sections, the two specific physical models of random waves will be discussed: randomly scattered speckle fields and ergodic modes.

4.3 Gaussian Random Waves from Scattered Speckle Patterns

Speckle patterns are a very common form of random fields in wave physics. They are the random two-dimensional interference patterns occurring in planes of propagating near-monochromatic waves, far from a scattering surface with roughness on the scale of the wavelength. Such fields have been studied in statistical wave acoustics (Ebeling 1984), although often in the context of modes of chambers or complex media with scatterers distributed throughout (Weaver 1982); I will concentrate here on the mathematically simpler case of a screen parallel to and far from the scattering surface. This is the case often studied in optical speckle patterns such as those formed by reflecting a laser from a rough wall. The randomness of speckle patterns comes simply from the single reflection from the rough wall: the random spatial structure of modes will be considered later in Section 4.4 and partially open chaotic enclosures in Chapter 6.

If the roughness has a sufficiently small correlation length, and the measurement plane is far from the scattering surface, the correlation function of the field in the plane is given by a Fourier-like transform of the amplitude of the incident field; in optical coherence theory, this result is known as the van Cittert–Zernike theorem (Born & Wolf 1959, Goodman 1985, 2007, Mandel & Wolf 1995). This is a randomized form of diffraction and can be justified using a Fresnel diffraction integral

(Born & Wolf 1959). Thus for this result to hold, the distribution of wavevectors of the reflected field must be almost uniformly propagating in the same direction, say, the z-direction, so that the Helmholtz equation (4.1) can be approximated by the paraxial equation $\partial_x^2\psi + \partial_y^2\psi = -2ik\partial_z\psi$.

The effect of the rough surface may be represented by an infinite set of secondary point sources in the scattering wall. These have a uniformly random phase (representing the random height of the scatterer) and independent identically distributed positions; the resulting field (given the paraxial approximations below) is Gaussian random by the central limit theorem, in the limit of infinitely many secondary sources. The spatial distribution of these secondary sources in the scattering plane S is represented by the (normalized) intensity distribution I_{in} of the wave incident on the wall (assuming the random roughness of the scattering surface is statistically stationary and isotropic) and neglecting any detailed treatment of absorption.

With these assumptions, the unnormalized two-point correlation function of the scattered wave ψ_0, infinitesimally close to the scattering wall S, containing points $\mathbf{s}_1$ and $\mathbf{s}_2$, can be written in terms of the incident intensity,

$$\langle\psi_0^*(\mathbf{s}_1)\psi_0(\mathbf{s}_2)\rangle = I_{\text{in}}(\mathbf{s}_1)\delta(\mathbf{s}_1 - \mathbf{s}_2) \tag{4.15}$$

because no correlation between neighboring scatterers on the scale of wavelength is assumed, and the scatterers only randomize the phases (not the intensity); there are no averages on the right-hand side of Equation (4.15). The scattered field $\psi_z(\mathbf{r})$ at $\mathbf{r} = (x, y)$ in a plane at distance z from the scattering plane S may be written as a diffraction integral depending on z and $\psi_0(\mathbf{s})$; within the paraxial approximation, this is given by the Fresnel integral (Born & Wolf 1959, Mandel & Wolf 1995):

$$\psi_z(\mathbf{r}) = -\frac{ik}{2\pi z}\int_S \psi_0(\mathbf{s})\exp\left(ik|\mathbf{r}-\mathbf{s}|^2/2z\right)\,\mathrm{d}^2\mathbf{s}. \tag{4.16}$$

(This propagator equation is the same as the Schrödinger propagator, as the paraxial equation is equivalent to the two-dimensional time-dependent Schrödinger equation with propagation in z.) Thus at points $\mathbf{r}_1, \mathbf{r}_2$ in the measurement plane with distance z from the scattering plane S, the correlation function $C(\mathbf{r}_1, \mathbf{r}_2)$ can be found by substituting Equations (4.15) and (4.16),

$$C(\mathbf{r}_1, \mathbf{r}_2) = \exp(ik(r_2^2 - r_1^2)/2z)\int_S I_{\text{in}}(\mathbf{s})\exp(ik(\mathbf{r}_2 - \mathbf{r}_1)\cdot\mathbf{s}/2z)\,\mathrm{d}^2\mathbf{s}. \tag{4.17}$$

Up to an unimportant phase factor, the two-point correlation function is given by the Fourier transform of the incident intensity distribution I_{in}.

An example of such a field is given in Figure 4.6. The scatterer is represented by many point sources with random positions within a disk of radius a in the plane S, with uniformly random phases. The resulting field in a far-field plane is isotropic and random, with correlation lengthscale $2z/ka$; its two-point correlation function is (ignoring the phase factor) the Fourier transform of a disk, $4az\,\mathrm{J}_1(kra/2z)/kr$.

The wavefield in the preceding argument has fixed wavenumber k, although the correlation function $C(r)$ is not restricted to have the form (4.11). This is because

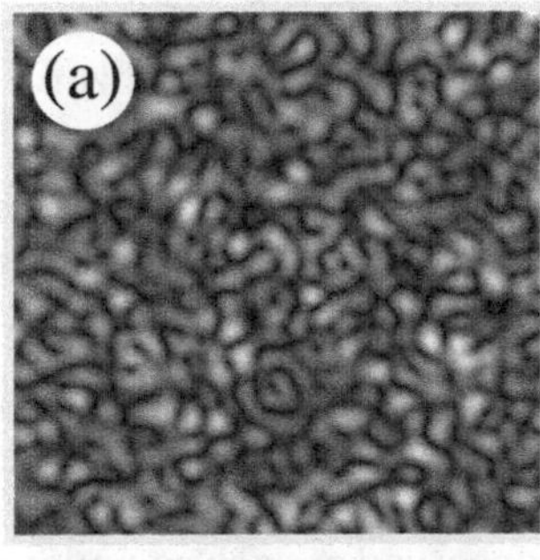

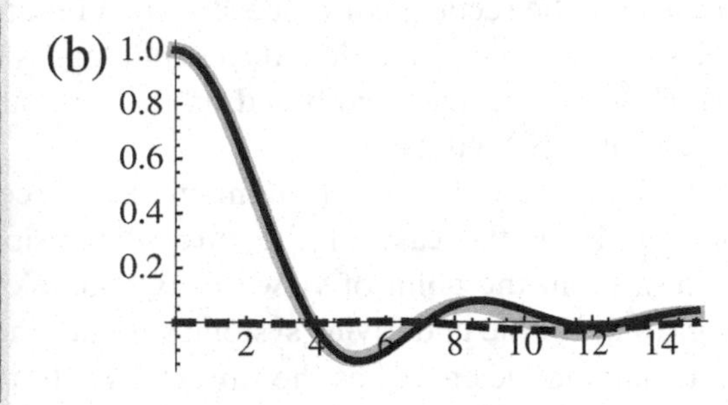

Figure 4.6. Isotropic random speckle pattern. (a) Intensity pattern $|\psi|^2$ of a random speckle field, simulated by 1,000 point sources uniformly distributed in a disk of radius a in the scattering plane S. The speckle pattern is plotted in the far-field zone (large z), with an area of 100 z/ak squared. The scaling is the inverse of previous plots: white represents high intensity, and dark represents low intensity. (b) Numerically computed correlation function $\Re\{C(r)\}$, plotted against kra/z (black). The imaginary part is the dashed line, and the grey line is the ideal theoretical correlation function $2z\,\mathrm{J}_1(kra/z)/kra$, corresponding to the Fourier transform of the original circular distribution $I(\mathbf{s})$ of sources.

the two-dimensional random scattered field is not isotropic in three dimensions; because paraxial propagation is assumed, the distribution of the transverse wavenumbers is determined by the initial intensity distribution I_{in}, by the Wiener–Khinchin theorem (Goodman 1985, Mandel & Wolf 1995), Frequently in optics, the distribution of transverse wavenumbers, and consequently $C(r)$ are taken to be a Gaussian function with zero mean; this is appropriate for speckle patterns occurring because of reflection of a laser spot whose intensity profile is Gaussian.

Although this model for randomly scattered fields does not necessarily work very close to the scattering plane S, its accuracy has been well established for z larger than a few wavelengths. Of course, the paraxial approximation requires that $|\mathbf{r}| \ll z$, whatever the propagation distance z. At transverse length scales beyond z, the scattered field appears increasingly to be the field from a point source rather than isotropic and random.

The randomness in the scattered fields described in this section purely comes from the uncorrelated rough scatterers in a single scattering plane S; the argument that modes of closed cavities are themselves examples of random functions is discussed in the following section.

4.4 The Gaussian Random Wave Hypothesis for Modes

Mathematically, the modes of a cavity are eigenfunctions of a partial differential operator corresponding to a certain discrete eigenvalue; in the case of the Helmholtz equation (4.1), the Laplacian operator ∇^2 has eigenvalue $-k^2$ for a characteristic frequency (wavenumber) k, which depends on the shape and boundary conditions of the cavity. In elementary treatments of this problem, the method of separation of variables is then used to break the problem into a set of ordinary differential equations, which are then solved separately. Examples of cavities separable in this way include the circle, ellipses, and rectangles – although this is not necessarily the case if the boundary condition is a function along the boundary (Berry & Dennis 2008).

This is the case with the rectangular mode shown in Figure 4.2(b). This function has features typical of this type of solution: its morphology is built on its pattern of zero contours, which form a regular grid based on the separating coordinate system (in this case Cartesian coordinates).

However, there are only a small number of such coordinate systems for any such operator – 11 in the case of the three-dimensional Laplacian (Morse & Feshbach 1953). From the point of view of ray dynamics, separability is often related to integrability of the underlying system: whether there exists a constant of the motion other than kinetic energy as the rays evolve. In the case of the rectangular billiard, the magnitude of linear momentum in x and y is conserved at each specular reflection, leading to a regular pattern of reflections made of only four directions (as Figure 4.2(a)). From the point of view of classical mechanics, integrability implies the motion is confined to an invariant torus (Arnol'd 1978), usually resulting in separability in the corresponding wave dynamics (but not always, as in the case of the quantum Toda lattice; Gutzwiller 1980).

Ergodicity is a property possessed by systems where averaging in different natural ways is equivalent. For the classical dynamics of rays, these averages are over the phase space of directions and positions and over long time evolution; integrable ray dynamics such as the example in Figure 4.2 does not possess this property. For ensembles of waves such as those discussed in Section 4.2, these averages are over the ensemble of choices of random phases and directions and over spatial positions $\mathbf{r}$. However, a given, specific mode of a cavity (whose classical ray dynamics is ergodic) is not chosen from an ensemble but is deterministically fixed given the frequency. The first two stadium modes of Figures 4.1, as discussed in Section 4.2, have properties suggestive of typical sample Gaussian random functions. Under the ergodic wave hypothesis, the spatial behaviour of typical modes such as this may be represented by Gaussian random functions.

The origin of this hypothesis lies in deep and abstract mathematical theorems on "quantum ergodicity"[†] due to Shnirelman (1974), Colin de Verdière (1985), and Zelditch (1987) (also see Zelditch & Zworski 1996 for an explicit proof for billiards). In the language of modes of cavities, the theorem states that when the ray dynamics is ergodic, then in the semiclassical limit $k \to \infty$, the modes φ_k have a uniform distribution over the domain; that is, for a subset $A \subseteq D$ (whose volumes/areas are denoted $\mathrm{vol}(A)$, $\mathrm{vol}(D)$),

$$\lim_{k\to\infty} \frac{\int_A |\varphi_k|^2}{\int_D |\varphi_k|^2} = \frac{\mathrm{vol}(A)}{\mathrm{vol}(D)}. \tag{4.18}$$

The mathematical statement of this result is subtle and allows the possibility of excluding a subset of modes (of negligible size in the high-k limit); these excepted modes are understood to include the "scarred eigenfunctions," discussed next.

The formal idea of wave ergodicity from the viewpoint of semiclassical physics was approached by physicists at a similar time. Using ideas from statistical mechanics, Voros (1979) made a conjecture about the ergodicity of the phase-space representation of the semiclassical behaviour of a wave corresponding to an ergodic ray

[†] As elsewhere in this book, "quantum" signifies a statement about wave behavior, rather than "classical" or ray dynamics.

system in d dimensions: the Wigner function

$$W_\varphi(\mathbf{r}, \mathbf{k}) \equiv \int \varphi\left(\frac{\mathbf{r}+\mathbf{r}'}{2}\right) \varphi\left(\frac{\mathbf{r}-\mathbf{r}'}{2}\right) \exp(-\mathrm{i}\mathbf{k}\cdot\mathbf{r}')\, \mathrm{d}^d\mathbf{r}', \tag{4.19}$$

in the limit of high frequency k, is given by

$$W_\varphi(\mathbf{r}, \mathbf{k}) \sim \frac{\delta(k^2 - |\mathbf{k}|^2)}{\int_D \int \delta(k^2 - |\mathbf{k}|^2)\, \mathrm{d}^d\mathbf{k}\, \mathrm{d}^d\mathbf{r}}. \tag{4.20}$$

According to this conjecture, all wavevectors $\mathbf{k}$ of magnitude $|\mathbf{k}| = k$ have equal weight, and the shape of the cavity D does not contribute. In the language of statistical mechanics, Voros conjectured that the phase-space distribution is microcanonical.

Berry (1977) took this conjecture further, proposing that Voros's arguments apply not only in the limit but to a typical mode φ at high frequency k. Thus the right-hand side of Equation (4.20) is taken to represent $\langle W_\varphi \rangle_{\text{space}}$, the Wigner function spatially averaged over an area large compared with $1/k$ but small compared with vol(D), thus removing local fluctuations. Since the integral of the Wigner function $W_\varphi(\mathbf{r}, \mathbf{k})$ over $\mathbf{k}$ represents the local intensity $|\varphi(\mathbf{r})|^2$, the average intensity $\langle |\varphi(\mathbf{r})|^2 \rangle_{\text{space}}$ is independent of $\mathbf{r}$, and the correlation function representing the average correlation at position $\mathbf{r}$ between points separated by $\mathbf{r}'$ follows from the definition (4.19) of the Wigner function,

$$\begin{aligned} C(\mathbf{r}, \mathbf{r}') &= \frac{1}{\langle |\varphi(\mathbf{r})|^2 \rangle_{\text{space}}} \left\langle \int \exp(\mathrm{i}\mathbf{k}\cdot\mathbf{r}') W_\varphi(\mathbf{r}, \mathbf{k})\, \mathrm{d}^d\mathbf{k} \right\rangle_{\text{space}} \\ &= \frac{\int \delta(k^2 - |\mathbf{k}|^2) \exp(\mathrm{i}\mathbf{k}\cdot\mathbf{r}')\, \mathrm{d}^d\mathbf{k}}{\int \delta(k^2 - |\mathbf{k}|^2)\, \mathrm{d}^d\mathbf{k}}. \end{aligned} \tag{4.21}$$

The correlation function is thus independent of $\mathbf{r}$ and is equal to the Fourier transform of a delta function at fixed wavenumber k, corresponding to a circular ring in two dimensions and a spherical shell in three, whose solution is given in terms of the Bessel functions (4.9). As a statistical statement, this is the Wiener–Khinchin theorem (Goodman 1985, Mandel & Wolf 1995) for a random function whose power spectrum is a delta function concentrated on the $(d-1)$-dimensional spherical shell centered on the origin with radius k.

Berry's ergodic mode hypothesis follows from the conjecture that for an ergodic mode with no other structure, this is the strongest statistical statement that can be made: The phases for each wavevector $\mathbf{k}$ of fixed magnitude semiclassically contributing to the mode φ at $\mathbf{r}$ are independent and random, giving a Gaussian random function with the prescribed correlation function and allowing a representation of the form of Equation (4.3). It is isotropic because all wavevectors contribute equally by (4.20), and stationary because averages are independent of position $\mathbf{r}$. Gaussianity gives that all statistics (such as multipoint correlations or statistics of derivatives) are determined purely by the isotropic, two-point correlation function $C(r)$, and equivalence between spatial and ensemble averages follows from dynamic ergodicity and stationarity. This is supported (for the modes shown in Figures 4.1(b) and (c)) by the illustrations of the distribution of values and correlation functions in Section 4.2; this has been supported by many studies at higher frequencies in the literature. It should be noted that the preceding arguments have been paraphrased

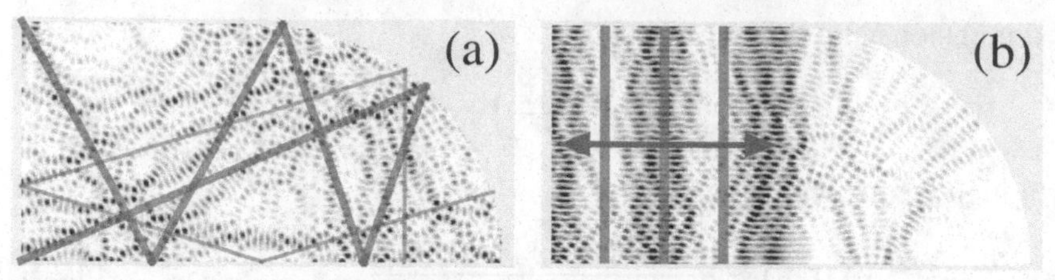

Figure 4.7. Examples of modes from Figure 4.1 and some periodic orbits. (a) Weakly scarred mode, together with some short unstable orbits. (b) Bouncing ball state, with three examples of the corresponding continuous family of periodic orbits.

from those in the original papers to apply simply to the case of cavities and billiards, and the original arguments are more general.

Thus the justification that a typical ergodic eigenfunction in the high-frequency limit is a Bessel-correlated Gaussian random function rests on two assumptions: to the mode at each point, the wavevector directions, corresponding semiclassically to the ergodic ray directions, are uniformly distributed and independent. Without a deeper understanding of any correlations in the contributing semiclassical rays, it is natural to assume the behavior with the least structure, which is Gaussian randomness. This argument may be made more formal (Jarzynsky 1997): for a mode about which nothing is known except its frequency, it is natural to choose from a statistical ensemble of possible functions in the least biased way, and this ensemble may be shown to be the Gaussian random functions. Attempts to justify the random wave model have also emerged from scattering theory approaches and random matrix theory: in the limit of large matrix dimension, the probability distribution of the coefficients of the eigenvectors approaches independent, identically distributed Gaussians (Izrailev 1990, Blum et al. 2002; see also Chapter 3), which gives an ensemble of random functions with Gaussian distribution, regardless of the basis set of functions (i.e., superpositions of solutions of the Helmholtz equation other than plane waves).

The assumption of Gaussian randomness is far from the semiclassical trace formula approach to ergodic waves described elsewhere in this book because the periodic orbits of the underlying dynamics do not play any role. However, as first appreciated by Heller (1984) and subsequently studied and justified (Bogomolny 1988, Berry 1989), a small number of modes (in the technical sense of formal wave ergodicity above) have a spatial morphology rather different from isotropic and homogeneous Gaussian randomness and appear spatially localized near the periodic orbits of rays in the same cavity. This is certainly strongly the case for the mode shown in Figure 4.1(d): This "bouncing ball" mode, within the square part of the quarter stadium shape, looks similar to a separable mode of the square, with a higher frequency in the separated y component than x. The existence of such modes is associated with the continuous family of periodic orbits between the parallel walls in the x-direction; cavities without such symmetries do not have bouncing ball modes.

More common are scars of isolated, unstable periodic orbits: Many modes, to a greater or lesser extent, have stronger concentrations along short periodic orbits. Two such possible scarring periodic orbits for one of the example stadium modes are shown in Figure 4.7(a). I will not discuss the subtle and much-studied phenomenon of scarring in ergodic modes here; however, as observed by O'Connor et al. (1987), isotropic superpositions of plane waves with random directions and phases themselves are localized on scar-like filamentary structures, despite there being no underlying classical dynamics. These structures are clear in Figure 4.3, at approximately 30° and 60°. They are not observed in isotropic random superpositions with correlation functions (such as that of the speckle pattern shown in Figure 4.6). This has been argued to be a consequence of the slow decay of $J_0(kr)$ (O'Connor et al. 1987). Methods have been proposed to construct the modes of a cavity purely based around scarring around periodic orbits, using so-called scar functions (see Chapter 5 of the current book and references therein).

In the preceding discussion and arguments for Gaussian randomness, isotropy, and stationarity, it was assumed that ergodic modes have no overall structure: on average, the mode at every point in space is the same as every other point. However, real modes satisfy boundary conditions, and modes (and open cavities) close to boundaries have been confirmed to behave similar to the appropriate boundary-adapted random wave model (e.g., Berry & Ishio 2005, Höhmann et al. 2009). In fact, in the literature, this approach is typical of the way the ergodic mode hypothesis has been investigated: a particular property of Gaussian random waves, related to an analytic property of modes – such as the distribution of nodal domains in two dimensions, which is known to satisfy Courant's theorem (Blum et al. 2002) or the maximum intensity of a normalized mode (Aurich et al. 1999) – is both studied for a chosen relevant random wave model and compared with numerically computed modes in an ergodic cavity. Numerical approaches are necessary, as analytic determination of modes is beyond present techniques in mathematical analysis, and the proof of general theorems in quantum ergodicity is difficult and highly technical.

Further semiclassical arguments have been made to modify the basic isotropic random wave model for ergodic modes, on the basis of the Green's function for the cavity. These include a modification of the J_0 correlation function for long distances (Hortikar & Srednicki 1998) and an approach to include the effects of boundary walls and corners (Urbina & Richter 2004) or elasticity (Akolzin & Weaver 2004). However, space limits further discussion of these topics here. I will conclude this brief review with an acoustic application of the random wave model: a justification of Sabine's formula.

4.5 Ergodic Wavefunctions and Sabine's Formula for Reverberation Time

The ergodic mode hypothesis can be used to derive one of the most basic results in acoustics, *Sabine's law* for the decay time (reverberation time) in an ergodic enclosure. This approach was originally suggested by M. V. Berry in 1980† and proved in a different way than here by Legrand and Sornette (1991). According to Sabine's law, in an ergodic cavity D with weakly absorbing walls, the decay of sound intensity

† I am grateful to Michael Berry for discussions on this topic, as well as showing me his unpublished notes.

is exponential $\exp(-t/\tau)$, with decay time τ given by

$$\tau_{2D} = \frac{\pi c A}{\alpha P}, \qquad \tau_{3D} = \frac{4cV}{\alpha S}, \tag{4.22}$$

where α is a constant absorption coefficient, and in two dimensions, D has area A and perimeter length P and in three dimensions, volume V and surface area S. Of course, this law represents an ideal from which real auditoriums deviate. Nevertheless, the interplay between auditorium ergodicity (ensuring echoes become uniformly distributed in space and direction) and weak absorption (ensuring the reverberation time is neither too long nor too short) has driven architectural acoustics since Sabine himself (Morse & Ingard 1968).

Sabine's law was derived in the framework of geometrical acoustics by Joyce (1975), by assuming ergodicity of ray reflections in a closed cavity with weakly absorbing walls. He characterized the geometric rays as the trajectories of "phonons" reflecting specularly from the walls.† The phonons are assumed to travel at speed c, with α representing the probability of absorption at the wall. Joyce's derivation follows from the fact that the mean time between reflections is the mean path length of the phonon divided by the speed v. Assuming ergodicity, this mean path length is simply the mean chord length in the cavity (in the phase space measure), that is, the average straight line length within the cavity between two points on the boundary. From kinetic theory of gases (as argued by Joyce) or geometric probability (Solomon 1978) this mean chord length is $4V/S$ in three dimensions and $\pi A/P$ in two. The decay time (4.22) follows as the mean time (mean chord length/speed), divided by the absorption α.

As discussed earlier, the heart of the ergodic wave hypothesis is the correspondence principle – ray and wave properties are equivalent in the semiclassical limit – and chaos. It is therefore not surprising that wave ergodicity should be able to reproduce the decay times τ (4.22) and with a similar degree of simplicity to the above-given argument for ray ergodicity. The following argument is given here in entirety for completeness, although the first part (at least up to Equation (4.31)) is rather standard in the acoustics literature (for example, Morse & Ingard 1968).

The decay time of the sound wavefield ψ is related to a complex angular frequency $\eta = \omega' - \mathrm{i}\delta$; so the intensity $|\psi|^2$ is proportional to $\exp(-\mathrm{i}\eta t) \propto \exp(-2\delta t)$, and τ is the reciprocal of two times δ, the imaginary part of the angular frequency. The proof is based on a perturbation argument: the physical complex wavefield ψ is a perturbation of a real ergodic mode φ by small absorption, represented by small specific admittance (reciprocal impedance)

$$\beta = \rho c \left(\frac{u_n}{p}\right)_{\text{on wall}} = -\mathrm{i}\left.\frac{c\partial_n \psi}{\omega \psi}\right|_{\partial D}. \tag{4.23}$$

The unperturbed mode φ, being a pure mode of the ideal nonabsorbing cavity D, satisfies Neumann boundary conditions on the boundary ∂D and has characteristic frequency ω,

$$\nabla^2 \varphi + \frac{\omega^2}{c^2}\varphi = 0 \quad \text{inside } D, \qquad \partial_n \varphi = 0 \quad \text{on } \partial D. \tag{4.24}$$

† Not the sense of a phonon as a quantized lattice vibration in quantum condensed matter, but rather as the acoustic analog to a Newtonian light corpuscle that follows an optical ray.

The problem is thus to find the imaginary frequency correction for a non-Hermitian perturbation of a Hermitian boundary problem and follows the spirit of Morse and Feshbach (1953, part 2, chapter 9) and Morse and Ingard (1968).

Further, ψ itself satisfies the perturbed equation

$$\nabla^2\psi + \frac{\eta^2}{c^2}\psi = 0 \quad \text{inside } D, \qquad \partial_n\psi = \mathrm{i}\beta\frac{\omega}{c}\psi \quad \text{on } \partial D, \tag{4.25}$$

where β is assumed small. To first order in the perturbation, the imaginary part of the eigenvalue η^2 is $2\omega\delta$. Using standard perturbation theory of the boundary condition, this is

$$\Im\left\{\frac{\oint_{\partial D}\psi^*\partial_n\psi}{\int_D|\psi|^2}\right\} \approx \frac{\omega}{c}\Re\{\beta\}\frac{\oint_{\partial D}|\varphi|^2}{\int_D|\varphi|^2}, \tag{4.26}$$

where the integral in the denominator is over the area/volume of D and the numerator the integral over the its boundary ∂D. Therefore

$$\frac{1}{\tau} = 2\delta = \frac{\Re\{\beta\}}{c}\frac{\oint_{\partial D}|\varphi|^2}{\int_D|\varphi|^2}. \tag{4.27}$$

The reflection coefficient for a wall of D, depending on incident direction θ, is

$$\alpha(\theta) = 1 - \left|\frac{\cos\theta - \beta}{\cos\theta + \beta}\right| \approx \frac{4\Re\{\beta\}}{\cos\theta}, \tag{4.28}$$

assuming $\beta \ll \cos\theta$ (i.e., non-glancing incidence). The absorption coefficient α is the average of $\alpha(\theta)$ over incident projection directions,

$$\alpha = \frac{\int\cos\theta\,\alpha(\theta)\,\mathrm{d}\Omega}{\int\cos\theta\,\mathrm{d}\Omega}, \tag{4.29}$$

where $\mathrm{d}\Omega$ is the (solid) angle differential element over half-space: in two dimensions, $\mathrm{d}\Omega = \mathrm{d}\theta$, for $-\pi/2 \le \theta \le \pi/2$, and in three, $\mathrm{d}\Omega = \sin\theta\,\mathrm{d}\phi\,\mathrm{d}\theta$ for $0 \le \phi \le 2\pi, 0 \le \theta \le \pi/2$. Thus (Morse & Ingard 1968)

$$\alpha = 2\pi\Re\{\beta\} \quad \text{in two dimensions,} \qquad \alpha = 8\Re\{\beta\} \quad \text{in three dimensions.} \tag{4.30}$$

Thus (Morse & Ingard 1968, Legrand & Sornette 1991)

$$\tau_{2\mathrm{D}} = \frac{2\pi c}{\alpha}\frac{\int_D|\varphi|^2}{\oint_{\partial D}|\varphi|^2}, \qquad \tau_{3\mathrm{D}} = \frac{8c}{\alpha}\frac{\int_D|\varphi|^2}{\oint_{\partial D}|\varphi|^2}. \tag{4.31}$$

To justify the exact Sabine expressions, we turn to the ergodic wave hypothesis. Assuming that the mode φ has the properties discussed in previous sections, the volume integral of the intensity, $\int_D|\varphi|^2$, is proportional by ergodicity to the volume of the cavity (or area in two dimensions) times the ensemble average of the intensity in the interior (far from the boundary), $\langle\varphi^2\rangle_{\text{space}}$. The ergodic mode φ satisfies Neumann conditions at the boundary, on which $\langle|\varphi|^2\rangle_{\text{boundary}} = 2\langle|\varphi|^2\rangle_{\text{space}}$ by Equation (4.14). Thus the ratio of integrals in Equation (4.31) can be replaced by 1/2 times V/S or A/P, yielding Sabine's decay times (4.22). The assumptions are the same as those of Joyce: the absorption is weak (and so may be treated as a perturbation of a Hermitian problem), and the dynamics is ergodic.

Legrand and Sornette (1991) justified this by explaining the factor of 1/2 as originating from correct consideration of the reflection coefficient. Further, they numerically studied Sabine-type reverberation in the stadium billiard, by substituting explicit values of the average intensity on the boundary to average intensity inside the cavity; for the stadium example modes shown in Figure 4.1, this ratio is (b) 0.505, (c) 0.524, and (d) 0.549. The mean ray length of trajectory in Figure 4.1(a) is 1.026, compared with the mean chord length of 1.007.

5 Short Periodic Orbit Theory of Eigenfunctions

Eduardo G. Vergini
Departamento de Física, E.T.S.I. Agrónomos, Universidad Politécnica de Madrid, Madrid, Spain; Departamento de Física, Comisión Nacional de Energía Atómica, Buenos Aires, Argentina

Gabriel G. Carlo
Departamento de Física, Comisión Nacional de Energía Atómica, Buenos Aires, Argentina

The short periodic orbit (PO) approach was developed in order to understand the structure of stationary states of quantum autonomous Hamiltonian systems corresponding to a classical chaotic Hamiltonian. In this chapter, we will describe the method for the case of a two-dimensional chaotic billiard where the Schrödinger equation reduces to the Helmholtz equation; then, it can directly be applied to evaluate the acoustic eigenfunctions of a two-dimensional cavity. This method consists of the short-wavelength construction of a basis of wavefunctions related to unstable short POs of the billiard, and the evaluation of matrix elements of the Laplacian in order to specify the eigenfunctions.

5.1 Introduction

The theoretical study of wave phenomena in systems with irregular motion received a big impetus after the works by Gutzwiller (summarized in Gutzwiller 1990). He derived a semiclassical approach providing the energy spectrum of a classically chaotic Hamiltonian system as a function of its POs. This formalism is very efficient for the evaluation of mean properties of eigenvalues and eigenfunctions (Berry 1985, Bogomolny 1988), but it suffers from a very serious limitation when a description of individual eigenfunctions is required: the number of used POs proliferates exponentially with the complexity of the eigenfunction. In this way, the approach loses two of the common advantages of asymptotic techniques: simplicity in the calculation and, more important, simplicity in the interpretation of the results.

Based on numerical experiments in the Bunimovich stadium billiard (Vergini & Wisniacki 1998), we have derived a short PO approach (Vergini 2000), which was successfully verified in the stadium billiard (Vergini & Carlo 2000): the first 25 eigenfunctions were computed by using five periodic orbits. This formalism allows us to obtain all wave mechanical information of a chaotic Hamiltonian system in terms of a very small number of short POs. The essence of the method consists of the construction of wave functions related to short unstable POs and the evaluation of matrix elements between these wavefunctions. Moreover, in 2001 we have improved the wavefunction construction by the inclusion of transverse excitations (Vergini & Carlo 2001).

When the dynamical system is a billiard, quantum eigenfunction problems require nontrivial solutions of the Helmholtz equation

$$(\Delta + k^2)u = 0, \tag{5.1}$$

satisfying certain boundary conditions. These solutions, characterized by the wavenumber k, also correspond to the eigenfunctions of an acoustic cavity. In this respect, it is worth noticing that even though the ray method (Babic & Buldyrev 1991) provides asymptotic expansions of solutions to boundary value problems for (5.1), it cannot be applied in the case of chaotic billiards. For these billiards, the motion is irregular, and the only geometrical objects building some structure are the unstable POs. So, the first step of an asymptotic approach in chaotic billiards should be in connection with POs.

5.2 Description of Trajectories in the Vicinity of a Periodic Orbit

The objective of this section is to decompose the evolution of trajectories (or rays) in the neighborhood of an unstable PO γ of the billiard, in terms of a periodic evolution and a hyperbolic one; this is the content of the Floquet theorem.

Let x be a local coordinate along γ. Then, a neighboring trajectory δ is characterized at each value of x by the transverse coordinate y and its conjugated variable $p_y = \sin\phi$, with ϕ the angle between γ and δ so that p_y is the transverse momentum in units of the momentum of the PO. The evolution of y and p_y from x_1 to x_3, with a bounce at an intermediate point x_2, is described in the first approximation by the symplectic matrix $\mathbf{M}(x_3, x_1)^\dagger$ as follows:

$$\begin{pmatrix} y(x_3) \\ p_y(x_3) \end{pmatrix} = \mathbf{M}(x_3, x_1) \begin{pmatrix} y(x_1) \\ p_y(x_1) \end{pmatrix}, \tag{5.2}$$

with $\mathbf{M}(x_3, x_1) = \mathbf{M}_f(x_3 - x_2)\mathbf{M}_b(\rho_2, \theta_2)\mathbf{M}_f(x_2 - x_1)$. The elementary matrix $\mathbf{M}_f$ describes a free evolution, whereas $\mathbf{M}_b$ considers a bounce with the boundary

$$\mathbf{M}_f(l) = \begin{pmatrix} 1 & l \\ 0 & 1 \end{pmatrix}, \quad \mathbf{M}_b(\rho, \theta) = \begin{pmatrix} -1 & 0 \\ \frac{2}{\rho\cos(\theta)} & -1 \end{pmatrix}, \tag{5.3}$$

with ρ the radius of curvature of the billiard boundary at the bounce‡ and θ the angle between the incoming ray of the PO and the normal to the boundary (see Figure 5.1).

Let $\mathbf{M}(x)$ be the matrix describing the evolution of neighboring trajectories defined at the origin $x = 0$. Then, the evolution along the complete orbit is given by

† A 2×2 matrix $\mathbf{M}$ is symplectic if and only if det $(\mathbf{M}) = 1$; in such a case, $\mathbf{M}$ preserves symplectic areas.

‡ By convention, we take ρ positive when the curvature of the boundary is like the one shown in Figure 5.1, whereas for the opposite curvature ρ is negative.

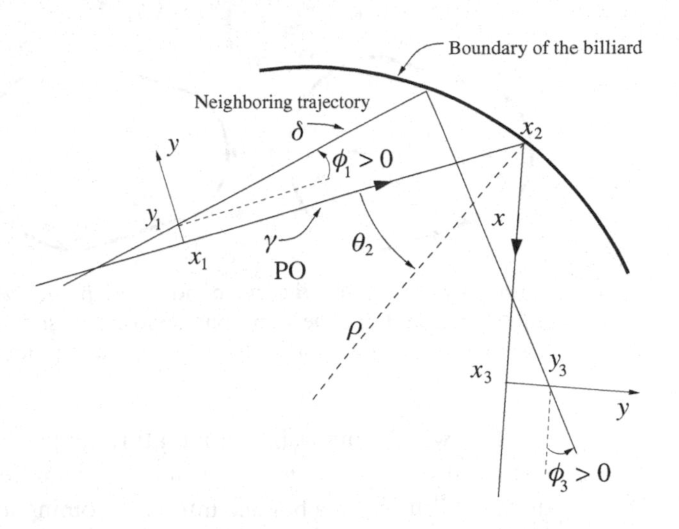

Figure 5.1. Evolution of a given orbit γ and its neighbor δ. We indicate the coordinates sets before, during, and after the bounce by the subscripts 1, 2, and 3. Then, θ is the reflection angle with respect to the normal to the boundary, ρ is the radius of curvature at the hitting point, and ϕ is the angle formed by the neighbouring trajectory with the orbit.

$\mathbf{M}(L)$, with L the length of γ; notice that x being periodic along γ, $x = 0$ and $x = L$ are the same point. This matrix is called the Poincaré return map or transverse monodromy matrix. Its eigenvectors $\boldsymbol{\xi}_u$ and $\boldsymbol{\xi}_s$, which are real vectors, define the unstable and stable directions, respectively. The corresponding eigenvalues are real numbers with absolute value $\exp(\lambda L)$ and $\exp(-\lambda L)$, where the positive number λ, which is measured in units of [length^{-1}], is the stability index of the PO.

The evolution of the unstable direction along the PO is described by $\tilde{\boldsymbol{\xi}}_u(x) = \mathbf{M}(x)\boldsymbol{\xi}_u$. While evolving, this vector dilates and rotates clockwise in the $y - p_y$ plane; in particular, it rotates discontinuously by around half a turn at a bounce (see Figure 5.6). Let μ be the total number of half turns made by $\tilde{\boldsymbol{\xi}}_u(x)$ during its evolution along the orbit. For even μ, the vector returns to the initial point $x = 0$ on the trajectory with the same direction and sense, and the eigenvalue of $\boldsymbol{\xi}_u$ is positive. For odd μ, its sense changes and the eigenvalue is negative; in this case the motion is called hyperbolic with reflection. We can write

$$\tilde{\boldsymbol{\xi}}_u(L) = (-1)^\mu e^{\lambda L} \boldsymbol{\xi}_u. \tag{5.4}$$

On the other hand, the vector $\tilde{\boldsymbol{\xi}}_s(x) = \mathbf{M}(x)\boldsymbol{\xi}_s$ contracts and rotates following $\tilde{\boldsymbol{\xi}}_u(x)$. This rotation is not rigid; the angle between the stable and unstable vectors varies during the evolution. However, the final cumulative rotation of the stable manifold is also μ, and we have

$$\tilde{\boldsymbol{\xi}}_s(L) = (-1)^\mu e^{-\lambda L} \boldsymbol{\xi}_s. \tag{5.5}$$

Because the stable and unstable manifolds rotate during their evolution along the trajectory, there is some value $x = x_0$ where these directions are symmetrical with respect to the axis, that is, where their slopes in the $y - p_y$ plane are the same in absolute value but with opposite sign. A simple condition in order to find x_0 is to require that the two diagonal elements of the return map starting at $x = x_0$ are the same, with this map easily derived from $\mathbf{M}(x)$ as follows:

$$\mathbf{M}_{x_0} = \mathbf{M}(x_0)\mathbf{M}(L)\mathbf{M}(x_0)^{-1}. \tag{5.6}$$

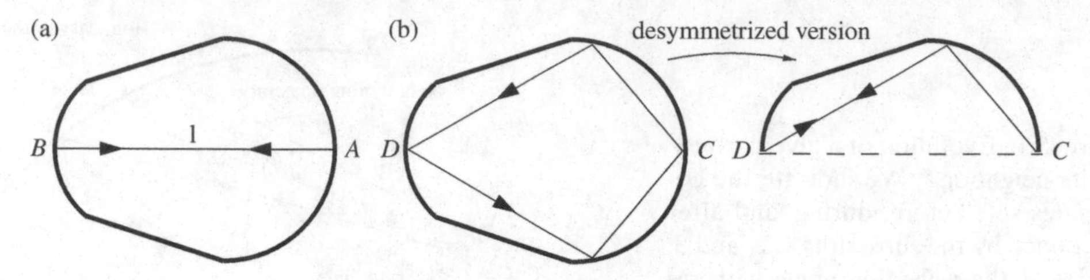

Figure 5.2. (a) A and B correspond to the points where the slopes of the unstable and the stable manifold are the same but opposite in sign. (b) For the case of orbits that after a desymmetrization become self-retracting these points correspond to C and D.

It is worth emphasizing that a turning point satisfies the previous requirement. Let us verify this statement for the two examples of Figure 5.2. We first decompose the contribution of a bounce into an incoming and outgoing contribution,

$$\mathbf{M}_b(\rho, \theta) = \mathbf{M}_b^{(\mathrm{out})}(\rho, \theta)\mathbf{M}_b^{(\mathrm{in})}(\rho, \theta), \tag{5.7}$$

with

$$\mathbf{M}_b^{(\mathrm{in})}(\rho, \theta) = -\mathbf{M}_b^{(\mathrm{out})}(\rho, \theta) = \begin{pmatrix} -1 & 0 \\ \frac{1}{\rho\cos(\theta)} & -1 \end{pmatrix}. \tag{5.8}$$

Then, the evolution for going from x_A to x_B is given by

$$\mathbf{M}(x_B, x_A) = \mathbf{M}_b^{(\mathrm{in})}(\rho_B, 0)\mathbf{M}_f(l)\mathbf{M}_b^{(\mathrm{out})}(\rho_A, 0) = \begin{pmatrix} -1 + \frac{l}{\rho_A} & -l \\ \frac{1}{\rho_A} + \frac{1}{\rho_B} - \frac{l}{\rho_A\rho_B} & -1 + \frac{l}{\rho_B} \end{pmatrix}, \tag{5.9}$$

whereas $\mathbf{M}(x_A, x_B)$ takes the same matrix elements but with the diagonal ones interchanged. In the same way, for the starting point C, which is a turning point of the desymmetrized PO, matrices $\mathbf{M}(x_D, x_C)$ and $\mathbf{M}(x_C, x_D)$ are the same but with the diagonal elements interchanged. Generalizing these observations, when the evolution along the PO can be decomposed into two contributions, one being the time reversal of the other, the return map takes the form

$$\mathbf{M}_{x_0} = \begin{pmatrix} d & b \\ c & a \end{pmatrix}\begin{pmatrix} a & b \\ c & d \end{pmatrix} = \begin{pmatrix} ad + bc & 2bd \\ 2ac & ad + bc \end{pmatrix}. \tag{5.10}$$

We redefine the origin $x = 0$ at one of these points x_0. With this choice and using the eigenvalues given in Equations (5.4) and (5.5), it is not difficult to see that the monodromy matrix $\mathbf{M}(L)$ acquires the form

$$\mathbf{M}(L) = (-1)^{\mu}\begin{pmatrix} \cosh(\lambda L) & \sinh(\lambda L)/\tan(\varphi) \\ \sinh(\lambda L)\tan(\varphi) & \cosh(\lambda L) \end{pmatrix}, \tag{5.11}$$

where $\tan(\varphi)$ represents the slope of the unstable manifold in the plane $y - p_y$ (whereas the slope of the stable manifold is $-\tan(\varphi)$). In particular, if the starting

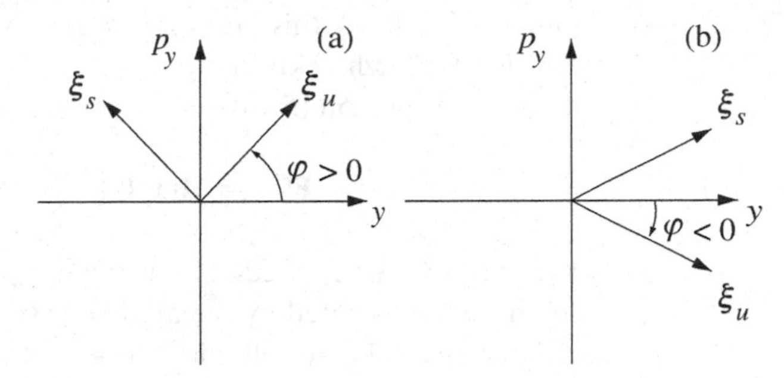

Figure 5.3. Unstable and stable directions ($\boldsymbol{\xi}_u$ and $\boldsymbol{\xi}_s$, respectively) at points x_0. The two possible signs of their slope are shown: $\varphi > 0$ and $\varphi < 0$.

point is a turning point we can use (5.10) in order to obtain

$$\lambda = \frac{1}{L}\ln\left(|A| + \sqrt{A^2 - 1}\right) \quad \text{and} \quad \tan(\varphi) = \text{sign}\,(acA)\sqrt{\left|\frac{ac}{bd}\right|}, \tag{5.12}$$

with $A \equiv ad + bc$.

In general it is impossible to compare vectors living in the plane $y - p_y$ because the axes have different units; of course, it is possible to introduce a norm, but in such a case the comparison depends on the arbitrary selected norm. However, when the directions of the vectors to be compared are symmetrical with respect to the axes it is only necessary to compare one component; this is the reason for the selection of x_0 as the origin. Then, we define the vectors $\boldsymbol{\xi}_u$ and $\boldsymbol{\xi}_s$ on the unstable and stable directions in such a way that their components are equal in absolute value, whereas the symplectic area defined by them is equal to the unity of length† (Figure 5.3 shows the two possibilities depending on the sign of φ):

$$\boldsymbol{\xi}_u \wedge \boldsymbol{\xi}_s = J. \tag{5.13}$$

The symplectic matrix B transforming coordinates along the new axes $\{\boldsymbol{\xi}_u, \boldsymbol{\xi}_s\}$ into the old ones and its inverse are

$$\mathbf{B} = (\boldsymbol{\xi}_u \boldsymbol{\xi}_s) = \frac{1}{\sqrt{2}}\begin{pmatrix} 1/\alpha & -s/\alpha \\ s\alpha & \alpha \end{pmatrix}, \quad \mathbf{B}^{-1} = \frac{1}{\sqrt{2}}\begin{pmatrix} \alpha & s/\alpha \\ -s\alpha & 1/\alpha \end{pmatrix} \tag{5.14}$$

with $\alpha = \sqrt{|\tan(\varphi)|}$ and $s = \text{sign}(\varphi)$; we took $\boldsymbol{\xi}_u$ as the first vector of the new basis by convention. In fact, a straightforward calculation using Equations (5.11) and (5.14) shows that

$$\mathbf{B}^{-1}\mathbf{M}(L)\mathbf{B} = (-1)^{\mu}\begin{pmatrix} \mathrm{e}^{\lambda L} & 0 \\ 0 & \mathrm{e}^{-\lambda L} \end{pmatrix} \tag{5.15}$$

in accordance with the eigenvalues of $\mathbf{M}(L)$ (see Equations (5.4) and (5.5)). Now, we decompose $\mathbf{M}(x)$ into a periodic matrix $\mathbf{F}(x)$ describing the directions of the manifolds and a matrix (depending in a simple way of x) describing the exponential contraction-dilation along the manifolds,

$$\mathbf{M}(x) = \mathbf{F}(x)\mathbf{B}\begin{pmatrix} \mathrm{e}^{\lambda x} & 0 \\ 0 & \mathrm{e}^{-\lambda x} \end{pmatrix}\mathbf{B}^{-1} = \mathbf{F}(x)\begin{pmatrix} \cosh(\lambda x) & s\sinh(\lambda x)/\alpha^2 \\ s\alpha^2\sinh(\lambda x) & \cosh(\lambda x) \end{pmatrix}. \tag{5.16}$$

† The unit of length is the only one relevant in this chapter; for instance, $J = 1$ cm.

We emphasize that this decomposition represents the Floquet theorem (Yakubovich & Starzhinskii 1975).

Of course, Equation (5.16) defines $\mathbf{F}(x)$ in terms of $\mathbf{M}(x)$,

$$\mathbf{F}(x) \equiv \mathbf{M}(x)\mathbf{B}\begin{pmatrix} \mathrm{e}^{-\lambda x} & 0 \\ 0 & \mathrm{e}^{\lambda x} \end{pmatrix}\mathbf{B}^{-1}, \tag{5.17}$$

and by using (5.15) we immediately see that $F(L) = (-1)^{\mu}\mathbf{I}$, with $\mathbf{I}$ the identity matrix. The motion described by the modified symplectic matrix $\mathbf{F}(x)$ is very unusual: it is neither hyperbolic nor elliptic. It is a neutral motion where all the trajectories, which are not given in terms of straight lines, are also periodic with period L for μ even and $2L$ for μ odd.

We finish this section by defining the following complex vector whose components play a key role in the next section:

$$\begin{pmatrix} Q(x) \\ P(x) \end{pmatrix} \equiv \mathbf{F}(x)(\boldsymbol{\xi}_u + \mathrm{i}\boldsymbol{\xi}_s) = \mathbf{F}(x)\mathbf{B}\begin{pmatrix} 1 \\ \mathrm{i} \end{pmatrix} = \mathbf{M}(x)\mathbf{B}\begin{pmatrix} \mathrm{e}^{-\lambda x} \\ \mathrm{i}\mathrm{e}^{\lambda x} \end{pmatrix}, \tag{5.18}$$

where the last two relations result from Equations (5.14) and (5.17), respectively. Using (5.18), the fact that $\mathbf{F}(x)$ preserves symplectic areas and (5.13), we obtain the normalization

$$Q^*(x)P(x) - Q(x)P^*(x) = 2\mathrm{i}\mathbf{F}(x)\boldsymbol{\xi}_u \wedge \mathbf{F}(x)\boldsymbol{\xi}_s = 2\mathrm{i}\boldsymbol{\xi}_u \wedge \boldsymbol{\xi}_s = 2\mathrm{i}J, \tag{5.19}$$

and from this it results

$$\frac{P(x)}{Q(x)} = g(x) + \mathrm{i}\frac{J}{|Q(x)|^2}, \tag{5.20}$$

where $g(x) = [Q(x)P^*(x) + Q^*(x)P(x)]/2|Q(x)|^2$ is a real function. Moreover, the derivative of the complex vector with respect to x is given by

$$\frac{\mathrm{d}}{\mathrm{d}x}\begin{pmatrix} Q(x) \\ P(x) \end{pmatrix} = \begin{pmatrix} 0 & 1 \\ 0 & 0 \end{pmatrix}\mathbf{M}(x)\mathbf{B}\begin{pmatrix} \mathrm{e}^{-\lambda x} \\ \mathrm{i}\mathrm{e}^{\lambda x} \end{pmatrix} + \mathbf{M}(x)\mathbf{B}\begin{pmatrix} -\lambda\mathrm{e}^{-\lambda x} \\ \mathrm{i}\lambda\mathrm{e}^{\lambda x} \end{pmatrix}, \tag{5.21}$$

with the derivative of $\mathbf{M}(x)$ obtained from $\mathbf{M}(\mathrm{d}x + x) = \mathbf{M}_f(\mathrm{d}x)\mathbf{M}(x)$. By using (5.18) and its complex conjugate it takes the simple form

$$\frac{\mathrm{d}}{\mathrm{d}x}\begin{pmatrix} Q(x) \\ P(x) \end{pmatrix} = \begin{pmatrix} P(x) \\ 0 \end{pmatrix} - \lambda\begin{pmatrix} Q^*(x) \\ P^*(x) \end{pmatrix}. \tag{5.22}$$

5.3 Short-Wavelength Construction of Resonances of Periodic Orbits

In this section we are going to construct a family of wavefunctions associated with an unstable PO, γ. The members of this family are labeled with the number n of excitations along the orbit, and this number defines the wavenumber of the resonance through a rule of the Bohr–Sommerfeld type.

In the previous section we have shown that the transverse motion in the neighborhood of a PO can be decomposed into a periodic one specified by $\mathbf{F}(x)$ and a pure hyperbolic motion (see Equation (5.16)). For the present construction, we will drop the pure hyperbolic motion. Then, we propose as a general recipe for the construction of resonances of γ the product of two functions. One is given by a plane wave e^{ikx}, whereas the other describes the evolution of a transverse wave packet

$$\psi(x, y) = \frac{1}{\sqrt{Q(x)}} \exp\left[ikx + ik\frac{y^2}{2}\frac{P(x)}{Q(x)}\right], \tag{5.23}$$

with $Q(x)$ and $P(x)$ defined in (5.18).

In first place we note that according to (5.20) the real part of the exponent is given by $-kJy^2/2|Q(x)|^2$; so the range of the transverse variable y is of order $k^{-1/2}$. With this in mind, we evaluate partial derivatives of ψ up to order k,

$$\frac{\partial^2}{\partial x^2}\psi(x, y) = -\left[k^2 + k^2y^2\frac{\mathrm{d}}{\mathrm{d}x}\frac{P(x)}{Q(x)} + \frac{ik}{Q(x)}\frac{\mathrm{d}}{\mathrm{d}x}Q(x) + O(1)\right]\psi(x, y), \tag{5.24}$$

$$\frac{\partial^2}{\partial y^2}\psi(x, y) = -\left[k^2y^2\frac{P^2(x)}{Q^2(x)} - ik\frac{P(x)}{Q(x)}\right]\psi(x, y), \tag{5.25}$$

and using Equation (5.22) the action of the Laplacian in leading order results

$$(\Delta + k^2)\psi(x, y) = -i\lambda k\frac{Q^*(x)}{Q(x)}\left(\frac{2ky^2J}{|Q(x)|^2} - 1\right)\psi(x, y) + O(1). \tag{5.26}$$

This result shows that non-diagonal matrix elements of the Laplacian are of order k in the case of unstable POs, whereas it is not difficult to show that they are of order 1 for stable POs. As a consequence of this fact, although wavefunctions of stable POs are able to support eigenfunctions (Babic & Buldyrev 1991) (the so-called eigenmodes in the literature), in the case of unstable POs it is necessary to include a linear combination of resonances of different POs in order to describe an eigenfunction. This point is clarified at the end of the section.

Resonances are constructed by associating the short-wavelength expression of Equation (5.23) to each straight line that constitute the PO. In Figure 5.4 a straight line specified by (x_1, y_1), with $x_1 = x$ inside the billiard, crosses the boundary at point A. Then, the continuation of the PO is provided by a second straight line with coordinates (x_2, y_2); also $x_2 = x$ inside the billiard. In a vicinity of order $k^{-1/2}$ around the point A, as the two lines contribute to the wavefunction we include the corresponding expressions as follows:

$$\psi_1(x_1, y_1) + e^{i\beta}\psi_2(x_2, y_2), \tag{5.27}$$

where the relative phase β between them should be selected to verify boundary conditions.

In order to find β let us write the coordinates of a point B in terms of the coordinate q along the boundary (with origin at A). To do that we use an intermediate Cartesian coordinate (t, n), tangent and normal to the boundary at A. Then, as $q_B = \alpha\rho_A$, $t_B = \sin\alpha\rho_A$ and $n_B = (\cos\alpha - 1)\rho_A$, it results

$$t_B = q_B + O(k^{-3/2}) \quad \text{and} \quad n_B = -\frac{q_B^2}{2\rho_A} + O(k^{-2}). \tag{5.28}$$

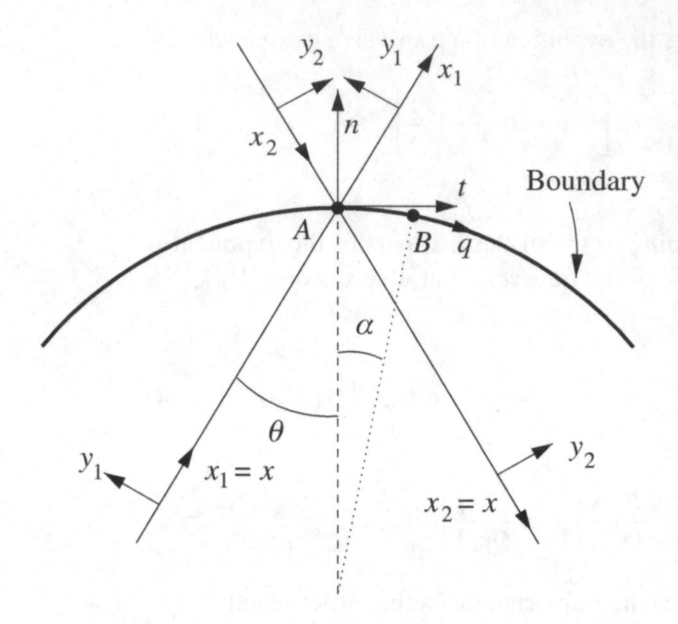

Figure 5.4. The straight lines specified by (x_1, y_1), with $x_1 = x$ inside the billiard, and (x_2, y_2) (also $x_2 = x$ inside the billiard) construct the PO. They hit the boundary at A, but the function can also be calculated outside of it. The tangential and normal coordinates at A are t and n respectively. The reflection angle is indicated by θ with respect to the normal to the boundary, and α is the angle defining a point B with respect to the normal. Finally, q is a coordinate along the boundary.

Later, coordinates (x_1, y_1) and (x_2, y_2) at B are given by

$$x_1 = x_A + \sin\theta q_B - \cos\theta \frac{q_B^2}{2\rho_A} + O(k^{-3/2}), \quad y_1 = -\cos\theta q_B + O(k^{-1}), \tag{5.29}$$

$$x_2 = x_A + \sin\theta q_B + \cos\theta \frac{q_B^2}{2\rho_A} + O(k^{-3/2}), \quad y_2 = \cos\theta q_B + O(k^{-1}). \tag{5.30}$$

Moreover, taking into account that we are going to compute (5.27) up to order 1, it is sufficient to evaluate Q and P at A. In particular, as $Q_2(x_A)$ and $P_2(x_A)$ are related with $Q_1(x_A)$ and $P_1(x_A)$ by the bounce at A, it results

$$Q_2(x_A) = -Q_1(x_A) \quad \text{and} \quad P_2(x_A) = \frac{2}{\rho_A \cos\theta} Q_1(x_A) - P_1(x_A). \tag{5.31}$$

We notice that the vector $(Q_1(x_1), P_1(x_1))$ is continuous along the first straight line, which is not limited to the segment inside the billiard; likewise for $(Q_2(x_2), P_2(x_2))$ along the second line.

Then, by substituting (5.29)–(5.31) in (5.27) we obtain

$$\frac{(1 - \mathrm{i}e^{\mathrm{i}\beta})}{\sqrt{Q_1(x_A)}} \exp\left[\mathrm{i}kx_A + \mathrm{i}k \left(\sin\theta q_B - \frac{\cos\theta q_B^2}{2\rho_A} + \frac{\cos^2\theta q_B^2 P_1(x_A)}{2Q_1(x_A)} \right) \right] + O(k^{-1/2}). \tag{5.32}$$

So, in order to cancel the first term and to provide Dirichlet boundary conditions up to order $k^{-1/2}$, we have to select $\beta = -\pi/2$.

For the evaluation of the relative phase in the case of Neumann boundary conditions it is necessary first to evaluate the boundary normal derivative of (5.27) and then to require that the leading order cancels. Working as previously, it is possible to verify that in this case $\beta = \pi/2$.

Now, we are in condition to specify the Bohr–Sommerfeld rule providing the possible wavenumbers k. Taking into account that the variable x is periodic with period L, the resonance at $x = 0$ should be continuous. But, as the complex numbers $Q(x)$ and $P(x)$ are periodic up to a global phase by construction $[(Q(0), P(0)) = (-1)^{\mu}(Q(L), P(L))]$, the continuity condition is satisfied if the accumulated phase after one period is a multiple of 2π. So, the admitted wavenumbers are labeled by the nonnegative integer n according to the following rule:

$$k_n L - \mu\frac{\pi}{2} - (N_D - N_N)\frac{\pi}{2} = 2\pi n, \tag{5.33}$$

with N_D (N_N) the number of bounces satisfying Dirichlet (Neumann) boundary conditions. The term kL is the phase accumulated by the first term of the exponent in Equation (5.23); the second term of the exponent does not accumulate phase. The term $-\mu\pi/2$ is the phase accumulated by the factor $1/\sqrt{Q(x)}$; although manifold directions rotate clockwise in the plane $y - p_y$, $Q(x)$ rotates counterclockwise in the complex plane. The last term of (5.33) is the phase accumulated by boundary conditions.

In conclusion, a resonance with n wavelengths (or excitations) along γ has a wavenumber defined by (5.33), and it consists of a sum of expressions like (5.23) with relative phases between consecutive contributions specified by boundary conditions. The normalization is simply obtained in leading order by integration along the transverse direction y (using Equation (5.20)) and then along x as follows:

$$\int |\psi(x, y)|^2 \,\mathrm{d}x\,\mathrm{d}y \simeq \int_0^L \mathrm{d}x \int_{-\infty}^{\infty} \mathrm{e}^{-\frac{k_n J y^2}{|Q(x)|^2}} \frac{\mathrm{d}y}{\sqrt{|Q(x)|^2}} = L\sqrt{\frac{\pi}{k_n J}}. \tag{5.34}$$

Moreover, by using (5.26) and normalized resonances the following diagonal matrix elements (which also reduce to Gaussian integrals) result:

$$\int \psi^*(x, y)(\Delta + k_n^2)\psi(x, y)\,\mathrm{d}x\,\mathrm{d}y = O(1) \tag{5.35}$$

and

$$\int [(\Delta + k_n^2)\psi(x, y)]^*(\Delta + k_n^2)\psi(x, y)\,\mathrm{d}x\,\mathrm{d}y \simeq 2\lambda^2 k_n^2. \tag{5.36}$$

In order to understand the meaning of the last two equations, let us consider the set of eigenvalues $\{-k_\nu^2\}$ and normalized eigenfunctions $\{\varphi_\nu\}$ of the billiard. Let $\{c_\nu\}$ be the set of amplitudes of the resonance in the basis of eigenfunctions, that is, $\psi = \sum c_\nu \varphi_\nu$. Then, as a consequence of Equations (5.35) and (5.36) the following relations are simply derived:

$$\sum k_\nu^2 |c_\nu|^2 = k_n^2 + O(1) \tag{5.37}$$

and

$$\sigma_n^2 \equiv \sum (k_\nu^2 - k_n^2)^2 |c_\nu|^2 \simeq 2\lambda^2 k_n^2, \tag{5.38}$$

where σ_n is the dispersion of the resonance in the spectrum of eigenvalues. Now, taking into account that the mean density of eigenvalues of a planar billiard of area

$\mathscr{A}$ is $\rho_{k^2} \simeq \mathscr{A}/4\pi$, the number of eigenfunctions contributing to the resonance is around

$$2\sigma_n \rho_{k^2} \simeq \frac{\mathscr{A}\lambda k_n}{\sqrt{2\pi}}. \tag{5.39}$$

In the same way, it is possible to establish that this number is the mean number of resonances of POs contributing to an eigenfunction (Vergini 2004).

We finish this section by noticing that even though the proposed resonances are complex functions, they can be reduced to real functions by using time reversal. For instance, as the PO of Figure 5.2(a) is invariant with respect to time reversal, its corresponding resonances directly result in real functions. On the other hand, in the case of Figure 5.2(b) a real resonance is constructed by the sum of the clockwise and the anticlockwise resonance. Of course, the same idea is applicable if the billiard satisfies a spatial symmetry.

5.4 Resonances with Minimum Dispersion: The Scar Function

In the previous section we have constructed resonances of POs with n excitations along the orbit but without excitations in the transverse direction y. In this section we construct resonances with m transverse excitations, obtaining a set of resonances of a given PO, all of them with the same wave number. Then, a linear combination of that resonance set, the so-called scar function (Vergini & Carlo 2001), is specified by requiring minimum dispersion.

A generalization of Equation (5.23), with m excitations in the transverse direction is given by Vergini & Carlo (2001):

$$\psi^{(m)}(x, y) = \frac{(kJ/\pi)^{1/4}\,\mathrm{H}_m\left[y\sqrt{kJ}/|Q(x)|\right]}{\sqrt{2^m m!\,L|Q(x)|}} \mathrm{e}^{\mathrm{i}kx+\mathrm{i}k(y^2/2)(P(x)/Q(x))-\mathrm{i}(m+\frac{1}{2})\phi(x)}, \tag{5.40}$$

with H_m the Hermite polynomial of order m ($\mathrm{H}_0(z) = 1$, $\mathrm{H}_1(z) = 2z$, $\mathrm{H}_2(z) = 4z - 2$, and so on) and $\phi(x)$ the angle swept in the anticlockwise sense by the complex number $Q(x)$ while evolving from 0 to x. This resonance is normalized in leading order, and its global phase convention is selected in such a way that $\psi^{(m)}(x = 0, y)$ is a real function.

The action of the Laplacian on $\psi^{(m)}$ gives, after a straightforward but tedious calculation (Vergini & Carlo 2001), the following result (the argument (x, y) is omitted):

$$(\Delta + k^2)\psi^{(m)} = sk\lambda\left[a_m\psi^{(m+2)} + b_m\psi^{(m-2)}\right] + O(1), \tag{5.41}$$

with $a_m = \sqrt{(m+1)(m+2)}$, $b_m = \sqrt{(m-1)m}$, $\psi^{(m-2)} = 0$ for $m < 2$, and s defined in Equation (5.14).

The accumulated phase of a resonance with n excitations along γ and m transverse excitations is given by

$$kL - \mu(m + \frac{1}{2})\pi - (N_D - N_N)\frac{\pi}{2} = 2\pi n. \tag{5.42}$$

So, for $m\mu$ even we can rewrite this as follows:

$$k_{n_0}L - \mu\frac{\pi}{2} - (N_D - N_N)\frac{\pi}{2} = 2\pi n_0 \quad \text{with} \quad n_0 = n + \frac{m\mu}{2}. \tag{5.43}$$

This means that there is a set of resonances of γ with wave number k_{n_0}.

In the following, we will construct the so-called even scar function (we refer to Vergini & Carlo 2001 for the odd one), which uses the set of resonances characterized by the following couple of excitations:

$$(n = n_0 - 2\mu j, m = 4j) \quad \text{for} \quad j = 0, 1, \ldots, N = \left[\frac{n_0}{2\mu}\right], \tag{5.44}$$

where $[\cdot]$ indicates integer part. The idea is to find the linear combination

$$\psi_{scar}(x, y) = \sum_{j=0}^{N} c_j \psi^{(4j)}(x, y) \tag{5.45}$$

that minimizes the dispersion σ_{n_0}, with $\sigma_{n_0}^2$ defined on the left-hand side of Equation (5.36).

Using Equation (5.41) it is possible to obtain expressions for σ_{n_0} and the amplitudes. However, as the calculation (developed in Vergini & Carlo 2001, Vergini & Schneider 2005) is cumbersome, we only give here the final results,

$$\Gamma \equiv \frac{\sigma_{n_0}}{k\lambda} \simeq \frac{2\pi}{A - 6\pi^2/A^2}, \tag{5.46}$$

with $A = 6.06 + \ln(1 + N/1.35)$, and

$$c_j \simeq \frac{(-1)^j}{\sqrt{j}}\left(1 - \frac{1}{16j}\right)\sin\left[\frac{\Gamma}{4}\ln\left(\frac{N+2}{j+1}\right)\right], \tag{5.47}$$

for $j = 1, 2, \ldots, N$.† Moreover, $c_0 = -c_1\sqrt{6}/(1 - \Gamma^2/2)$ for $N > 0$. It is worth noticing that the dispersion of a scar function is smaller than the corresponding dispersion of the resonance without transverse excitations (see Equation (5.36)) by the factor $\Gamma/\sqrt{2}$, which goes logarithmically to zero with N.

In the rest of this section we apply this formalism to a simple case. Let us consider the horizontal PO of the Bunimovich stadium billiard. Taking into account the symmetry of the PO with respect to the vertical axis, we can use the point A of Figure 5.5 as our starting point; by doing so, local coordinates x, y coincide with the global coordinates X, Y for the first line. Then, the symplectic matrix for going from $x_A = 0$ to $x_B = 2$‡ is given by

$$\mathbf{M}(2, 0) = \mathbf{M}_b^{(\text{in})}(1, 0)\mathbf{M}_f(2)\mathbf{M}_b^{(\text{out})}(\infty, 0) = \begin{pmatrix} -1 & -2 \\ 1 & 1 \end{pmatrix}, \tag{5.48}$$

and using Equation (5.12) it results

$$\lambda = \tfrac{1}{4}\ln(3 + \sqrt{8}) \quad \text{and} \quad \tan(\varphi) = \frac{1}{\sqrt{2}}. \tag{5.49}$$

† To normalize the scar function, normalize the coefficients in such a way that $\sum_{j=0}^{N} c_j^2 = 1$.

‡ From now on, we will omit an explicit reference to the unit J of length.

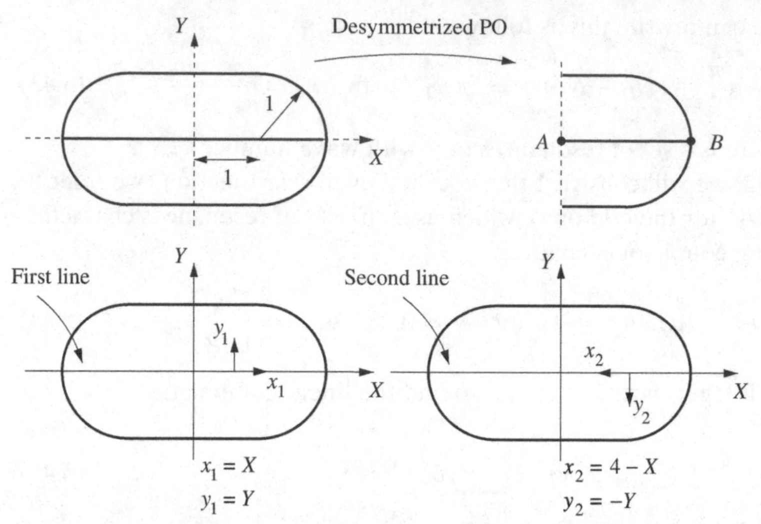

Figure 5.5. On the top we specify the stadium billiard and the horizontal PO. Moreover, the desymmetrized horizontal PO with its turning points A and B are indicated. On the bottom, the two lines required for the construction of resonances, the corresponding local coordinates, and their relation with the global coordinates are given.

Moreover, the manifold directions evolved by $\mathbf{F}(x)$ result

$$(\boldsymbol{\xi}_u(x)\boldsymbol{\xi}_s(x)) \equiv \mathbf{F}(x)\,(\boldsymbol{\xi}_u\boldsymbol{\xi}_s) = \mathbf{M}(x)\mathbf{B}\begin{pmatrix} \mathrm{e}^{-\lambda x} & 0 \\ 0 & \mathrm{e}^{\lambda x} \end{pmatrix} \tag{5.50}$$

with $\mathbf{B}$ given by Equation (5.14), and the symplectic matrix $\mathbf{M}(x)$ along the desymmetrized PO given by $\mathbf{M}(0) \equiv \mathbf{I}$, $\mathbf{M}(0^+) = \mathbf{M}_b^{(\mathrm{out})}(\infty, 0)\mathbf{M}(0) = \mathbf{I}$ where $(+)$ means after the bounce, $\mathbf{M}(x) = \mathbf{M}_f(x)\mathbf{M}(0^+)$ for $0^+ \leq x \leq 2^-$ where $(-)$ means before the bounce, $\mathbf{M}(2) = \mathbf{M}_b^{(\mathrm{in})}(1, 0)\mathbf{M}(2^-)$, $\mathbf{M}(2^+) = \mathbf{M}_b^{(\mathrm{out})}(1, 0)\mathbf{M}(2)$, $\mathbf{M}(x) = \mathbf{M}_f(x - 2)\mathbf{M}(2^+)$ for $2^+ \leq x \leq 4^-$, and finally $\mathbf{M}(4) = \mathbf{M}_b^{(\mathrm{in})}(\infty, 0)\mathbf{M}(4^-)$.

The evolved vectors $\boldsymbol{\xi}_u(x)$ and $\boldsymbol{\xi}_s(x)$ are plotted in Figure 5.6 for several relevant values of x. At $x = 0$, $x = 2$, and $x = 4$ the vectors are symmetrical with respect to the axes in accordance with the fact that these points are turning points of the desymmetrized PO. At $x = \sqrt{2}$ and $x = 4 - \sqrt{2}$ (they are the same point $X = \sqrt{2}$ and $Y = 0$ in the billiard) one of the vectors is vertical; this point, which is called a self-conjugated point, is characterized because the scar function diverges in the limit of short wavelengths (Bogomolny 1988). The quantities θ_u and θ_s indicate the angle swept by the unstable and stable vectors, respectively, as they evolve along the PO. After one period $\theta_u = \theta_s = 3\pi$, and this indicates that $\mu = 3$. In general, μ is given by the sum of the number of self-conjugated points (one for this example) and the number of bounces with the boundary (two in this case).

Resonances related to this PO are constructed with two straight lines (see Figure 5.5). The first one is specified by the coordinates $x_1 = X$ and $y_1 = Y$, with X and Y the coordinates on the plane, whereas the complex vector on the line is (see (5.18))

$$\begin{pmatrix} Q_1(x_1) \\ P_1(x_1) \end{pmatrix} = \mathbf{M}_f(x_1)\mathbf{M}(0^+)\mathbf{B}\begin{pmatrix} \mathrm{e}^{-\lambda x_1} \\ \mathrm{i}\mathrm{e}^{\lambda x_1} \end{pmatrix} \tag{5.51}$$

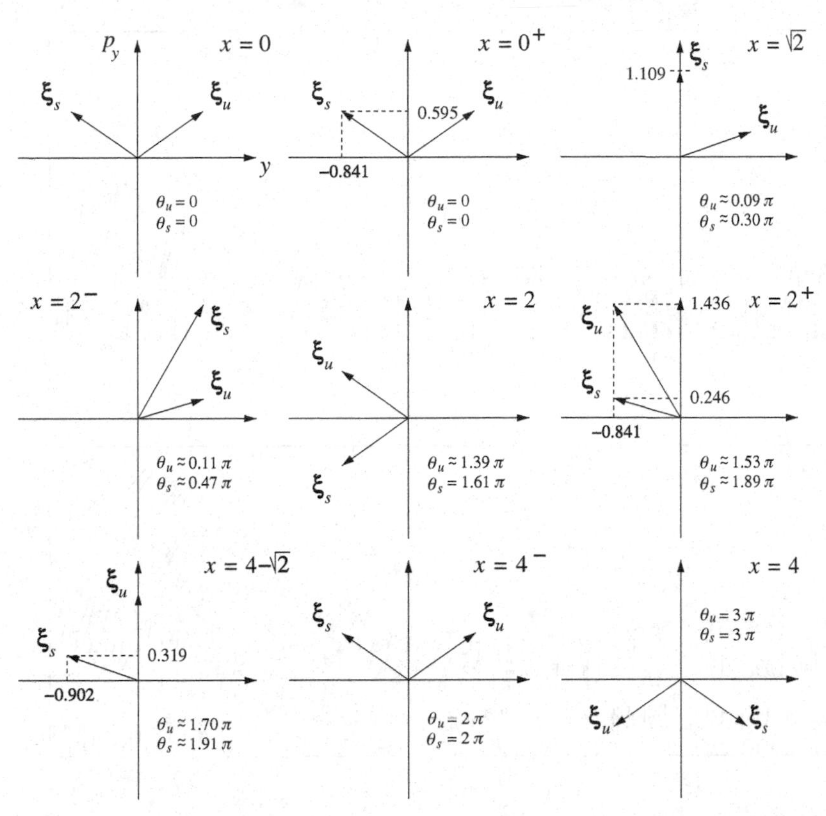

Figure 5.6. For representative values of x we show the evolved manifold directions as specified by Equation (5.50). The description of the different situations is explained in the text.

for $-\infty < x_1 < \infty$. On the other hand, the second straight line is given by the coordinates $x_2 = 2 - X$ and $y_2 = -Y$, and the corresponding complex vector results

$$\begin{pmatrix} Q_2(x_2) \\ P_2(x_2) \end{pmatrix} = \mathbf{M}_f(x_2 - 2)\mathbf{M}(2^+)\mathbf{B} \begin{pmatrix} \mathrm{e}^{-\lambda x_2} \\ \mathrm{i}\mathrm{e}^{\lambda x_2} \end{pmatrix} \tag{5.52}$$

for $-\infty < x_2 < \infty$. Then, normalized resonances are given by

$$\psi_1^{(m)}(x_1, y_1) + \mathrm{e}^{\mathrm{i}\beta}\psi_2^{(m)}(x_2, y_2), \tag{5.53}$$

with $\beta = -\pi/2$ $(\pi/2)$ for Dirichlet (Neumann) boundary conditions on the stadium billiard. When the wavenumber satisfies the Bohr–Sommerfeld rule given by Equation (5.42), resonances of Equation (5.53) are (up to a global phase) real functions. In fact, as $\psi^{(m)}(x_2 = 2 - X, y_2 = -Y)$ is to minus a global phase the complex conjugate of $\psi^{(m)}(x_1 = X, y_1 = Y)$, it is not necessary to evaluate explicitly the second line. Vergini and Carlo (2000) and Carlo et al. (2002) have provided recipes in order to directly construct real resonances in the stadium billiard with the right spatial symmetry.

In the Appendix we have computed explicitly Equation (5.53); in particular, Equation (5.61) provides resonances of the horizontal PO with Dirichlet conditions on the stadium boundary and Neumann conditions on the vertical axis. Figure 5.7 displays some representative resonances and also the scar function computed with

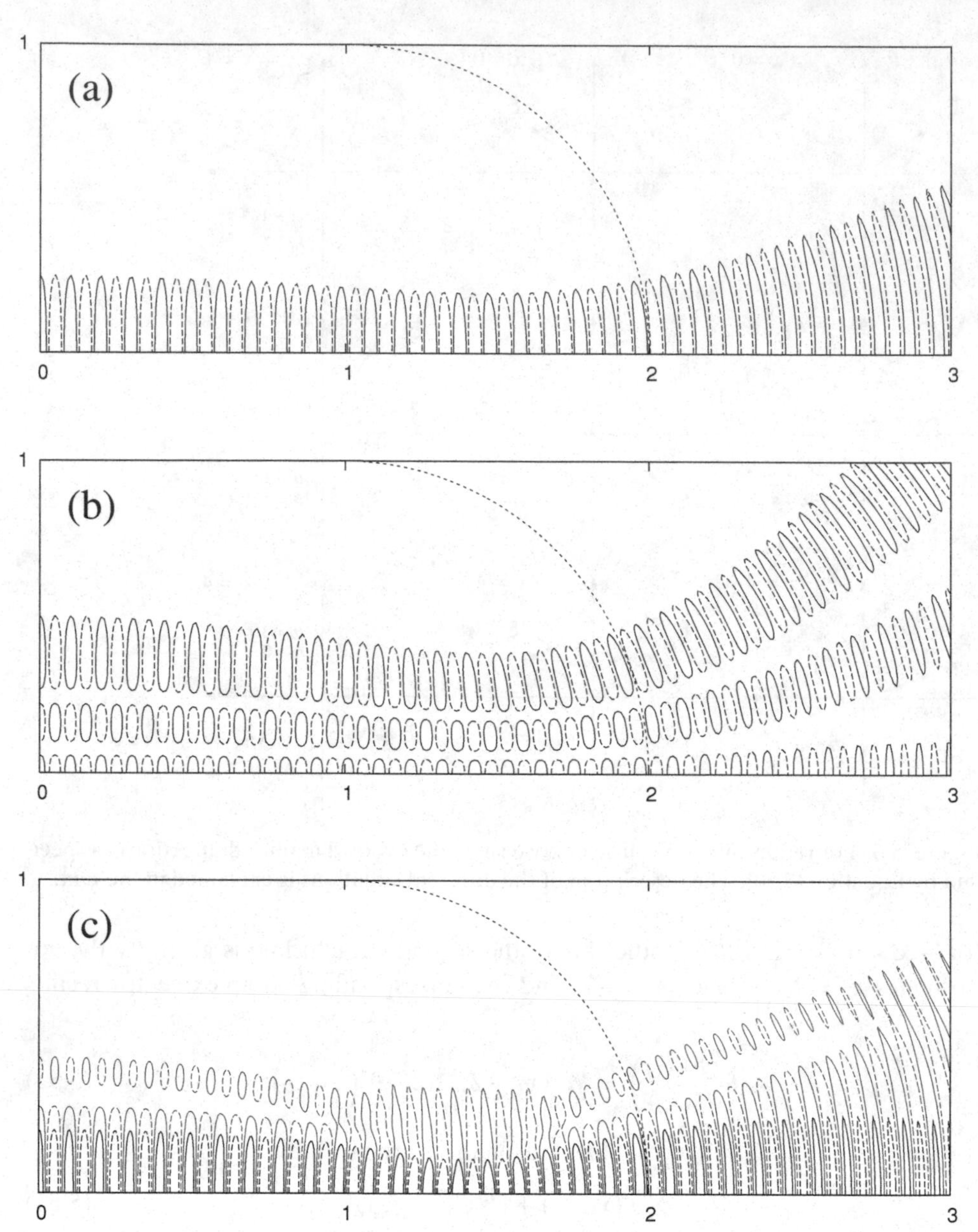

Figure 5.7. Contour lines at levels −0.35 and 0.35 for the resonances, (a) $n_0 = 40, m = 0$ and (b) $n_0 = 40, m = 4$, according to Equation (5.61). In (c), contour lines are at levels −0.5, −0.2, 0.2, and 0.5 for the scar function with $n_0 = 40$.

the assistance of (5.45). Contour lines of the scar function with levels −0.5 and 0.5 show a neck at $X = \sqrt{2}$, which demonstrates the accumulation of density at self-focal points predicted by Bogomolny (1988).

In order to appreciate the advantages of these constructed wave functions as a basis for the description and evaluation of eigenfunctions, Figure 5.8 displays the decomposition of the resonance $n_0 = 40$, $m = 0$ and the corresponding scar function on the basis of even–even eigenfunctions of the Bunimovich stadium billiard. For comparison, we also show the decomposition of the symmetrized plane wave

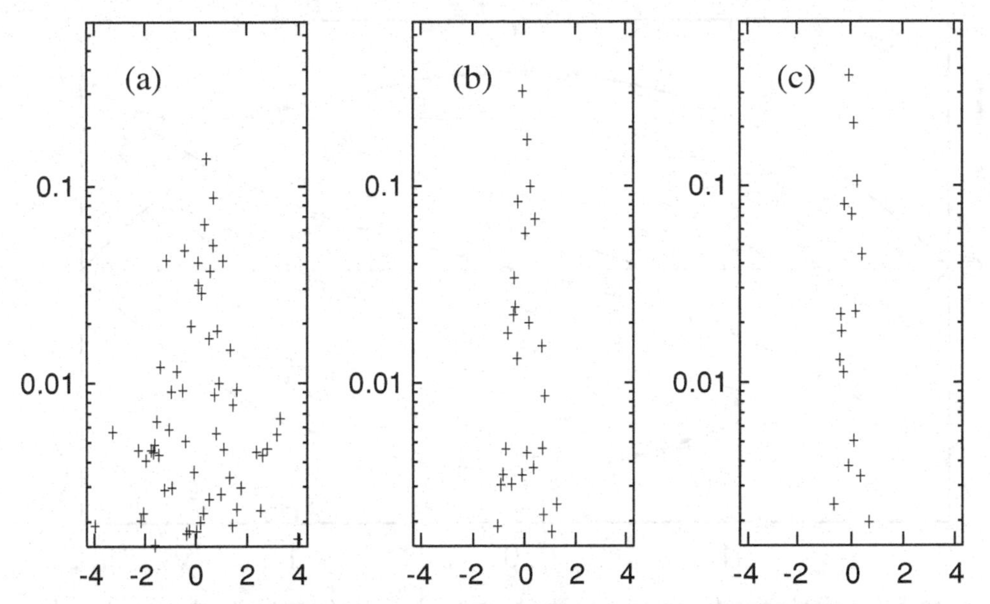

Figure 5.8. Square modulus of the overlap between even–even eigenfunctions of the Bunimovich stadium billiard and (a) a symmetrized plane wave, (b) the resonance $n_0 = 40, m = 0$, and (c) the scar function with $n_0 = 40$. The horizontal axis indicates the corresponding eigenwavenumber minus $k_{n-0} \simeq 64.01$.

$\cos(k_x X)\cos(k_y Y)$, with $k_x = 3k_{n_0}/5$ and $k_y = 4k_{n_0}/5$. In this respect, the figure demonstrates that even though plane waves are very powerful for the evaluation of eigenfunctions of the stadium (see the scaling method of Vergini & Saraceno 1995), the constructed wavefunctions on POs are much more localized in the spectrum.

We finish the section with a didactic exercise. We will show that trajectories in the neighborhood of the horizontal PO are closed when the evolution is governed by $F(x)$. Let us write $Q(x)$ and $P(x)$ in terms of their real and imaginary parts: $Q(x) = y_u(x) + \mathrm{i}y_s(x)$ and $P(x) = p_{y_u}(x) + \mathrm{i}p_{y_s}(x)$, with $(y_u(x), p_{y_u}(x))$ the components of $\xi_u(x)$ (the other functions being the components of $\xi_s(x)$). Then, taking into account that $\xi_u(x)$ and $\xi_s(x)$ evolve governed by $F(x)$ (see (5.50)), neighboring trajectories result: $y(x) = \epsilon[y_u(x)\cos\delta + y_s(x)\sin\delta]$ and $p_y(x) = \epsilon[p_{y_u}(x)\cos\delta + p_{y_s}(x)\sin\delta]$, with the pair ϵ, δ characterizing a particular trajectory. By using the expressions provided in (5.54) and (5.58), we compute in Figure 5.9 neighboring trajectories in configuration space for fixed ϵ and $-\pi/4 \leq \delta \leq \pi/4$; the curve labeled with a (b) corresponds to $\delta = \pi/4$ ($\delta = -\pi/4$), whereas thin curves indicate trajectories with $-\pi/4 < \delta < \pi/4$. The vertical axis is given at arbitrary scale because we are using a linear approximation†; of course, the accuracy increases for small ϵ.

5.5 Eigenfunctions Calculation

In this section we are going to sketch how the resonances that we have previously constructed can be applied to a specific example, the stadium billiard, in order to

† In this approximation, semicircles of the stadium boundary are replaced by vertical straight lines, as is the case for concave mirrors in geometrical optics

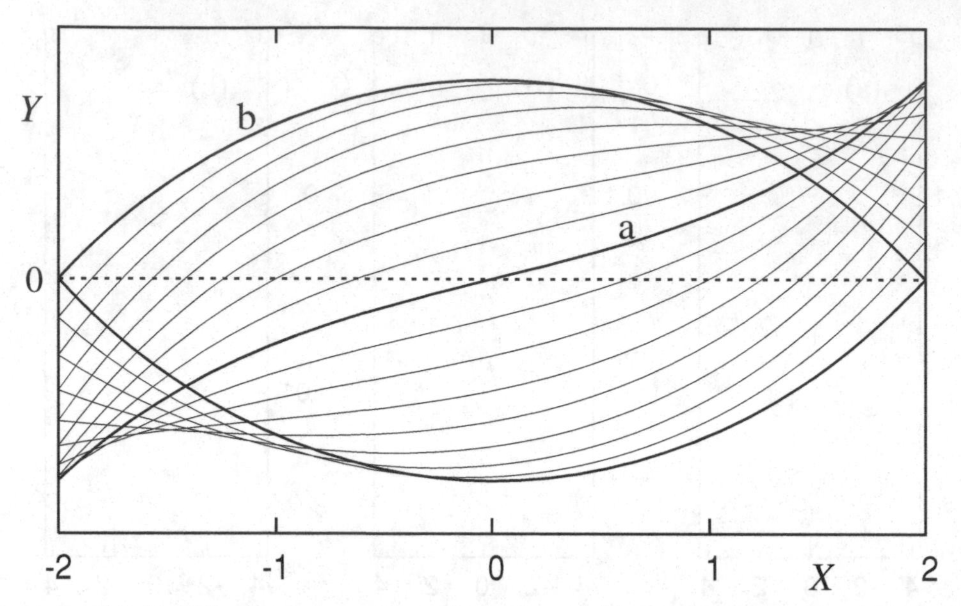

Figure 5.9. Neighboring trajectories to the horizontal PO obtained with the modified dynamics introduced by $F(x)$ with δ varying from $\pi/4$ (a) to $-\pi/4$ (b). As can be seen, all neighboring trajectories result POs for the modified dynamics; see the text for details.

calculate its eigenfunctions (we use its desymmetrized version). This is a very well-known system in the quantum chaos literature and provides the essential nontrivial features that make it one of the simplest realistic examples of chaos.

In the previous section we have constructed a resonance associated with a periodic orbit of this system. Because this is the first step toward our goal, we now briefly explain the construction for another orbit in order to further illustrate the procedure. This orbit corresponds to Figure 5.10(c). Resonances are constructed with straight lines by associating a semiclassical expression with each one. The first line is defined by the segment of γ starting at $x_1 = 0$. Let x_2 ($> x_1$) be the value of x such that the path reaches the stadium boundary while evolving along γ, in this case corresponding to $\sqrt{3}/2$. The path going out of x_2 defines the second line, up to $x = L/2 = 3\sqrt{3}/2$. The transformation from local coordinates (x, y) on some

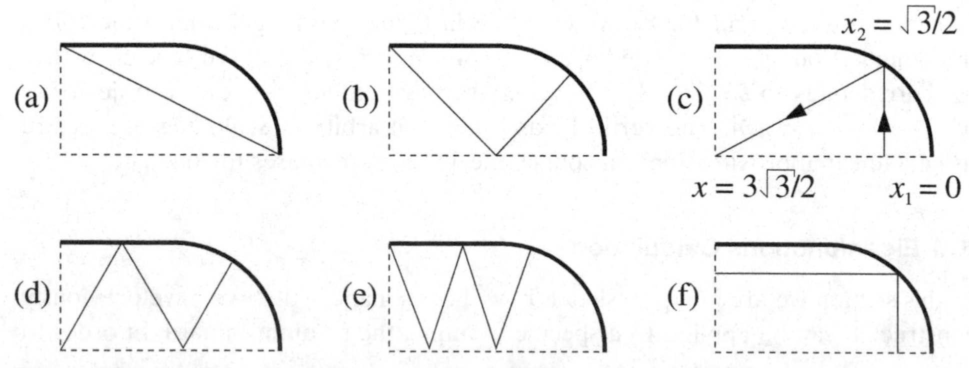

Figure 5.10. Set of shortest POs of the stadium billiard (desymmetrized version) used in the eigenfunctions calculation. Straight lines needed to construct the resonances are indicated.

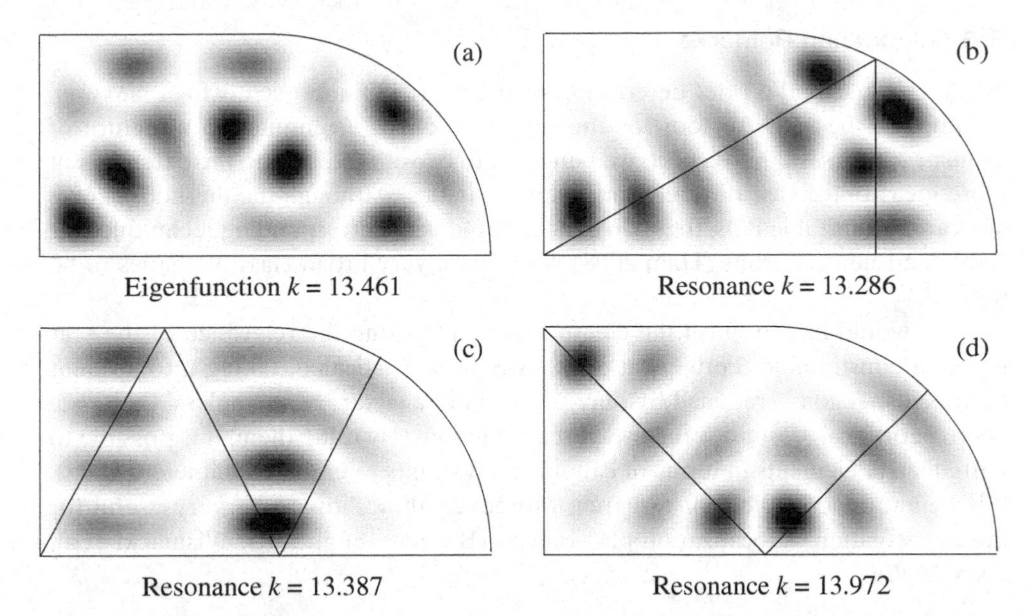

Figure 5.11. Density plot (black corresponds to high intensity) of (a) the 21st odd–odd squared eigenfunction of the stadium billiard; (b)–(d) resonances contributing strongly to this eigenfunction; superimposed, the corresponding periodic orbits. Resonance (c) reproduces 46% of the eigenfunction, (c) plus (d) 82%, and a linear combination of the three resonances reproduces 93% of (a) (by which we mean that the square modulus of the overlap with (a) is 0.93). The eigenvalue and the Bohr–Sommerfeld quantized wave numbers are also shown.

line to global coordinates (X, Y) (horizontal and vertical directions in the plane respectively) is obtained through a local rotation. Finally, the family of resonances is constructed with all the lines including symmetries (see Vergini & Carlo 2000 for details).

The next step consists of selecting the set of resonances defining an appropriate basis. We restrict to odd symmetry on the horizontal and vertical directions. There is a special kind of resonances that are of the bouncing ball type. These are characteristic of this system, and they are only a fraction of the mean number of states specified by Weyl's law, but we have to take care of them (for details see Vergini & Carlo 2000). Looking at Figure 5.10 we can see the set of orbits used for the resonance basis construction. We take as many of them as necessary so that the density of resonances agrees with the semiclassical mean density. Orbit (f) is not considered because the associated resonances do not satisfy boundary conditions with sufficient accuracy for low energies. Though the eigenfunctions are obtained with this set of wavefunctions, only a limited number of them are required for each one, as can be seen for example in Figure 5.11.

The last step consists of evaluating the overlaps and the matrix elements using the prescription given in (5.41). It is possible to obtain them by direct integration on the domain (the quarter of billiard in this case). Then, by solving a generalized eigenvalue problem, the set of eigenfunctions is obtained. Figure 5.11 shows linear density plots of the 21st squared eigenfunction and its main components (where black corresponds to maximum intensity).

5.6 Concluding Remarks

We have presented a brief description of the short PO approach, which, anyway, we hope to be a useful tool for the investigation of eigenfunction structure of acoustic cavities. In this respect, we believe it to be very interesting the extension of these ideas to the case of three-dimensional cavities. Although the methods shown here are for unstable POs, they can be integrated with corresponding techniques for stable and neutral orbits (Ham 2008), allowing a very broad class of shapes to be treated.

We would like to finish the chapter by emphasizing the relevance of the scar function construction. Perhaps, it appears a bit sophisticated, taking into account that its dispersion is reduced by only a logarithmic factor. Nevertheless, there is a deep theoretical reason for the use of scar functions. By using them, it is possible to write matrix elements in terms of canonical invariants (Vergini & Schneider 2005), this being an active field of research. Moreover, they show to be very useful for the investigation of localization phenomena (Borondo et al. 2005, Wisniacki et al. 2005, 2006).

5.7 Appendix

In this Appendix we are going to evaluate explicitly Equation (5.53) in terms of the global coordinates X and Y. Even though explicit expressions are not required for the calculation of resonances, it is useful to obtain them in a simple case in order to have some intuition about the construction.

By replacing (5.14) and (5.49) in (5.51), we obtain

$$\begin{pmatrix} Q_1(x_1) \\ P_1(x_1) \end{pmatrix} = \frac{1}{\sqrt{2}} \begin{pmatrix} (1/\alpha + \alpha x_1)\mathrm{e}^{-\lambda x_1} + \mathrm{i}(-1/\alpha + \alpha x_1)\mathrm{e}^{\lambda x_1} \\ \alpha(\mathrm{e}^{-\lambda x_1} + \mathrm{i}\mathrm{e}^{\lambda x_1}) \end{pmatrix}. \tag{5.54}$$

And using the global coordinate X, with $X = x_1$, this yields

$$|Q_1(X)|^2 = (1/\alpha^2 + \alpha^2 X^2)\cosh(2\lambda X) - 2X\sinh(2\lambda X) \tag{5.55}$$

and (see (5.20))

$$g_1(X) = [\alpha^2 X\cosh(2\lambda X) - \sinh(2\lambda X)]/|Q_1(X)|^2. \tag{5.56}$$

Moreover, the angle ϕ_1 swept by Q_1 in the complex plane results (see (5.40))

$$\phi_1(X) = \mathrm{sign}(X)\left\{\arctan\left[\left(\frac{\alpha^2|X| - 1}{\alpha^2|X| + 1}\right)\mathrm{e}^{2\lambda|X|}\right] + \frac{\pi}{4}\right\}, \tag{5.57}$$

with the function $\arctan(\cdot)$ being in the range $(-\pi/2, \pi/2)$.

Using the previous expressions, the contribution $\psi_1^{(m)}(X,Y)$ of the first line to (5.53) is given by (5.40). Furthermore, working in the same way with the second line, we obtain from (5.52) the following relations:

$$Q_2(X) = Q_1(-X) = -\mathrm{i}Q_1^*(X) \quad \text{and} \quad P_2(X) = P_1(-X) = \mathrm{i}P_1^*(X), \tag{5.58}$$

with $X = 4 - x_2$. And from these expressions one obtains

$$|Q_2(X)| = |Q_1(X)|, \quad g_2(X) = -g_1(X), \quad \text{and} \quad \phi_2(X) = 2\pi - \phi_1(X). \tag{5.59}$$

Finally, in order to establish a relation between $\psi_1^{(m)}(X,Y)$ and $\psi_2^{(m)}(X,Y)$, we have to specify the value of the wavenumber, and therefore the boundary conditions. For instance, let us consider Dirichlet conditions on the stadium boundary (at the turning point B in Figure 5.10) and Neumann conditions on the vertical axis (at the turning point A). Then, $N_D = N_N = 1$ and from (5.43) one obtains

$$k_{n_0} = \frac{\pi}{2}\left(n_0 + \frac{3}{4}\right) \quad \text{with} \quad n_0 = n + \frac{3}{2}m; \tag{5.60}$$

the integers n and m (which is even) are the number of excitations along and transverse to the periodic orbit, respectively. We have named $n = n_0 - \mu m/2$ the number of excitations along the orbit according to the quantization rule (5.43); however, it is worth mentioning that the number of nodal lines along the orbit is $n' = n_0 - \nu m/2$, with ν the number of self-focal points along the periodic orbit (for the desymmetrized horizontal periodic orbit (see Figure 5.10), there is one self-focal point at $X = \sqrt{2}$, $Y = 0$). In any case, the couple (n_0, m) unequivocally defines the resonance.

For $k = k_{n_0}$, one can verify the relation $\psi_2^{(m)}(X,Y) = i\psi_1^{(m)*}(X,Y)$, and then, Equation (5.53) reduces to

$$\psi_1^{(m)}(X,Y) + e^{-i\pi/2}\psi_2^{(m)}(X,Y) = \frac{(k_{n_0}/\pi)^{1/4}}{\sqrt{2^m\, m!\, |Q_1(X)|}} H_m\left[\frac{Y\sqrt{k_{n_0}}}{|Q_1(X)|}\right]$$
$$\times \exp\left[-\frac{Y^2 k_{n_0}}{2|Q_1(X)|^2}\right] \cos\left[k_{n_0}X + Y^2 k_{n_0} g_1(X)/2 - (m+1/2)\phi_1(X)\right]. \tag{5.61}$$

This expression is normalized in leading order to unity inside the desymmetrized billiard of Figure 5.10.

6 Chaotic Wave Scattering

Jonathan P. Keating
Marcel Novaes
School of Mathematics, University of Bristol, Bristol, UK

We give an overview of wave scattering in complex geometries, where the corresponding rays are typically chaotic. In the high-frequency regime, a number of universal (geometry-independent) properties that are described by random matrix theory emerge. Asymptotic methods based on the underlaying rays explain this universality and are able to go beyond it to account for geometry-specific effects. We discuss in this context statistics of the scattering matrix, scattering states, the fractal Weyl law for resonances, and fractal resonance wavefunctions.

6.1 Introduction

Our purpose here is to give an introductory overview of wave scattering in complex geometries, where the corresponding rays are typically chaotic. For simplicity, we focus our discussion on domains with lossless walls, inside which the wave speed is constant. The rays then are straight, with specular reflections at the boundaries. This situation is often encountered in experiments (Stöckmann 1999, Kuhl et al. 2005, Tanner & Søndergaard 2007) and in acoustic applications. However, many of the features we shall identify occur much more generally. Indeed, most recent developments in the subject have taken place in the context of quantum wave scattering, where the underlying rays are the classical trajectories of Newtonian mechanics. Much of this review will be devoted to translating quantum results into the language of classical wave scattering.

One of the main observations we wish to make is that many of the essential mathematical features of wave scattering in complex geometries can be found in certain very simple discrete models, which we here call *wave maps*. These wave maps therefore allow one to explore many of the fundamental problems of the theory without the distraction of inessential mathematical details.

Our review is organized as follows: In Section 6.2 we recall, for purposes of later comparison with open systems, some fundamental short-wavelength asymptotic properties of waves in closed systems in which the ray dynamics is chaotic. These include *Weyl's law*, which relates the locally averaged eigenfrequency counting function to the volume of the enclosing domain, and *wave ergodicity* of the stationary states. We conclude with a discussion of *wave maps* in the context of closed geometries.

In Section 6.3 we extend the discussion to waves in open geometries. We introduce the scattering matrix description and outline the connection with random matrices. The short-wavelength approximation in terms of scattering rays is introduced and briefly discussed. We also mention the modeling of chaotic scattering states by random waves.

Finally, we consider in Section 6.4 the description of *resonances*, that is, eigenstates of the wave equation with outgoing boundary conditions, which correspond to transient states with complex eigenfrequencies. A fundamental role is played in this context by the *trapped sets* of the ray dynamics: the generalization of the Weyl law to resonances is related to the fractal dimension of the intersection of the trapped sets, and they support the long-lived resonance wavefunctions.

6.2 Closed Systems

Let us consider waves in a closed domain of arbitrary shape in d dimensions. We denote by D the interior of the domain and by ∂D its boundary. The waves inside D are solutions of the wave equation

$$\left(\Delta - \frac{1}{c^2}\frac{\partial^2}{\partial t^2}\right)\psi = 0, \tag{6.1}$$

where Δ is the Laplace operator and c is the wave speed. We assume Dirichlet boundary conditions,

$$\psi(\mathbf{r}, t)\big|_{\mathbf{r}\in\partial D} = 0, \tag{6.2}$$

corresponding to fixed membrane edges (or pressure-release boundaries in acoustics, such as the sea surface seen from below). The stationary states $\{\Psi_n(\mathbf{r})\mathrm{e}^{\mathrm{i}\omega_n t}\}$ associated with a discrete set of eigenfrequencies $\{\omega_n = k_n c\}$ satisfy the Helmholtz equation

$$\Delta\Psi_n(\mathbf{r}) = -k_n^2\Psi_n(\mathbf{r}). \tag{6.3}$$

We assume that the spectrum is ordered, $k_n \geq k_{n-1}$, and that the eigenfunctions are orthonormal, $\int_D \Psi_n(\mathbf{r})\Psi_m(\mathbf{r})\,\mathrm{d}\mathbf{r} = \delta_{nm}$. In the following, we shall sometimes refer to the set $\{k_n\}$ as eigenfrequencies (i.e., assume implicitly that $c = 1$).

We denote the Green's function by $G(z; \mathbf{r}', \mathbf{r})$. This is the integral kernel of the resolvent operator,

$$(z - \Delta)^{-1}\Psi(\mathbf{r}') = \int G(z; \mathbf{r}', \mathbf{r})\Psi(\mathbf{r})\,\mathrm{d}\mathbf{r}, \tag{6.4}$$

where $\Im\{z\} > 0$. The Green's function can be expanded in the eigenbasis formed by the eigenfunctions $\Psi_n(\mathbf{r})$,

$$G(z; \mathbf{r}', \mathbf{r}) = \sum_n \frac{\Psi_n(\mathbf{r}')\Psi_n(\mathbf{r})}{z - k_n^2}, \tag{6.5}$$

and so may be seen to have poles at $z = k_n^2$ with residues related to the eigenfunctions. Since the eigenfunctions are orthonormal, the distribution of the eigenfrequencies k_n is determined by the trace of the Green's function (i.e., the integral of

$G(z; \mathbf{r}, \mathbf{r})$ over positions $\mathbf{r}$ in D). For example, the eigenfrequency counting function,

$$N(k) = \#\{k_n : 0 < k_n \leq k\} \tag{6.6}$$

(the "modecount" function of Chapter 1), is determined in this way by tr G.

The ray dynamics inside D consists of free propagation (straight rays) combined with specular reflection at ∂D. In the short-wavelength limit, it is natural to expect solutions of the wave equation to inherit the structure of the rays. The counting function can be decomposed into a smooth (locally averaged) part and a fluctuating part, $N(k) = N_{\text{sm}}(k) + N_{\text{fl}}(k)$. The latter is given in terms of a sum over all periodic rays by the trace formula (Gutzwiller 1970, 1971, Balian & Bloch 1972, 1974; see Chapter 1 of this book). The former is given by Weyl's law:

$$N_{\text{sm}}(k) \sim \frac{V(D)k^d}{(4\pi)^{d/2}\Gamma(\frac{d}{2}+1)}, \quad \text{as } k \to \infty, \tag{6.7}$$

where $V(D)$ is the volume of D. (Note that as $k \to \infty$, $N_{\text{fl}}(k)/N_{\text{sm}}(k) \to 0$, and so $N(k)$ itself satisfies Weyl's law.)

According to a theorem of Schnirelman, when the underlying rays are ergodic, the densities $|\Psi_n|^2$ become uniformly distributed over D as $n \to \infty$ (Schnirelman 1974, Berry 1977, Colin de Verdière 1985). We call this *wave ergodicity* (in the quantum mechanics literature it is called *quantum ergodicity*). Three important remarks are in order. First, this is to be understood in a weak sense; that is, the wavefunction will have oscillations on the scale of the wavelength k^{-1}, but these are washed out if $|\Psi_n|^2$ is smoothed over a k-independent range, in the limit as $n \to \infty$. In other words, if A denotes a subregion of D with volume $V(A)$, then

$$\lim_{n\to\infty} \int_A |\Psi_n(\mathbf{r})|^2 \, d\mathbf{r} = \frac{V(A)}{V(D)}. \tag{6.8}$$

Second, the previous equation is true for almost all possible subsequences of the mode labels n as $n \to \infty$ (strictly speaking, it is true for subsequences of density 1), but in principle one may find particular exceptional sequences for which it fails (i.e., for which $|\Psi_n|^2$ has a nonuniform limit). It has been suggested that states in these exceptional sequences may show localization around periodic rays, a phenomenon called *scarring* (Heller 1984, Bogomolny 1988, Berry 1989, Keating & Prado 2001). Scarring certainly occurs when stationary states are averaged over a range of wavenumbers containing a large number of eigenfrequencies. This is sometimes called *weak scarring* and is important in applications where the individual modes cannot be resolved. Third, in time-reversal-symmetric situations (which includes most acoustic applications) the wavefunctions Ψ are real, and so $|\Psi|^2 = \Psi^2$. Fluctuations around the ergodic average are believed to be described statistically by the *random wave model* Berry 1977; see Chapter 4 of the current book).

Wavefunctions can also be considered from a phase space point of view. This is usually done in terms of the Wigner function, which is a function of both position and wavevector defined as

$$W_\psi(\mathbf{r}, \mathbf{k}) = \int \psi^*(\mathbf{r} - \mathbf{r}')\psi(\mathbf{r} + \mathbf{r}')e^{-2\pi i \mathbf{k}\cdot\mathbf{r}'} \, d\mathbf{r}'. \tag{6.9}$$

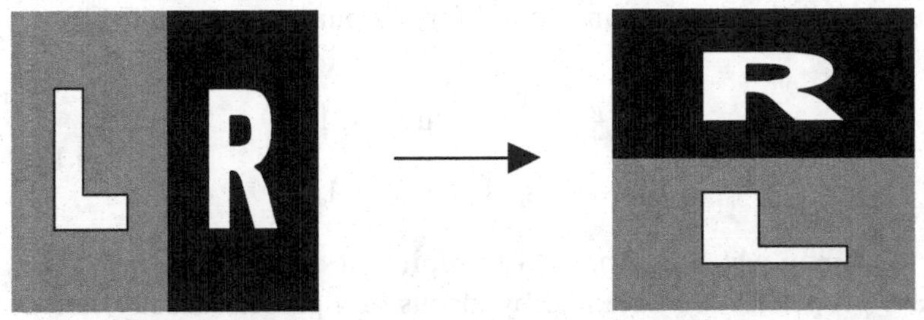

Figure 6.1. The baker map acting on a square of unit side. The rectangles are compressed by a factor of 2 in the vertical direction, are stretched by the same factor in the horizontal direction, and are repositioned (they are not rotated). The only fixed points are $(0, 0)$ and $(1, 1)$. The dynamics is uniformly hyperbolic with Lyapunov exponent $\log 2$.

The Wigner function is real and satisfies the relations

$$\int W_\psi(\mathbf{r}, \mathbf{k})\,\mathrm{d}\mathbf{k} = |\psi(\mathbf{r})|^2, \quad \int W_\psi(\mathbf{r}, \mathbf{k})\,\mathrm{d}\mathbf{r} = |\widetilde{\psi}(\mathbf{k})|^2, \tag{6.10}$$

where $\widetilde{\psi}(\mathbf{k})$ is the wavefunction in $\mathbf{k}$ space (the Fourier transform of $\psi(\mathbf{r})$). The random wave model for Ψ_n is equivalent to assuming the corresponding $W_{\Psi_n}(\mathbf{r}, \mathbf{k})$ to be concentrated on the surface defined by $|\mathbf{k}| = k = k_n$, that is, as $n \to \infty$,

$$W_{\Psi_n}(\mathbf{r}, \mathbf{k}) \sim \frac{2\pi^{d/2} k^{d-1}}{\Gamma(\frac{d}{2}) V(D)} \delta(k - k_n). \tag{6.11}$$

We finish this section by presenting an alternative description of the ray dynamics of Dirichlet-condition systems in $d = 2$. This is the bounce map. It is essentially a way of removing the trivial propagation between reflections (or bounces) at the boundary. We introduce the variable q, the position along ∂D at which a reflection occurs measured from an arbitrary reference point, and the variable p, the cosine of the angle between the ray and the normal to ∂D at q; q can be measured in units such that the perimeter has unit length. A succession of reflections is then a succession of points (q, p). The space spanned by these variables, also known as Birkhoff variables, is the cylinder $X = S^1 \times [-1, 1]$ (the variable q is periodic). The dynamics is now replaced by a map $\mathcal{M} : X \to X$, which is area preserving. The time it takes to travel between successive reflections is lost in this description – the dynamics is now discrete. On the other hand, the bounce map description offers a considerable simplification because the dimensionality of the phase space is reduced.

Given that the ray dynamics inside the domain D can be represented as an area-preserving map, it is natural to investigate the typical properties of waves in situations where the ray dynamics is chaotic by analyzing the corresponding properties of chaotic maps. Examples of such maps are the baker map, Arnold's cat map, and Chirikov's standard map (Ott 1993). We illustrate in Figure 6.1 the action of the baker map, which is a prototype of the "Smale horseshoe," a stretch-and-fold procedure typical of chaotic dynamics. The map is most naturally considered as acting on the torus T, which is a square of unit side with opposite sides identified, rather than a cylinder, but this distinction is unimportant for our purposes. Analytically, it takes the form

$$\mathcal{M}_{\text{baker}}(q,p) = \begin{cases} \left(2q, \frac{p}{2}\right) & \text{if } q < 1/2, \\ \left(2q-1, \frac{p+1}{2}\right) & \text{if } q > 1/2. \end{cases} \tag{6.12}$$

Once we have a torus map which is supposed to represent our ray dynamics, the corresponding wave properties are obtained by means of a process usually called "quantization" (Hannay & Berry 1980), in which an $N \times N$ unitary matrix $\mathbf{U}$ is associated with the map $\mathcal{M}$. The matrix $\mathbf{U}$ is a representation of the discrete wave time-evolution operator. The parameter N, which is given by $N = kL$, where L is the length of ∂D, is required to be an integer. The limit $N \to \infty$ replaces $k \to \infty$. For example, in the case of the baker map (Balazs & Voros 1989, Hannay et al. 1994)

$$\mathbf{U} = \mathbf{F}_N^{-1} \begin{pmatrix} \mathbf{F}_{N/2} & 0 \\ 0 & \mathbf{F}_{N/2} \end{pmatrix}, \tag{6.13}$$

where $\mathbf{F}_N$ is the discrete Fourier transform in N (assumed even) dimensions,

$$(\mathbf{F}_N)_{nm} = \frac{1}{\sqrt{N}} \mathrm{e}^{-2\pi i nm/N}. \tag{6.14}$$

The unitary matrix $\mathbf{U}$ may be thought of as a *chaotic wave map*. It is believed to exhibit all of the generic properties of complex wave systems; for example, there is a random wave model and a trace formula, and it displays random matrix statistics. On the other hand, its analysis is simpler because we deal with a finite-dimensional matrix rather than a differential operator. The analog of Weyl's law, for example, simply states that the phases of the eigenvalues of $\mathbf{U}$ (which lie on the unit circle in the complex plane because $\mathbf{U}$ is unitary) are uniformly distributed on the interval $(0, 2\pi]$.

6.3 Open Systems: Scattering

After this brief review of closed chaotic systems, we turn now to our main topic, which is chaotic wave scattering. In a typical scattering problem there is a certain region of space from which incident waves may be scattered. Far away from this region the waves propagate freely. Frequently, detailed information about the scatterer is not available, and it is convenient to express the theory in terms of asymptotic quantities only. This is the purpose of the $\mathbf{S}$ matrix formulation. Assume that the incoming wave can be written in terms of a complete set of functions like $\Psi^{\text{in}}(\mathbf{r}) = \sum_j a_j \phi_j^{\text{in}}(\mathbf{r})$, where a_j are complex coefficients and ϕ_j may be plane waves, cylindrical harmonics, and the like. Assume that the same is true for the outgoing wave, possibly in terms of a different set, $\Psi^{\text{out}}(\mathbf{r}) = \sum_j b_j \phi_j^{\text{out}}(\mathbf{r})$. The linearity of the wave equation implies that the whole scattering process can be characterized by a linear operator $\mathbf{S}$, which relates the incoming and outgoing coefficients,

$$\mathbf{b} = \mathbf{S}\mathbf{a}. \tag{6.15}$$

With appropriate normalization of the basis, this operator is unitary, reflecting conservation of total flux in the process.

If the scatterer is sufficiently complicated, the scattering process can display properties that are typical of chaos (Gaspard 1998, Smilansky 1991). One might hope that this situation could be described by random matrix theory. This is indeed the case, as was first proposed by Blümel and Smilansky (1988, 1990). If the dynamics is invariant under the time-reversal operation, the **S** matrix is symmetric. In analogy to closed systems, it is assumed that the **S** matrix of a system with sufficiently complicated ray dynamics is statistically equivalent to a random matrix. The appropriate ensemble in this situation is called the circular orthogonal ensemble. This is the ensemble of symmetric unitary matrices endowed with a probability measure that is invariant under all orthogonal transformations (i.e., changes of basis).

For definiteness, we shall focus our attention on a particular (representative) example of a scattering system: a Dirichlet-condition domain to which a number of waveguides are attached (see Figure 6.2). This has been used in a considerable number of experiments (Stöckmann 1999, Kuhl et al. 2005, Tanner & Søndergaard 2007). When the sum of the widths of the leads is very small compared with the perimeter, one expects the chaotic dynamics of the closed system to influence markedly the transport properties of the open problem. If a waveguide is assumed to have constant cross-sectional width W, the wave equation is separable inside it, and there are $N = [kW/\pi]$ scattering channels, corresponding to the transverse modes (here $[\cdot]$ denotes the integer part). Let us assume that a wave enters the scattering domain through this waveguide and that there is another waveguide that also has, for simplicity, N channels. The dimension of the **S** matrix is then $2N \times 2N$. We assume $N \gg 1$ because we are interested in the high-k regime. Because of unitarity, the eigenvalues $\{e^{i\theta_n}, 1 \le n \le 2N\}$ of **S** lie on the unit circle in the complex plane. According to random matrix theory predictions for the circular orthogonal ensemble, they have the joint probability density (Mehta 1991)

$$P(\{\theta_n\}) = \frac{1}{(4\sqrt{\pi})^N \Gamma(\frac{N}{2}+1)} \prod_{j<m} |e^{i\theta_j} - e^{i\theta_m}|. \tag{6.16}$$

In the situation described above the **S** matrix assumes a block form,

$$\mathbf{S} = \begin{pmatrix} \mathbf{R} & \mathbf{T}' \\ \mathbf{T} & \mathbf{R}' \end{pmatrix}, \tag{6.17}$$

where $\mathbf{R}$ and $\mathbf{R}'$ are reflection matrices whereas $\mathbf{T}$ and $\mathbf{T}'$ are transmission matrices. The unitarity of **S** implies certain restrictions. For example, the Hermitian matrices $\mathbf{T}\mathbf{T}^\dagger, \mathbf{T}'\mathbf{T}'^\dagger, 1 - \mathbf{R}\mathbf{R}^\dagger$, and $1 - \mathbf{R}'\mathbf{R}'^\dagger$ all have the same set of eigenvalues, $\{T_1, \ldots, T_N\}$. Each of these is a real number between 0 and 1, which quantifies how much of the corresponding eigenfunction is transmitted through the cavity. Within random matrix theory they are distributed according to Forrester (2006) and Beenakker (1993, 1997):

$$\mathscr{P}(\{T_n\}) = \mathscr{N} \prod_{i<j} |T_i - T_j| \prod_m T_m^{-1/2}, \tag{6.18}$$

where $\mathscr{N}$ is a normalization constant. For many applications it is enough to know the probability density, which for $N \gg 1$ is given by

$$\rho(T_1 \equiv T) = \int_0^1 \cdots \int_0^1 \mathscr{P}(\{T_n\})\, dT_2 \cdots dT_N = \frac{1}{\pi} \frac{1}{\sqrt{T(1-T)}}. \tag{6.19}$$

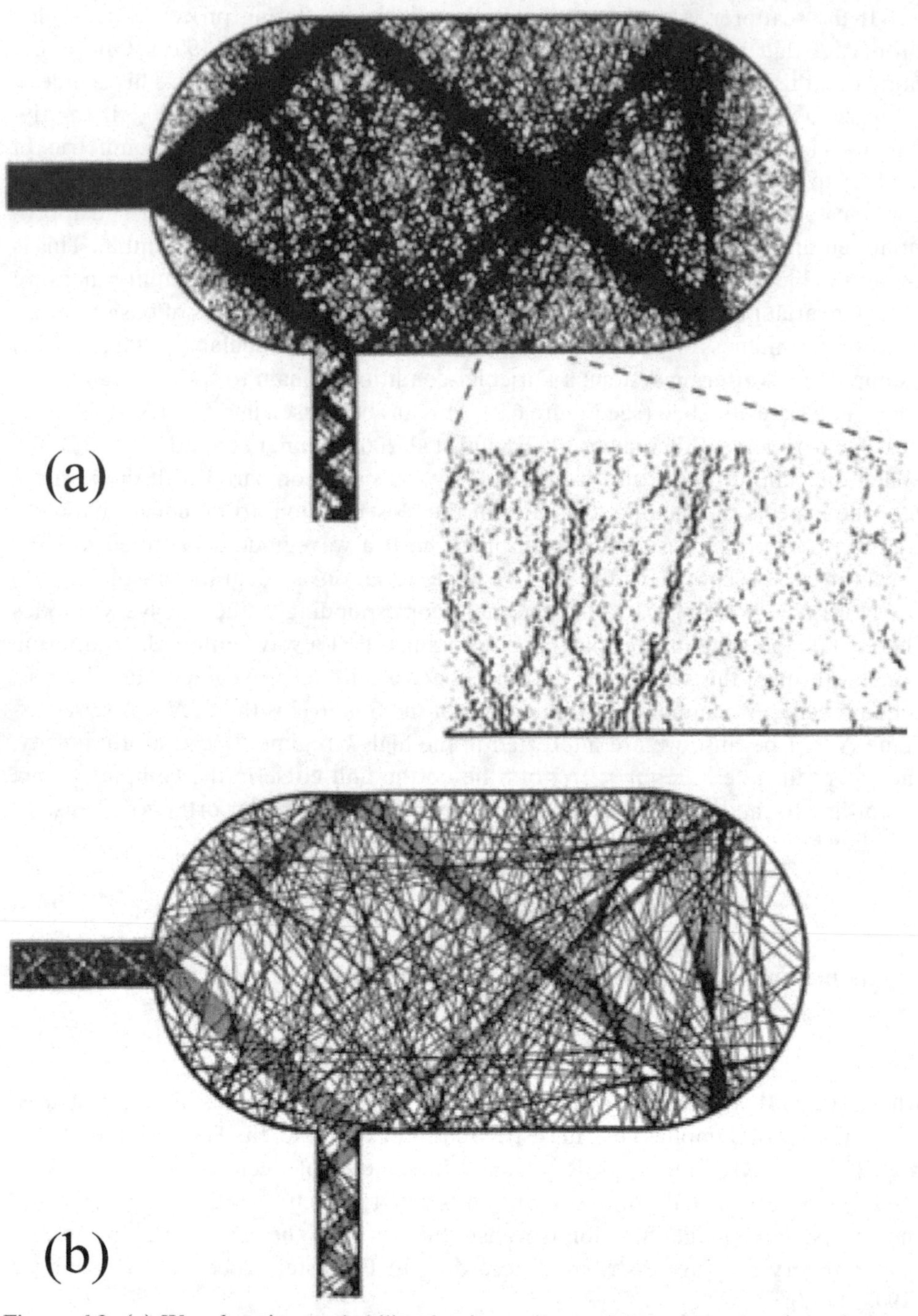

Figure 6.2. (a) Wavefunction probability density and a partial magnification (zoom), for a stadium-shaped domain of area A and wavenumber such that $kA \approx 3046$. There are 90 open channels in each lead, and a wave is entering the cavity from the left in mode $n = 16$. (b) A set of 22 rays enter, with the appropriate angle corresponding to this mode, from different points in the left lead and are plotted until they escape. Taken from Ishio and Keating (2004).

The quantity $g = \mathrm{tr}(\mathbf{TT}^{\dagger}) = \sum_m T_m$ measures what is called the conductance of the cavity (in appropriate units). The average value of g over all possible realizations of the $\mathbf{S}$ matrix is $\langle g \rangle = N/2$, whereas the variance is $\langle g^2 \rangle - \langle g \rangle^2 = 1/8$. This last result implies that conductance fluctuations are universal (independent of both geometry and wavenumber) (Beenakker 1993, 1997, Jalabert et al. 1994, Baranger & Mello 1994).

Scattering states can be described by a random wave model (Ishimaru 1978, Pnini & Shapiro 1996, Ishio et al. 2001, Kuhl et al. 2005, Tanner & Søndergaard 2007). Let us denote by $\Psi(\mathbf{r}, t)$ a wavefunction which satisfies the wave equation inside our cavity with scattering boundary conditions; that is, it is asymptotically equal to incoming or outgoing waves in the corresponding leads. This function can be written, in the interior, as a superposition of traveling waves (Pnini & Shapiro 1996):

$$\Psi(\mathbf{r}, t) = \sum_n \cos(\varphi_n + \mathbf{k}_n \cdot \mathbf{r} - kct), \tag{6.20}$$

where the phases are random and again the wavevectors $\mathbf{k}_n$ are uniformly distributed on a $(d-1)$-dimensional sphere of radius k. The intensity averaged over a period is

$$I(\mathbf{r}) = \frac{kc}{2\pi} \int_0^{\frac{2\pi}{kc}} \Psi^2(\mathbf{r}, t)\, \mathrm{d}t = \sum_{nm} \cos(\varphi_n - \varphi_m + (\mathbf{k}_n - \mathbf{k}_m) \cdot \mathbf{r}). \tag{6.21}$$

Noticing that the same expression arises for a complex field $\psi(\mathbf{r}) = \sum_n \mathrm{e}^{\mathrm{i}(\varphi_n + \mathbf{k}_n \cdot \mathbf{r})}$ with $I(\mathbf{r}) = |\psi(\mathbf{r})|^2$, we see that by the central limit theorem both $\Re\{\psi\}$ and $\Im\{\psi\}$ are independent Gaussian variables with zero mean and equal variances, and the intensity distribution is thus

$$P(I) = V(D)\mathrm{e}^{-V(D)I}, \tag{6.22}$$

where $V(D)$ is the volume of the cavity.

The element S_{ab} of the $\mathbf{S}$ matrix represents the amplitude of the $a \to b$ scattering process. An approximation in terms of rays exists (Jalabert et al. 1990, Baranger et al. 1993):

$$S_{ab} \approx \sum_{\gamma : a \to b} A_\gamma \mathrm{e}^{\mathrm{i}kL_\gamma}, \tag{6.23}$$

where γ is a ray that enters the cavity through channel a, scatters inside for some time, and exits through channel b. A natural question, which arises in analogy with the case of closed systems, is whether this approximation can be used, together with an assumption about the effective "randomness" of the rays, to derive the random matrix theory predictions. A number of encouraging results have been recently obtained in this direction (Richter & Sieber 2002, Heusler et al. 2006, Berkolaiko et al. 2008). The main ingredients are classical rays that are correlated, in the sense that they have similar lengths. For instance, the conductance

$$g = \sum_{\gamma, \rho} A_\gamma A_\rho^* \mathrm{e}^{\mathrm{i}k(L_\gamma - L_\rho)} \tag{6.24}$$

involves a double sum over pairs of scattering trajectories. If we consider the average of this quantity over a small window in k, the rapidly oscillating phase factor will average to zero, except if $L_\gamma - L_\rho \approx k^{-1}$.

It is important to note that the relationship between the asymptotic limit $k \to \infty$, in which the ray-based approximation is expected to be valid, and random matrix theory is not as straightforward for scattering as it is in the case of closed systems. This is due to the presence of two characteristic timescales. One of them is the average dwell time τ_D, given basically by the relative volume of the apertures. The other is the Ehrenfest time $\tau_E = \lambda_L^{-1} \log(kW)$, where λ_L is the Lyapunov exponent of the ray dynamics. This is essentially the time it takes for two rays initially only one wavelength apart to become separated by a distance W. For both closed and open systems one can defined a "chaotic time" τ_0, roughly the time it takes for the ray dynamics to be "randomized." In terms of these timescales, the regime of applicability of random matrix theory is expected to be when $\tau_D \gg \tau_0$, so that the incoming rays spend sufficient time inside the cavity to randomize effectively, but also $\tau_D \gg \tau_E$, so that waves also become effectively random. This last relation is nontrivial because τ_D remains fixed in the limit $k \to \infty$, while τ_E slowly diverges. Therefore, if k is large enough that $\tau_E \gtrsim \tau_D$ the universality predicted by random matrix theory breaks down and geometry-specific effects become important (Jacquod & Sukhorukov 2004, Whitney & Jacquod 2005, Rahav & Brouwer 2006, Brouwer & Rahav 2006). The wavefunctions of scattering states are then closely related to scattering rays that are far from being uniformly distributed. They are therefore not described by the random wave model. An example is shown in Figure 6.2, taken from Ishio and Keating (2004).

6.4 Open Systems: Resonances

Resonances, sometimes called quasibound states or Gamow states, are poles of the scattering matrix (and of the Green's function) corresponding to complex wavenumbers. Resonance wavefunctions, which form part of a natural basis set in scattering theory (Tolstikhin et al. 1997, 1998, Ching et al. 1998), are related to the corresponding residues in the same way that the eigenfunctions are in closed systems. The poles lie in the lower half-plane, $\widetilde{k}_n = k_n - \mathrm{i}\Gamma_n$, and the negative imaginary part $\Gamma_n > 0$ is called the decay rate. If a system is only weakly open, the resonances are close to the real axis, in such a way that the real parts $\widetilde{k}_n$ approximate the original spectrum of the closed system. Another way to look at resonances is in terms of wavefunctions satisfying purely outgoing boundary conditions: from the $\mathbf{S}$ matrix relation (6.15) we see that an outgoing wave $\mathbf{b}$ can exist with no incoming wave ($\mathbf{a} = 0$) only when $\mathbf{S}$ has a pole.

Resonances may be detected indirectly, by measuring their effects on observable quantities. When resonances are close to the real axis and nonoverlapping, they lead to sharp and isolated features in the scattering cross section. The width of such a resonant peak is then proportional to the corresponding decay rate. We mention in passing that resonances are of utmost importance in the optics of microlasers. Here a cavity is filled with an optically active medium, which amplifies a quasibound state leading to laser action (the optical amplification essentially brings the pole up to the real line). The spatial profile of the resonance wavefunction crucially determines the direction of laser emission (Nockel & Stone 1997, Gmachl et al. 1998).

There exists a large body of literature dedicated to the random matrix theory approach to resonances. One way to model non-Hermitian Hamiltonians is by writing $\mathbf{H} = \mathbf{H}_0 - \mathrm{i}\mathbf{V}$, where both $\mathbf{H}_0$ and $\mathbf{V}$ are Hermitian. The matrix $\mathbf{H}_0$ is supposed to model the closed system, whereas $\mathbf{V}$ is responsible for coupling this with the outside. It is then natural to assume $\mathbf{H}_0$ to be a random element of some random matrix ensemble, whereas $\mathbf{V}$ can be taken as either random or constant. Another possibility is to implement the model at the level of a non-unitary propagator $\mathbf{U}$, which may be taken as $\mathbf{U} = \mathbf{U}_0\mathbf{V}$ with $\mathbf{U}_0$ a random unitary matrix. Interesting results exist for both approaches, in different combinations of ensembles and choices of coupling (see the review by Fyodorov & Sommers 2003 and references therein).

A useful tool in studying resonances in the high-frequency regime is the "semiclassical" (ray-based) zeta function, which in two dimensions can be written

$$1/\zeta_{\mathrm{sc}}(k) = \prod_{p}\prod_{j=0}^{\infty}\left(1 - \mathrm{e}^{\mathrm{i}kL_p - \lambda_L L_p(j+1/2)}\right)^{j+1}, \tag{6.25}$$

in terms of periodic rays p. Here L_p denotes the length of p and again λ_L is the Lyapunov exponent. The function $1/\zeta_{\mathrm{sc}}(k)$ is an approximation to the spectral determinant $\det(k^2 - \Delta)$ corresponding to (6.3), and its zeros $1/\zeta_{\mathrm{sc}}(z_j) = 0$ approximate the resonances, $z_j \approx \widetilde{k}_j$. The semiclassical zeta function was employed in the study of scattering by a set of fixed disks in Cvitanovic and Eckhardt (1989, 1993), Gaspard and Rice (1989), Gaspard, Alonso, Okuda and Nakamura (1994), Wirzba (1999), Gaspard and Ramirez (1992), and Gaspard and Baras (1995). In particular, it has been observed experimentally using microwaves (Lu et al. 1999) that the peaks in the transmission function $T(k)$ between two antennas inside a chaotic cavity are well approximated by Lorentzian contributions from $z_j = s_j - \mathrm{i}s'_j$,

$$T(k) \approx \sum_j \frac{c_j s'_j}{(k - s_j)^2 + s'^2_j}. \tag{6.26}$$

In the high-frequency regime resonances are intimately related to *trapped sets* of the corresponding ray dynamics. The *forward trapped set* is the set of initial conditions inside the scattering region that remain inside for all times when propagated forward. We call this set K_+. The *backward trapped set* K_-, on the other hand, is the set of initial conditions inside the scattering region that remain inside for all times when propagated backward. If the dynamics is chaotic these two sets have fractal dimensions. Their intersection, $K_0 = K_+ \cap K_-$, is an invariant set called the repeller. The analog of Weyl's law (6.7) for resonances is the so-called fractal Weyl law (Lin & Zworski 2002, Lu et al. 2003), which relates the number of resonances up to a height k in the real axis with decay rates less than some constant γ to the fractal dimension d_R of the repeller,

$$\#\{\widetilde{k}_n : \Im\{\widetilde{k}_n\} > -\gamma, \quad \Re\{\widetilde{k}_n\} < k\} \sim k^{d_R/2}. \tag{6.27}$$

This result has not been rigorously proven, but it is consistent with a rigorous upper bound (Sjöstrand & Zworski 2007) and is supported by extensive numerical evidence (Lin & Zworski 2002, Lu et al. 2003, Schomerus & Tworzydlo 2004, Nonnenmacher & Zworski 2005, 2007a).

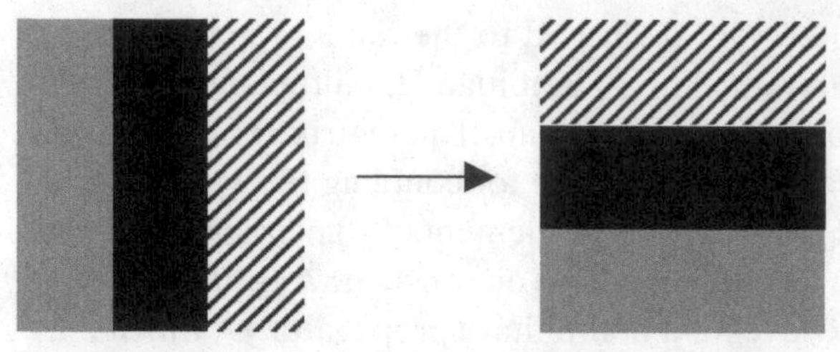

Figure 6.3. The triadic baker map acting on a square of unit side. The rectangles are compressed by a factor of 3 in the vertical direction, stretched by the same factor in the horizontal direction, and repositioned (they are not rotated). The only fixed points are (0, 0), (0.5, 0.5), and (1, 1). The dynamics is uniformly hyperbolic with Lyapunov exponent log 3.

Concerning the resonance wavefunctions, one must distinguish between the right and left eigenstates associated with a resonance $\widetilde{k}_n = k_n - \mathrm{i}\Gamma_n$. When we take the limit of large real part $k_n \to \infty$, keeping the decay rate bounded, $\Gamma_n < \gamma$, stationary states concentrate on trapped sets (Keating et al. 2006, Nonnenmacher & Rubin 2007). In this sense they may be thought of as "fractal wavefunctions" (Casati et al. 1999). How they concentrate on their support depends on the value of the decay rate. They are believed to be approximately uniformly distributed if the decay rate is equal to the classical escape rate of the system. On the other hand, they tend to concentrate on the repeller K_0 if the decay rate is very small. However, there may exist a minimum decay rate if the dynamics is sufficiently chaotic (Cvitanovic & Eckhardt 1989, 1993, Gaspard & Rice 1989, Gaspard et al. 1994, Wirzba 1999, Nonnenmacher & Zworski 2007b).

In order to illustrate the properties outlined above, we now discuss these questions in more detail for maps. Suppose $\mathscr{M}$ is a map acting on the torus T. An illustrative example is the triadic baker map,

$$\mathscr{M}_{\text{tribaker}}(q, p) = \begin{cases} (3q, \frac{p}{3}) & \text{if } 0 \leq q < \frac{1}{3}, \\ (3q-1, \frac{p+1}{3}) & \text{if } \frac{1}{3} \leq q < \frac{2}{3}, \\ (3q-2, \frac{p+2}{3}) & \text{if } \frac{2}{3} \leq q < 1, \end{cases} \tag{6.28}$$

which is similar to (6.12) but involves three different regions. Its action is shown in Figure 6.3. We obtain an open analog of the map $\widetilde{\mathscr{M}}$ by identifying some region of T as the opening $\mathscr{O}$; that is, points that fall in that region escape to infinity and are removed. In the bounce map discussed in Section 6.2 this would be the portion of the perimeter where a lead is attached. The construction of this type of map for scattering problems has been discussed at length in Vallejos and Ozorio de Almeida (1999) and Ozorio de Almeida and Vallejos (2000, 2001).

If $\mathscr{O}_m = \mathscr{M}^m(\mathscr{O})$ denotes the mth image of the opening under the action of the map, the forward-trapped and backward-trapped sets are defined, respectively, by removing the backward and forward images of $\mathscr{O}$,

$$K_+ = T \setminus \bigcup_{m=0}^{\infty} \mathscr{O}_{-m}, \quad K_- = T \setminus \bigcup_{m=1}^{\infty} \mathscr{O}_m. \tag{6.29}$$

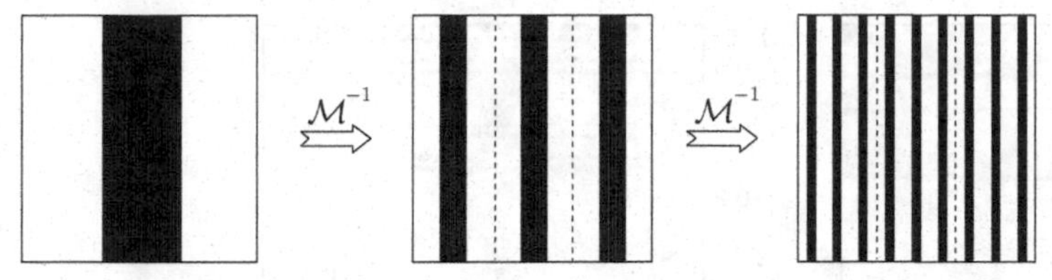

Figure 6.4. We depict in black the opening (left) and the regions $\mathcal{O}_{-n}$ for $n = 1$ (middle) and $n = 2$ (right), for the triadic baker map. The forward-trapped set K_+ is therefore a Cantor set of vertical lines.

Let us also define the set of points that fall into the opening after m steps, but not earlier,

$$\mathcal{R}_+^m = \{x \in \mathcal{O}_{-m},\ x \notin \mathcal{O}_{-n},\ 0 \leq n < m\}. \tag{6.30}$$

In addition, we define $\mathcal{R}_-^m = \{x \in \mathcal{O}_m,\ x \notin \mathcal{O}_n,\ 1 \leq n < m\}$ for $m > 1$ and $\mathcal{R}_-^1 = \mathcal{O}_1$. For (6.28) we can identify the middle region as $\mathcal{O}$. Its pre-images $\mathcal{O}_{-n}$ are vertical strips of exponentially decreasing width, as we show in Figure 6.4. It is easy to see that by removing them we generate K_+ as a Cantor set of vertical lines. On the other hand, K_- is a Cantor set of horizontal lines. The fractal dimension of these sets is $1 + \log(2)/\log(3)$. The dimension of their intersection is $d_R = 2\log(2)/\log(3)$.

Suppose now that $\mathbf{U}$ is an N-dimensional unitary matrix representing the wave map associated with the ray map $\mathcal{M}$. The system is then opened. If the opening $\mathcal{O}$ occupies a fraction M/N of the total area in phase space, then the open wave map corresponds to a non-unitary matrix $\widetilde{\mathbf{U}} = \mathbf{U}\mathbf{\Pi}$, where $\mathbf{\Pi}$ is a projector onto the complement of the opening. The result of multiplying by $\mathbf{\Pi}$ is to set M columns of $\mathbf{U}$ equal to zero. The open wave map corresponding to (6.28) is

$$\mathbf{U} = \mathbf{F}_N^{-1} \begin{pmatrix} \mathbf{F}_{N/3} & 0 & 0 \\ 0 & 0 & 0 \\ 0 & 0 & \mathbf{F}_{N/3} \end{pmatrix}, \tag{6.31}$$

where again $\mathbf{F}_N$ is the discrete Fourier transform in N (assumed to be a multiple of 3) dimensions. The fractal Weyl law predicts that the number of eigenvalues with modulus larger than a fixed value r_0 should depend on the dimension as $N^{-\log 2/\log 3}$. This has been confirmed by numerical computation (Nonnenmacher & Zworski 2005, Nonnenmacher & Zworski 2007a).

A heuristic physical explanation for the fractal Weyl law in an open chaotic wave map can be given as follows (Schomerus & Tworzydlo 2004): States localized on pre-images $\mathcal{O}_{-n}$ of the opening with small n have an almost-complete decay and thus correspond to generalized eigenvectors with almost null eigenvalue. This leads to an accumulation of the spectrum at the origin. The fraction of states with null eigenvalue is given roughly by the area occupied by the union of all pre-images up to $n = \tau_E$, which is roughly $1 - \mathrm{e}^{-\tau_E/\tau_D}$ (Schomerus & Tworzydlo 2004). The fraction of the spectrum that does not accumulate at the origin is then $N\mathrm{e}^{-\tau_E/\tau_D} \propto N^{1-(\lambda_L \tau_D)^{-1}}$, corresponding to a power-law dependence on N with fractional exponent. For small openings the exponent indeed agrees with the fractal dimension.

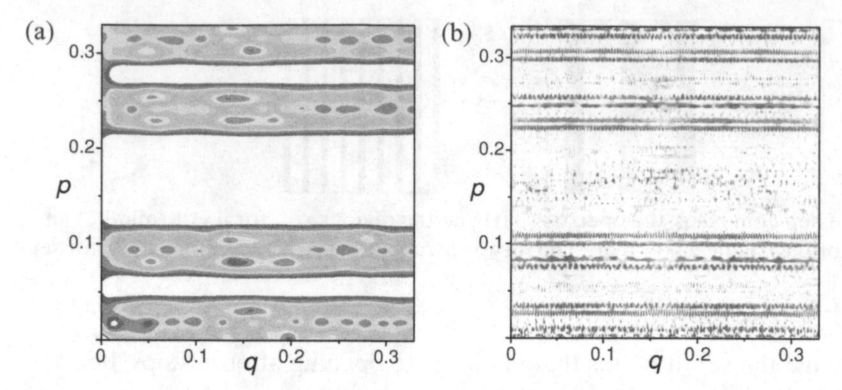

Figure 6.5. (a) The Wigner function and (b) its smoothing, both averaged over the 100 longest-lived right eigenstates of the open triadic baker map, for $N = 3^7$. In the white regions the Wigner function is non-positive. Both figures show the expected concentration on the backward-trapped set of the ray dynamics. Taken from Keating et al. (2006). Copyright 2006 by the American Physical Society. See http://www.cambridge.org/wrightandweaver for a colour version of this figure.

Since the matrix $\widetilde{\mathbf{U}}$ is not unitary, we must distinguish between its left and right eigenvectors,

$$\sum_j (\widetilde{\mathbf{U}})_{mj} (\Psi_n^R)_j = z_n (\Psi_n^R)_m, \quad \sum_j \left(\Psi_n^L\right)_j (\widetilde{\mathbf{U}})_{jm} = z_n \left(\Psi_n^L\right)_m. \tag{6.32}$$

We assume $\int_T |\Psi_n^L(q)|^2 \, dq = \int_T |\Psi_n^R(q)|^2 \, dq = 1$. The eigenvalues z_n lie inside the unit circle in the complex plane, $|z_n|^2 = e^{-\Gamma_n} \leq 1$, where $\Gamma_n \geq 0$ is the decay rate. The typical dwell time is $\tau_D = N/M$, whereas the Ehrenfest time is $\tau_E = \lambda_L^{-1} \ln M$.

It was shown in Keating et al. (2006) and Nonnenmacher and Rubin (2007) that as $N \to \infty$ the right eigenstates are vanishingly small outside K_- if the decay rate is bounded from above. Denoting the Husimi function, a Gaussian-weighted local average of the Wigner function, by

$$H_\psi(q, p) = N \int W_\psi(q', p') e^{-\pi N(q-q')^2 - \pi N(p-p')^2} \, dq' \, dp', \tag{6.33}$$

this corresponds to

$$H_{\Psi_n^R}(x) \approx 0 \quad \text{if } x \notin K_- \ (\Gamma_n < \gamma, N \to \infty). \tag{6.34}$$

Analogously, left eigenstates concentrate on K_+. The localized nature of the wavefunctions of the open triadic baker map can be observed in Figure 6.5 (from Keating et al. 2006), where we plot the averages over the 100 longest-lived right eigenstates of the Husimi and the Wigner function.

With regard to the weight of the resonance wavefunctions on different regions of phase space, a remarkably simple relation holds (Keating et al. 2006),

$$\int_{\mathscr{R}_+^m} H_{\Psi_n^R}(x) \, dx \approx e^{-m\Gamma_n}(1 - e^{-\Gamma_n}). \tag{6.35}$$

Notice that (6.35) vanishes for any fixed m as $\Gamma_n \to 0$; thus if the decay rate approaches zero as $N \to \infty$, the corresponding wavefunction becomes localized on the invariant set K_0 (again, sometimes a minimum decay rate exists). For the

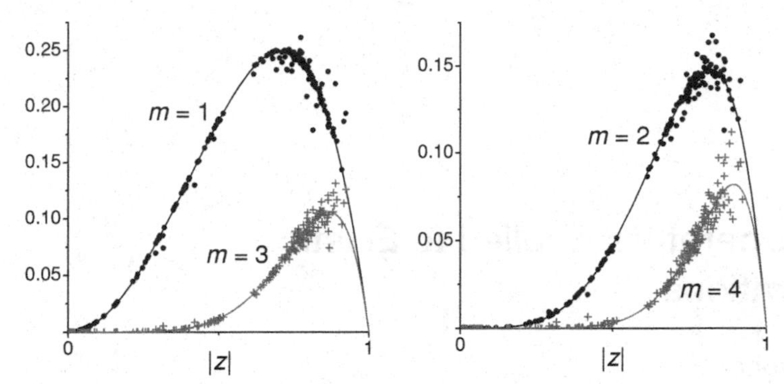

Figure 6.6. The weight $\int_{\mathscr{R}_+^m} H_{\Psi_n^R}(x)\,dx$ of right eigenstates on the regions $\mathscr{R}_+^m$, for the triadic baker map (6.31). The solid line is the approximation (6.35). Taken from Keating et al. (2006). Copyright 2006 by the American Physical Society.

tribaker, (6.35) is in good agreement with the numerical results, as may be seen from Figure 6.6.

It is interesting to note that the distribution of the eigenstates in phase space depends explicitly on the corresponding decay rate. This is an important difference with respect to the case of closed systems, when the large-wavenumber limit is generically believed to be unique (wave ergodicity). We also note that the areas of the regions $\mathscr{R}_+^m$ are proportional to $e^{-m\gamma_c}$, where γ_c is the classical escape rate. Therefore one may conjecture (as was done in Casati et al. 1999) that resonance wavefunctions with $\Gamma_n \approx \gamma_c$ should be constant (up to small fluctuations and in the weak sense) over the set K_-.

6.5 Summary

The main points in this chapter are the following.

- In closed geometries the eigenfrequencies corresponding to the stationary solutions of the wave equation are real. Their average density depends on the volume of the domain (Weyl's law). The associated eigenfunctions are, in the short-wavelength limit, uniform over the domain (wave ergodicity), with fluctuations on the scale of a wavelength.
- The scattering of waves can be characterized by a scattering matrix **S**. In complex geometries, many statistical properties of **S** are universal (geometry independent) and can be modeled by those of random matrices. This can be understood in terms of the chaotic properties of rays, which are trapped for a long time. Geometry-specific corrections arise at very high frequencies.
- In open geometries the analogs of the eigenfrequencies are resonances. These are complex poles of **S** and relate to transient states. The imaginary parts determine the lifetimes of the corresponding states. The density of the resonances in the complex plane is related to the fractal dimension of the set of rays that are trapped inside the scattering domain (the fractal Weyl law). The resonance wavefunctions are also supported on trapped sets and so again inherit their fractal structure.

7 Transfer Operators Applied to Elastic Plate Vibrations

Niels Søndergaard

Division of Mathematical Physics, Lund Technical University, Lund, Sweden

In ray approximations of boundary integral kernels, unitary transfer operators arise. The existence of transfer operators implies trace formulas for the fluctuating part of the spectral density and due to unitarity a reduction of the number of periodic orbits used in these formulas. We introduce the transfer operator method and discuss its applications in elasticity in the case of plate vibrations.

7.1 Introduction

In the construction of fluctuating spectral densities (Gutzwiller 1990, Brack & Bhaduri 1997, Stöckmann 1999), the physicist Gutzwiller took a trace of a unitary evolution operator, the Green's function in path integral form. The resulting *trace formulas* are expressed as a sum over periodic orbits. Later, in the setting of resonators it was discovered that the more conventional boundary integral kernels (Kitahara 1985, Bonnet 1995) could do the same job and that these kernels could also be made unitary.

One such approach is the transfer matrix of Bogomolny (Bogomolny 1992, Boasman 1994). Such ray kernels contain the essence of semiclassical approximations and can be used for numerics even without periodic orbits. Furthermore, these operators are unitary, leading to a truncation in the trace formula (called *resurgence*) of the infinitely many periodic orbits of the system in question (Berry & Keating 1990, Georgeot & Prange 1995). Another unitary approach is *inside–outside duality* based on scattering (Smilansky 1995, Smilansky & Ussishkin 1996).

Fortunately, these methods are not restricted to purely quantum mechanical problems. In elasticity, ray methods are standard, and furthermore, unitarity follows naturally from energy conservation (Achenbach et al. 1982). Thus, in elasticity of plates two cases have been treated with the transfer matrix: that of in-plane elasticity and that of bending elasticity.

This presentation is organized as follows: First the method of the transfer matrix is presented for the scalar Helmholtz equation. Next we continue with two-dimensional elasticity treating in-plane fields (plane strain) and bending motion respectively. We then briefly discuss possible extensions of these results.

7.2 Scalar Helmholtz Equation

7.2.1 Periodic Orbits: Classical Quantities

For simplicity, we first discuss the transfer matrix method for the Helmholtz partial differential equation with the Dirichlet boundary condition

$$\Delta\psi + k^2\psi = 0 \quad \text{with} \quad \psi = 0 \quad \text{at} \quad \partial\Omega. \tag{7.1}$$

The scalar field is denoted ψ. This scalar may represent pressure or in the case of quantum billiards the idealized case of an electronic quantum dot. The wavenumber k shall be taken as our spectral parameter.

The ray limit of (7.1) gives rise to a classical dynamics. In this dynamics, we consider polygonal orbits inscribed in the billiard region Ω. Certain orbits do not contain repeats of shorter orbits. These are referred to as *prime* orbits. A given prime orbit p has n_p impacts with the boundary. The impacts are labeled with i below.

The length of the segment between impact i and $i+1$ is denoted $l_{i,i+1}$ ($i+1$ is calculated modulo n_p). Then, the total length becomes

$$L_p = \sum_{i=1}^{n_p} l_{i,i+1}^{(p)}. \tag{7.2}$$

To calculate the stability of a given closed ray, we follow the procedure used in geometric optics. There, stabilities of optical setups are assessed by the use of ray matrices.

Our conventions for ray coordinates are x for the position in arc length of a ray and $p = k\sin\theta$ for the projected tangential momentum with respect to the boundary. We state without proof the corresponding matrices $\mathbf{J} = \mathbf{F}\cdot\mathbf{R}$ of flight and reflection. They act from the left as

$$\begin{pmatrix}\delta x_+\\ \delta p_+\end{pmatrix} = \mathbf{J}\cdot\begin{pmatrix}\delta x_-\\ \delta p_-\end{pmatrix}, \tag{7.3}$$

transforming initial infinitesimal variations of the ray to final.

In the coordinates mentioned, flight from impact i to impact $i+1$ is given by

$$\mathbf{F}_i = \mathbf{F}(i+1,i) = \begin{pmatrix}1 & \frac{L}{k\cos^2\theta}\\ 0 & 1\end{pmatrix} \tag{7.4}$$

and reflection at impact i by

$$\mathbf{R}_i = -\begin{pmatrix}1 & 0\\ \frac{\Delta k}{r} & 1\end{pmatrix} \tag{7.5}$$

with L the flight length, $\Delta k = 2k\cos\theta$ the change in normal momentum, and r the local radius of curvature of the boundary.

Therefore, the product of all ray matrices

$$\mathbf{J}_p = \prod_{i=1}^{n_p} \mathbf{F}_i\cdot\mathbf{R}_i \tag{7.6}$$

is the total stability matrix of an orbit.

7.2.2 Transfer Matrix

Consider the setting of the boundary integral equations (Kitahara 1985, Bonnet 1995). Denote points on the boundary x, y. There will be an integral equation of the form

$$\phi = K\phi \tag{7.7}$$

with $\phi = \partial\psi/\partial n$ and $K = K(k)$, whose solution depends on a corresponding determinental condition

$$\det(1 - K) = 0 . \tag{7.8}$$

When the boundary integral kernel is replaced with its ray approximation, the result is found to be equivalent to the following Van Vleck operator kernel typical of semiclassical approximations for the unitary evolution operator in quantum mechanics (Boasman 1994):

$$T(x, x') = \frac{1}{\sqrt{2\pi \mathrm{i}}}\sqrt{\frac{\partial^2(kL(x, x'))}{\partial x \partial x'}}\, \mathrm{e}^{\mathrm{i}kL(x,x') - \mathrm{i}\mu\pi/2} , \tag{7.9}$$

where μ is a phase shift from caustics. For convenience we also include possible phase shifts (e.g., -1 for Dirichlet conditions) from the boundary. By equivalent, we mean that the traces of this operator $\mathrm{tr}(T^n)$ are identical to those from the ray boundary integral kernel $\mathrm{tr}(K^n)$; namely, (7.8) can also be written using Fredholm's cumulant expansion

$$0 = \exp\left(-\sum_{n=1}^{\infty}\frac{\mathrm{tr}(K^n)}{n}\right). \tag{7.10}$$

If we denote an eigenvalue of K with λ, Fredholm's trace expansion of the infinite determinant in (7.8) is seen formally from the following calculation:

$$\begin{aligned}
\det(1 - K) &= \prod_{\lambda}(1 - \lambda) = \prod_{\lambda}\exp(\log(1 - \lambda)) \\
&= \prod_{\lambda}\exp\left(-\sum_{n=1}^{\infty}\frac{\lambda^n}{n}\right) = \exp\left(-\sum_{\lambda}\sum_{n=1}^{\infty}\frac{\lambda^n}{n}\right) \\
&= \exp\left(-\sum_{n=1}^{\infty}\frac{\mathrm{tr}(K^n)}{n}\right).
\end{aligned} \tag{7.11}$$

In conclusion, the two operators have the same spectrum in the ray limit.

7.2.2.1 Interpretations and Properties of the Transfer Operator

MOMENTA. We define a dimensionless action, similar to the eikonal of geometric optics,

$$S(x, x') = kL(x, x'), \tag{7.12}$$

and *tangential momenta* at x,

$$\partial S/\partial x = p, \tag{7.13}$$

respective at x',

$$\partial S/\partial x' = -p' . \tag{7.14}$$

The p, p' are strictly speaking tangential wavenumbers; however to connect with classical mechanics we have decided to call them momenta.

To get the interpretation of tangential momentum, we write the action in more detailed form:

$$S = S_2(x, y) + S_1(y, z) \equiv kL(x, y) + kL(y, z) , \tag{7.15}$$

where

$$L(x, y) \equiv |\mathbf{x} - \mathbf{y}| = \sqrt{(\mathbf{x} - \mathbf{y}) \cdot (\mathbf{x} - \mathbf{y})}. \tag{7.16}$$

As mentioned, we parametrize $\mathbf{y} = \mathbf{y}(s)$ with s the arc length in the counterclockwise direction. In particular,

$$\frac{d\mathbf{y}}{ds} \equiv \mathbf{t} \tag{7.17}$$

is the unit tangent on the boundary. We see that

$$\partial_y S_2(x, y) = -\mathbf{t} \cdot \left(k \frac{\mathbf{x} - \mathbf{y}}{|\mathbf{x} - \mathbf{y}|} \right) \equiv -p(x \leftarrow y), \tag{7.18}$$

whereas

$$\partial_x S_2(x, y) = \mathbf{t} \cdot \left(k \frac{\mathbf{x} - \mathbf{y}}{|\mathbf{x} - \mathbf{y}|} \right) \equiv +p(x \leftarrow y). \tag{7.19}$$

From (7.16), the condition for stationary phase is

$$\begin{aligned} 0 = \frac{\partial S}{\partial y} &= \frac{\partial S_2(x, y)}{\partial y} + \frac{\partial S_1(y, z)}{\partial y} \\ &\equiv -p(x \leftarrow y) + p(y \leftarrow z), \end{aligned} \tag{7.20}$$

which indeed is the condition for conservation of tangential momentum of a ray coming from z, reflecting at y, and ending at x. We shall call the stationary point y in (7.20): $\bar{y}(x, z)$. Note that in concave geometries, extra care is needed: this leads to the discussion of "ghost orbits" whose contributions can be shown to add up to zero.

Following Bogomolny (1992), we switch back to the actions S_1 and S_2 without using their specific form. Partial differentiation with respect to x in (7.20) gives

$$B^{(1)} + \frac{\partial \bar{y}}{\partial x} D = 0 , \tag{7.21}$$

where the amplitudes $B^{(i)}$ of $T(x, y)$, $T(y, z)$ respective to the Hessian coming from the stationary phase are defined as

$$B^{(1)} = \frac{\partial^2 S_2}{\partial x \partial y}, \qquad B^{(2)} = \frac{\partial^2 S_1}{\partial y \partial z}, \quad \text{and} \quad D = \frac{\partial S_2}{\partial y^2} + \frac{\partial S_1}{\partial y^2}. \tag{7.22}$$

The action of the operator $T(x, z)$ we set equal to

$$\bar{S}(x, z) = S_2(x, \bar{y}) + S_1(\bar{y}, z). \tag{7.23}$$

Partial differentiation of (7.23) with respect to x and z gives

$$L \equiv \frac{\partial^2 \bar{S}}{\partial x \partial z} = \frac{\partial \bar{y}}{\partial x}\frac{\partial \bar{y}}{\partial z} D + B^{(1)} \frac{\partial \bar{y}}{\partial z} + B^{(2)} \frac{\partial \bar{y}}{\partial x}. \tag{7.24}$$

Applying (7.21) to simplify (7.24) produces

$$L = -\frac{B^{(1)} B^{(2)}}{D}, \tag{7.25}$$

which proves the multiplicative property.

With these definitions, we can interpret the amplitude in (7.9) with

$$\frac{\partial^2 (S(x, x'))}{\partial x \partial x'} = \frac{\partial p}{\partial x'}, \tag{7.26}$$

that is, how the final tangential momentum p depends on the initial position x'.

MULTIPLICATIVITY. For the calculation of traces we need to compose T with itself. The composition law

$$T(x, z) \approx \int T(x, y) T(y, z) \, \mathrm{d}y \tag{7.27}$$

holds approximately in the ray limit. The left-hand side in (7.27) is understood for an open orbit consisting of two joined segments going from z to another point $\bar{y}$ at the boundary respective from $\bar{y}$ to x, where $\bar{y}$ is stationary with respect to action variations.

UNITARITY. For the integral kernel T, the unitarity condition reads

$$\int T(x, y) T^*(z, y) \approx \delta(x - z) \, \mathrm{d}y. \tag{7.28}$$

The action now changes to

$$S = k(L(x, y) - L(z, y)), \tag{7.29}$$

leading to a change of sign in front of $L(z, y)$ in the parenthesis of the right-hand side of (7.20):

$$0 = -p(x \leftarrow y) - p(y \leftarrow z). \tag{7.30}$$

Thus, the tangential momentum of the incoming ray has to equal the opposite tangential momentum of the outgoing ray. That condition can be fulfilled if the ray retraces its path, that is, ultimately if $z = x$. In the simple (no-"ghost") case, we expand the action $S(x, y)$ to first order in x around z:

$$(T \cdot T^\dagger)(x, z) = \frac{1}{2\pi} \int \left| \frac{\partial^2 S}{\partial z \partial y} \right| \mathrm{e}^{\mathrm{i}p(x-z)} \, \mathrm{d}y = \delta(x - z), \tag{7.31}$$

where $p = \partial S / \partial z$ was used in a change of variables.

7.2.3 Trace Formula from the Transfer Operator

Consider an *open* orbit returning to itself at position x via $n-1$ reflections at the boundary,

$$T^n(x,x) = \int \cdots \int T(x, y_{n-1})T(y_{n-1}, y_{n-2}) \cdots T(y_1, x)\, \mathrm{d}y_1 \cdots \mathrm{d}y_{n-1}$$
$$\approx \sum_{\text{paths}:x\to x} \sqrt{\frac{\partial p}{\partial x}} \exp\left(\mathrm{i}k L(x, y_{n-1}, \ldots, y_1, x) + \mathrm{i}\tilde{\mu}\frac{\pi}{2}\right). \tag{7.32}$$

The tangential momentum p is the returning momentum at the point x. To complete the calculation of the trace we need to evaluate

$$\mathrm{tr}(T^n) = \int T^n(x,x)\, \mathrm{d}x. \tag{7.33}$$

If we define the corresponding action in question as

$$S = kL(x_+, y_{n-1}, y_{n-2}, \ldots, y_1, x_-)\Big|_{x_-=x_+=x}, \tag{7.34}$$

then at the point x of stationary phase for this action

$$0 = \frac{\partial S}{\partial x} = \left(\frac{\partial S(x_+, \ldots, x_-)}{\partial x_+} + \frac{\partial S(x_+, \ldots, x_-)}{\partial x_-}\right)\Bigg|_{x_-=x_+=x} = p_+ - p_-. \tag{7.35}$$

Thus, the final tangential momentum equals the initial tangential momentum, precisely the condition for a reflection. Therefore, the law of reflection holds at each impact, and a particular stationary point $(x, y_{n-1}, \ldots, y_1)$ corresponds to a classical periodic orbit. Given such a stationary point, the cyclic permutations such as $(y_1, x, y_{n-1} \ldots, y_2)$ also are stationary points, all representing the same orbit. Thus a prime orbit with n_p impacts occurs n_p times.

We then proceed to the amplitude of this final stationary phase integral:

$$\frac{\partial^2 S}{\partial x^2} = \left(\frac{\partial^2 S}{\partial x_-^2} + \frac{\partial^2 S}{\partial x_+^2} + 2\frac{\partial^2 S}{\partial x_- \partial x_+}\right)\Bigg|_{x_-=x_+=x}. \tag{7.36}$$

Define

$$a = \frac{\partial^2 S}{\partial x_-^2}, \quad b = \frac{\partial^2 S}{\partial x_- \partial x_+}, \quad \text{and} \quad c = \frac{\partial^2 S}{\partial x_-^2}. \tag{7.37}$$

Then

$$\mathrm{d}p_- = -a\, \mathrm{d}x_- - b\, \mathrm{d}x_+ \quad \text{and} \quad \mathrm{d}p_+ = b\, \mathrm{d}x_- + c\, \mathrm{d}x_+. \tag{7.38}$$

We continue the calculation of $\mathrm{tr}(T^n)$. Multiplying with the prefactor in (7.32) and using (7.36) and (7.37) the effective amplitude $\mathscr{A}$ becomes

$$\mathscr{A} = \left|\frac{\frac{\partial^2 S}{\partial x_- \partial x_+}}{\partial^2 S/\partial x^2}\right|^{1/2} = \left|\frac{b}{a+c+2b}\right|^{1/2} = \left|2 + \frac{a+c}{b}\right|^{-1/2}. \tag{7.39}$$

Furthermore, expanding the differentials of the momenta in (7.3) via (7.38) reveals after a short calculation

$$J_{11} = -\frac{a}{b} \quad \text{and} \quad J_{22} = -\frac{c}{b}. \tag{7.40}$$

Therefore $\mathscr{A}$ may be expressed in terms of the stability matrix $\mathbf{J}$:

$$\mathscr{A} = \left|2 - \text{tr}(\mathbf{J})\right|^{-1/2} = |\mathbf{I} - \mathbf{J}|^{-1/2}, \tag{7.41}$$

where $|\mathbf{J}| = 1$ was used in the last equality. This fact is related to the symplectic nature of the mapping in phase space from (x_-, p_-) to (x_+, p_+).

Thus, consider the evolution of *oriented volume elements* in phase space $\{(x, p)\}$:

$$\mathrm{d}p_- \wedge \mathrm{d}x_- \mapsto ?$$

Here, the wedge product above is the two-dimensional outer product; it may be represented with the familiar cross product in the present case if we allow a third direction normal to the (x, p)-plane. In two dimensions $\mathrm{d}p \wedge \mathrm{d}x$ is also called the *symplectic form.*

If we use (7.3), then the oriented volume element is scaled by the determinant $|\mathbf{J}|$:

$$\mathrm{d}p_+ \wedge \mathrm{d}x_+ = |\mathbf{J}|\, \mathrm{d}p_- \wedge \mathrm{d}x_-. \tag{7.42}$$

For our explicit choice of coordinates leading to a particular representation of $\mathbf{J}$, Equations (7.4) and (7.5), we see that $|\mathbf{J}| = 1$. Thus, oriented phase-space volume elements are preserved by the map $\mathbf{J}$. In conclusion, the mapping from (x_-, p_-) to (x_+, p_+), $\mathbf{J}$, is by definition a two-dimensional *symplectic* map.

In general, ray dynamics arising from variational principles leads to symplectic maps. This holds in higher dimensions as well; the symplectic form is generalized to $\sum_i \mathrm{d}p_i \wedge \mathrm{d}x^i$ and is conserved.

We note that the amplitude blows up in the case of an eigenvalue of $\mathbf{J}$ equal to unity. That corresponds to regular and stable regions of phase space and shows up in for example the circle billiard or polygonal billiards. In those cases, orbits are not isolated but occur in families. These can also be handled using the method of the transfer matrix (Bogomolny & Hugues 1998).

Earlier, the calculation of the amplitude $\mathscr{A}$ was done in two dimensions. Similar results hold in higher dimensions for general hamiltonian flows with $\mathscr{A} = |\mathbf{1} - \mathbf{J}|^{-1/2}$ and $\mathbf{J}$ the corresponding stability (Gutzwiller 1990).

We find from (7.10) and our calculations for $\text{tr}(T^n)$ that

$$\text{tr}(T^n) = \sum_p \sum_{r:n=n_p r} \frac{1}{r} \frac{\mathrm{e}^{\mathrm{i}rkL_p - \mathrm{i}r\mu_p \frac{\pi}{2}}}{|\mathbf{I} - \mathbf{J}_p^r|^{1/2}}, \tag{7.43}$$

where the sum over r is a sum over repeats of a given prime orbit and μ_p is the *Maslov index* that takes into account all caustics of the orbit plus possible phase shifts from the boundary (Gutzwiller 1990). This gives

$$\det(1 - T) = \exp\left(-\sum_p \sum_{r=1}^{\infty} \frac{1}{r} \frac{\mathrm{e}^{\mathrm{i}rkL_p - \mathrm{i}r\mu_p \frac{\pi}{2}}}{|\mathbf{I} - \mathbf{J}_p^r|^{1/2}}\right), \tag{7.44}$$

the ray expansion of the spectral determinant.

To summarise, we have expressed the condition for an eigenmode, the vanishing of the secular determinant (7.8), which we have expanded in the ray limit using the transfer matrix T.

We turn to the problem of creating the spectral distribution from the resonance condition. As T is unitary, the condition (7.8) is fulfilled whenever one of its (k-dependent) eigenphases $\exp(\mathrm{i}\phi)$ of T equals unity. Hence, the number of modes in a given interval $[k, k+\mathrm{d}k]$,

$$\begin{aligned}
\mathrm{d}N &= d(k)\mathrm{d}k \\
&= \sum_{\phi}\sum_{n=-\infty}^{\infty} \delta(\phi - n2\pi)\mathrm{d}\phi \\
&= \sum_{\phi}\sum_{N=-\infty}^{\infty} \int \mathrm{d}n\, \mathrm{e}^{\mathrm{i}2\pi nN}\, \delta(\phi - n2\pi)\mathrm{d}\phi \\
&= \sum_{\phi}\sum_{N=-\infty}^{\infty} \mathrm{e}^{\mathrm{i}\phi N}\frac{\mathrm{d}\phi}{2\pi}.
\end{aligned} \tag{7.45}$$

In the preceding expression, we denoted the formal sum over eigenphases with $\sum_\phi$ and used Poisson summation at the third equality sign.

We now concentrate on all terms in (7.45) with $N \neq 0$, as these terms will lead to fluctuations. Dividing by $\mathrm{d}k$ gives the oscillatory density of states:

$$\begin{aligned}
d_{osc}(k) &\equiv \sum_{\phi}\sum_{N=1}^{\infty} \mathrm{e}^{\mathrm{i}\phi N}\frac{\phi'}{2\pi} + \text{c.c.} \\
&= \frac{\mathrm{d}}{\mathrm{d}k}\left(\sum_{\phi}\sum_{N=1}^{\infty} \frac{1}{2\pi\mathrm{i}N}\mathrm{e}^{\mathrm{i}\phi N} - \text{c.c.}\right) \\
&= \frac{1}{2\pi\mathrm{i}}\frac{\mathrm{d}}{\mathrm{d}k}\left(\sum_{N=1}^{\infty} \frac{1}{N}\,\mathrm{tr}(T^N) - \text{c.c.}\right) \\
&= \frac{1}{\pi}\frac{\mathrm{d}}{\mathrm{d}k}\sum_{p}\sum_{r=1}^{\infty} \frac{1}{r}\frac{\sin\left(rkL_p - r\mu_p\frac{\pi}{2}\right)}{|\mathbf{I} - \mathbf{J}_p^r|^{1/2}} \\
&= \frac{1}{\pi}\sum_{p} L_p \sum_{r=1}^{\infty} \frac{\cos(rkL_p - r\mu_p\frac{\pi}{2})}{|\mathbf{I} - \mathbf{J}_p^r|^{1/2}}.
\end{aligned} \tag{7.46}$$

In the derivation, we defined $\phi' \equiv \mathrm{d}\phi/\mathrm{d}k$ and used the expression for the traces (7.43).

In conclusion, the oscillatory spectral density is given by

$$d_{osc}(k) = \frac{1}{\pi}\sum_{p} L_p \sum_{r=1}^{\infty} \frac{\cos(krL_p - r\mu_p\frac{\pi}{2})}{|\mathbf{I} - \mathbf{J}_p^r|^{1/2}}. \tag{7.47}$$

We see that each orbit and its repeats lead to fluctuations in the spectral density. Very unstable orbits have large values of $|\mathbf{I} - \mathbf{J}_p^r|$, giving small weights. However, these orbits tend to be long, and the number of longer orbits usually grows exponentially.

7.3 Elastic In-Plane Modes

7.3.1 Wave Equation

There exist two-dimensional idealizations of elasticity used in the case of translational symmetry (plane strain) or very thin flat plates (plane stress). For such in-plane elasticity in its simplest form, *isotropic* plane strain (respectively plane stress with a slight change of constants), the underlying vectorial displacement field $\mathbf{u}$ has two components. The partial differential equation governing $\mathbf{u}$ written in vector form is

$$\mu\Delta\mathbf{u} + (\lambda + \mu)\nabla\nabla \cdot \mathbf{u} + \rho\omega^2\mathbf{u} = 0, \tag{7.48}$$

where μ, λ are the Lamé constants for the isotropic elastic material of density ρ (e.g., Auld 1973, Landau & Lifshitz 1959).

The *free* boundary condition is of particular interest,

$$\mathbf{t(u)} = \boldsymbol{\sigma}(\mathbf{u}) \cdot \mathbf{n} = \mathbf{0}, \tag{7.49}$$

for the displacement field at the boundary, where $\mathbf{n}$ denotes the normal to the boundary. The operator $\mathbf{t}$ refers to the traction coming from the stress tensor given in components by

$$\sigma_{ij} = \lambda\, \partial_k u_k \delta_{ij} + \mu\, (\partial_i u_j + \partial_j u_i)\,. \tag{7.50}$$

7.3.2 Ray Dynamics

The wave Equation (7.48) allows two different polarizations, longitudinal L and transverse T, each with their own wave speed:

$$c_{\mathrm{L}}^2 = \frac{\lambda + 2\mu}{\rho} \quad \text{and} \quad c_{\mathrm{T}}^2 = \frac{\mu}{\rho}. \tag{7.51}$$

As in the scalar case, each polarization mode propagates in the interior of a resonator in straight lines.

Because of the difference in wave speeds, the ray dynamics now also involves *refraction* and *ray splitting* (see Figure 7.1). Therefore, an incoming angle θ_- of a ray no longer has to equal the outgoing θ_+, but rather it obeys

$$\frac{c_{\mathrm{L}}}{c_{\mathrm{T}}} = \frac{\sin\theta_{\mathrm{L}}}{\sin\theta_{\mathrm{T}}}, \tag{7.52}$$

the law of refraction. At the level of plane waves, each impact is associated with certain reflection coefficients $\boldsymbol{\alpha}$ (Landau & Lifshitz 1959). It turns out that those entering the trace formula are those that are *unitary*. They can be found from the ordinary coefficients denoted $\mathbf{a}$ by putting for example

$$\alpha(L \leftarrow T) = \left(\frac{c_{\mathrm{L}}\cos\theta_{\mathrm{L}}}{c_{\mathrm{T}}\cos\theta_{\mathrm{T}}}\right)^{1/2} a(L \leftarrow T), \tag{7.53}$$

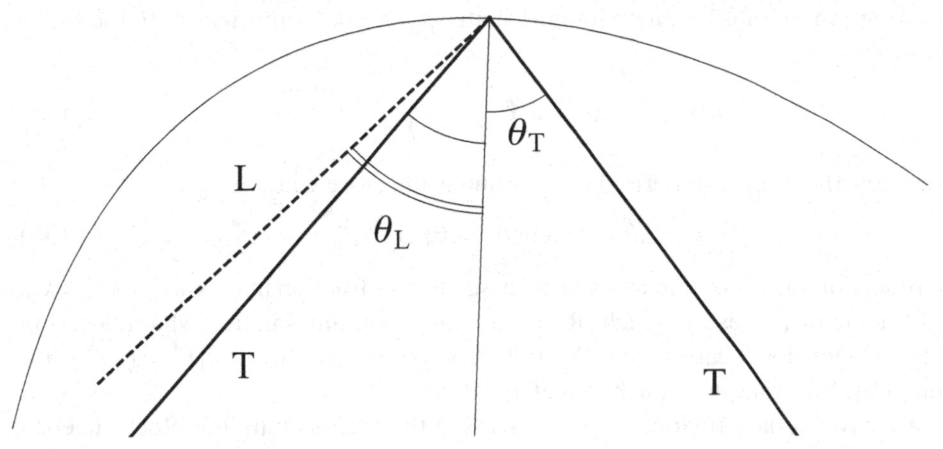

Figure 7.1. Ray splitting.

and likewise for the opposite conversion. These unitary coefficients then represent the conservation of the normal energy flux.

7.3.3 Transfer Matrix

The derivation of the transfer matrix in the elastic case is a longer affair, and so we shall refer to Tanner and Søndergaard (2007). However, what is crucial is its close resemblance to the scalar transfer operator (7.9):

$$T(x,x') = \frac{1}{\sqrt{2\pi \mathrm{i}}} \sqrt{\frac{\partial^2 (L(x,x'))}{\partial x \partial x'}} \boldsymbol{\alpha} \begin{pmatrix} \sqrt{k_\mathrm{L}} \mathrm{e}^{\mathrm{i} k_\mathrm{L} L(x,x')} & 0 \\ 0 & \sqrt{k_\mathrm{T}} \mathrm{e}^{\mathrm{i} k_\mathrm{T} L(x,x')} \end{pmatrix}. \tag{7.54}$$

Now, the transfer matrix is also a matrix in polarization indices, but otherwise it is identical to the scalar transfer matrix. Because of these similarities it is clear that the trace formula derived in the previous section generalizes with a few modifications.

7.3.4 Trace Formula

In the present case of elastodynamics, there are several wavenumbers in play, both the longitudinal and the transverse. As a spectral parameter we then choose the more convenient angular velocity ω. The parameter dual to the angular velocity is time. This then enters the trace formula via the total cycle time

$$T_p = \sum_{i=1}^{n_p} \frac{l_{i,i+1}^{(p)}}{c_{i,i+1}^{(p)}} \tag{7.55}$$

instead of the total length.

Define the polarization at stage i as $\sigma_i = L, T$. The mode conversion amplitude is the product over all mode conversion coefficients along the orbit:

$$\alpha_p = \prod_{i=1}^{n_p} \alpha_{\sigma_i \sigma_{i+1}}. \tag{7.56}$$

The ray splitting leads to a new stability matrix relevant for refraction: In the (x, p) coordinates,

$$\mathbf{R} = -\begin{pmatrix} 1 & 0 \\ \frac{\Delta k}{r} & 1 \end{pmatrix}, \tag{7.57}$$

where now the change in normal momentum is the more general

$$\Delta k = k_- \cos\theta_- + k_+ \cos\theta_+ . \tag{7.58}$$

The reflection matrix is seen as a special case of the refraction matrix when $k_- = k_+$.

Other coordinates for which $|\mathbf{R}| \neq 1$ are also sometimes in use; what matters in the two-dimensional case is that the full stability matrix (7.6) has $|\mathbf{J}| = 1$, which is fulfilled by, for example, Couchman et al. (1992).

We have in the previous section discussed the various building blocks needed for the trace formula. Because the reflection coefficient matrix $\boldsymbol{\alpha}$ is a full matrix, the traces $\mathrm{tr}(T^n)$ involve sums over orbits that contain segments of different polarization.

It can be checked that the stationary phase integrations with respect to the impact positions x_i of an orbit polygon force the tangential momenta to coincide – just as in the scalar case. In the case of ray splitting, this conservation

$$p_- = p_+ \iff k_- \sin\theta_- = k_+ \sin\theta_+ \iff \frac{\sin\theta_+}{\sin\theta_-} = \frac{k_-}{k_+} = \frac{c_+}{c_-} \tag{7.59}$$

is precisely the law of refraction. Therefore, a stationary polygonal orbit obeys the usual laws of geometric ray propagation. The total phase from the actions becomes

$$\sum_{i=1}^{n_p} k_{\sigma_i} l_{i,i+1} = \omega T_p , \tag{7.60}$$

when expressed via the orbit's cycle time.

The derivation of the geometric amplitude $\mathscr{A}$ is similar to the scalar case and leads again to the factor $|\mathbf{I} - \mathbf{J}|^{-1/2}$, where now the more general refraction ray matrix is used. Further contributions to the amplitude come from the product of the reflection coefficients, α_p.

As the matrix $\boldsymbol{\alpha}$ is full in general, orbits with segments of different polarizations are generated in the trace expansion $\mathrm{tr}(T^n)$. These orbits are new relative to the scalar case. Therefore with ray splitting, more orbits are generated. Likewise if we sort orbits after their cycle time, the number of orbits grows more rapidly than in the scalar case.

The effect of the ray-splitting orbits on the fluctuating spectral density can therefore be written as

$$d(\omega) = \frac{1}{\pi} \sum_p T_p \sum_{r=1}^{\infty} \alpha_p^r \frac{\cos(\omega r T_p - r\mu_p \pi/2)}{|\mathbf{I} - \mathbf{J}_p^r|^{1/2}} . \tag{7.61}$$

Thus, each orbit gives rise to fluctuations in the spectral density with a spacing controlled by the orbit period as $2\pi/T_p$: in general, very fast orbits give the information on the coarsest scale and furthermore tend to be more stable, leading to a higher weight in (7.61). A formally identical trace formula was written down in the quantum case with ray splitting by Couchman et al. (1992).

In the case of T-rays hitting the boundary almost tangentially, the rays are beyond the critical angle given by $\arcsin(c_T/c_L)$. Then there is no mode conversion or loss of energy. Thus, the accompanying reflection coefficient is just given by a phase shift, which also enters the phase of the trace formula (see Sondergaard & Tanner 2002 for details of the case of the elastic disk). As expected, it was found that purely T-polarized orbits having all incidence angles beyond the critical angle had high strength compared with the general ray-splitting orbits. Pure surface rays corresponding to Rayleigh waves are also expected to be important but will not be discussed here (Viktorov 1967).

7.4 Elastic Bending Modes

A plate's bending is the transverse field in the normal direction. In this case the field obeys the *biharmonic* equation

$$D\Delta^2 w + \rho h \omega^2 w = 0 \tag{7.62}$$

with $D = Eh^3/12(1-\nu^2)$ the flexural rigidity, E the modulus of extension, ν the Poisson ratio, and h the thickness. Flexural waves are dispersive:

$$k^4 = \frac{h\rho}{D}\omega^2. \tag{7.63}$$

The corresponding transfer matrix was found by Bogomolny and Hugues (1998). It has form similar to that of the scalar Helmholtz Equation (7.9). The difference is an additional phase shift ϕ depending on the particular boundary conditions in use. For example, in the case of clamped boundary conditions

$$w = 0 \quad \text{and} \quad \frac{\partial w}{\partial n} = 0, \tag{7.64}$$

the shift ϕ_c equals

$$\phi_c = -2\arctan\left(\frac{\cos\theta}{\sqrt{1+\sin^2\theta}}\right). \tag{7.65}$$

This leads to an additional factor $2\exp(i\phi_c)$ in front of the transfer matrix. This factor also enters the trace formula with the accumulated additional phase shifts

$$\Phi_p = \sum_{i=1}^{n_p} \phi_{c,i} \tag{7.66}$$

in (7.47) using the wavenumber k as a spectral parameter.

7.4.1 Smooth Term: Results

For completeness, we state the results for the *smooth* modal count $\bar{N}(\omega)$. These were originally derived in statistical physics to calculate heat capacities of elastic bodies. For a general discussion see Bogomolny and Hugues (1998), and see Bertelsen et al. (2000) for experiments. To the best of the author's knowledge, for the counting function of isotropic in-plane elasticity, only the first two terms in the asymptotic series

$$\bar{N} \sim \bar{N}_2 + \bar{N}_1 + \cdots \tag{7.67}$$

are known.

The leading term $\bar{N}_2$ equals the "available phase-space area":

$$\bar{N}_2 = \pi \left(c_L^{-2} + c_T^{-2}\right) |B| f^2 \tag{7.68}$$

with $|B|$ the surface area. The first two factors correspond to the area in slowness space ($s = 1/c$), whereas the third factor is the area in real space.

The next term $\bar{N}_1$ proportional to the boundary length $|\partial B|$ equals

$$\bar{N}_1 = \frac{1}{2c_T} \beta |\partial B| f, \tag{7.69}$$

where

$$\beta = \frac{4}{\gamma} - 3 + \sqrt{\alpha} + \frac{4}{\pi} \int_{\sqrt{\alpha}}^{1} \arctan\left(\frac{(2-\xi^{-2})^2}{\sqrt{(1-\alpha\xi^{-2})(\xi^{-2}-1)}} \right) \mathrm{d}\xi \tag{7.70}$$

and

$$\gamma = \frac{c_R}{c_T} \tag{7.71}$$

is the ratio between the Rayleigh and transverse wave speeds.

7.5 Discussion

We have discussed the transfer matrix method in two-dimensional elastic plate vibrations in analogy with its use in quantum billiards. Because this method also applies to the inhomogeneous setting of the general Schrödinger equation we expect it to generalize to shell vibrations as well. Thus, the fluctuations in the spectral density should be governed by the actions of the orbits, the orbit periods $T_p(\omega)$. The dispersion in thin plates and shells renders these periods frequency dependent. Yet locally in a narrow frequency band, the fluctuations should have fluctuations given by these time periods.

8 Mesoscopics in Acoustics

Richard Weaver

Department of Physics, University of Illinois at Urbana-Champaign, Urbana, IL, USA

The term *mesoscopics* has its origin in the electronics of small devices at low temperature. Ordinary electronics is well described by a picture of electrons as diffusing in an irregular geometry owing to scattering by impurities or rough surfaces or heterogeneous potentials. This picture of the electrons makes no reference to their wave nature. The electron wave is presumed incoherent and its phase irrelevant. In fact, though, the electron does have phase. But the phase is rendered incoherent by repeated scatterings in a time-varying medium, the time dependence being a consequence of thermally excited phonons or thermally excited other electrons. In the quantum literature such scattering is termed *inelastic* because scatterings that garble phase do not conserve the energy of the original electron.

At low temperature, however, where there are few thermal excitations to scatter off, the lifetime of an electron against inelastic scattering can be longer than other relevant timescales. In this case the electron maintains a degree of phase coherence for a period long enough to be manifest macroscopically (or at least *mesoscopically*). Constructive and destructive interferences then lead to a variety of behaviors that can be nonintuitive to a theoretician operating within the simple diffusion picture. The regime is described by the inequality

$$\tau_{\text{elastic}} \ll T < \tau_{\text{inelastic}} \tag{8.1}$$

in which the mean free time against elastic scattering is short compared with the dwell time T of a wave in the sample before measurement (otherwise the field would be more propagative than diffuse), in turn much less than the time for phase garbling. The period $2\pi/\omega$ is typically comparable to or less than τ_{elastic}.

The term has been adopted by acousticians and others to describe certain behaviors of classical waves in random multiply scattering or reflecting media. The issues arise in room acoustics, ultrasonics in composite materials, vibrations in complex structures (see Chapters 14 and 13), and seismic waves in a complex lithosphere (see Chapter 12). For these systems a theoretician often implicitly adopts a picture of waves or rays whose phases are garbled by the scatterings, as if they are scattered inelastically. It is often imagined that the wave energy density (or more generally some mean square field quantity related to energy) is the only important dynamical coordinate and that all field quantities like pressure or strain can be inferred from statistics and specified mean squares. But as with electrons, an acoustic wave carries phase. Any description in terms of energy density or intensity alone is manifestly incorrect. Mesoscopics is the study of the consequences of residual phase coherence in a nominally diffuse acoustic field. It is concerned with nonintuitive behaviors of such fields. As such, it is somewhat ill defined; one person's nonintuitive might be another's self evident.

It is rare for acoustics to be studied in rapidly time-varying media; the acoustic analog to electronic inelastic scattering is therefore not very important. Similarly, dissipation, of great importance in acoustics, is relatively unimportant for electrons. Although electrons do scatter, they tend not to escape except through leads explicitly provided for that purpose. For this reason, the inequality cited earlier is modified for applications in acoustics:

$$\tau_{\text{mfp}} \ll T < \tau_{\text{absorption}}. \quad (8.2)$$

Phenomena in acoustics can be *mesoscopic* when scattering is rapid compared with the timescale T of the dwell of the waves and the timescale of the measurements (otherwise the acoustics would be best described by a simple picture of ray propagation.) The second inequality may be relaxed if the experimenter has a good signal-to-noise ratio and can extract signals even after they are heavily attenuated. The effects of residual coherence sometimes require long dwell times to be appreciated; thus the second inequality ought not be strongly violated.

Depending on one's intuition or lack thereof, a list of mesoscopic behaviors might be long. Here we discuss a limited list that includes enhanced backscatter, quantum echo, localization, and power-law decay. We mention others and direct the reader elsewhere (see especially Chapters 3, 10, 12, and 14 of the current book, for more thorough discussions of them).

8.1 Enhanced Backscatter

Enhanced backscatter, also called the coherent backscatter effect or weak Anderson localization, is perhaps the simplest to understand and demonstrate in the lab and perhaps for that reason is one of the more striking. It was observed optically in backscatter from a multiply scattering half-space. Illumination from a specified incident angle results in diffuse backscatter whose intensity is smoothly distributed about that incident angle. At the precise angle of incidence, however, the backscattered intensity is roughly twice as strong (see Figure 8.1). The cone of enhancement tends to be narrow, with an opening angle comparable to the product of wavenumber and mean free path. In acoustics the phenomenon is more readily observed in a multiply reflecting or scattering cavity (Figure 8.2).

Consider a reverberant room or multiply scattering acoustic medium, with a transient source at a position **s** and a receiver at a position **r**. Propagation between these points is mediated by a large number of rays, each with a different multiply scattered or multiply reflected itinerary. A few of these are illustrated in Figure 8.2. The field at **r** is given by the superposition of all these rays. An intuitive diffuse field picture holds these rays to have uncorrelated random phases; hence the intensity received at **r** is the incoherent sum of the intensities of all the rays. The net intensity is thus, statistically, the number of rays times the intensity of a single ray. If $\mathbf{r} = \mathbf{s}$, the number of rays is much the same as it is if **r** is merely near **s**. But

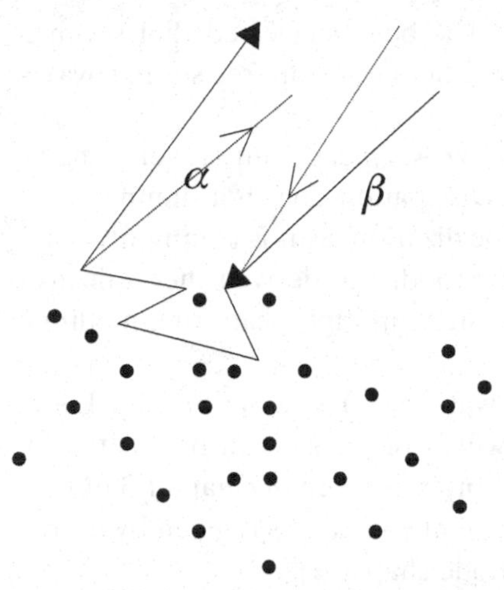

Figure 8.1. A retroreflectance measurement on a multiply scattering half-space. A source illuminates the specimen from an angle α. A detector at angle β receives the reflected multiple scattered re-emission. The bold lines indicate this process. The two cases illustrated are for $\beta \neq \alpha$ and $\beta = \alpha$. If the angles α and β are close, there are an $\approx$ equal number of distinct multiply scattered sequences contributing to the two pictures. If the ray paths are incoherent, the intensity is merely the incoherent sum of these contributions and hence on average proportional to the number of rays. But in the case $\beta = \alpha$ the time-reversed sequence (illustrated by the dashed line on the left) is one of these sequences and has the same phase as the direct sequence, and the many rays are not incoherent. The intensity at $\beta \neq \alpha$ is then inferred to be half that at $\beta = \alpha$.

if $\mathbf{r} = \mathbf{s}$, or within a half wavelength, then the many rays are not uncorrelated; for each ray there is a time-reversed version with identical phase. These two rays interfere constructively; so their superposition carries four times the intensity of one ray. The net result is that the intensity at $\mathbf{r} = \mathbf{s}$ is twice the intensity a half wavelength away. The enhancement requires that the backscattered field is not dominated by single scattering. de Rosny et al. (2000) and Weaver and Lobkis (2000b) have confirmed this effect using laboratory ultrasonics. Larose, Margerin, van Tiggelen and Campillo (2004) have done so with seismic waves, as described by Campillo and

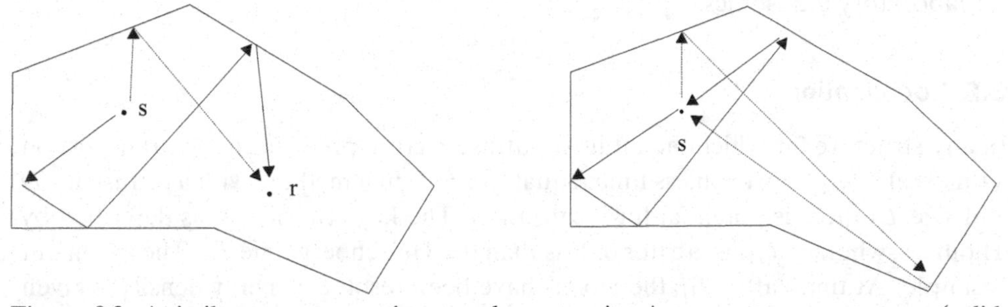

Figure 8.2. A similar measurement in a reverberant cavity. A source at **s** generates rays (solid lines), which, according to the usual diffuse ray picture, soon fill the volume uniformly. If all rays are incoherent, the intensity at all points **r** is merely proportional to the density of rays and equal everywhere. But if the receiver is at **s**, then the rays (dashed lines) occur in path-reversed pairs with the same phase. The rays are not incoherent, and the average intensity at **s** is twice that at **r**.

Margerin in Chapter 12 of the current book. It has been suggested that seismic wave enhanced backscatter will permit the measurement of mean free seismic wave paths.

Enhanced backscatter fails if the medium is *gyroscopic*, or time-reversal non-invariant, such that a wave cannot be reversed. Dissipation does not diminish the effect. The effect is also called weak Anderson localization, as it is commonly conjectured that Anderson localization can be understood, and derived theoretically, from repeated enhanced backscatterings in a strongly multiply scattering medium (Cherroret & Skipetrov 2008).

A different kind of enhanced backscatter was long ago described by Lyon (1969), for the case of a reverberant cavity in which responses can be written in terms of discrete normal modes. A modern description is given in Chapter 3 of this book. In this case the signal at $\mathbf{r}$ because of a transient source at $\mathbf{s}$ is given by a sum over modes, n, with natural frequencies ω_n and mode shapes $u^{(n)}$:

$$\psi_r(t) = \sum_n \frac{u_r^{(n)} u_s^{(n)} \sin(\omega_n t)}{\omega_n}. \tag{8.3}$$

On making an assumption of no degeneracies $\omega_n = \omega_m$ and of uncorrelated mode shapes u, one can square and time-average the above-given expression and make estimates for the mean square signal at late times. If t is great enough, the cross terms in the square of (8.3) average to zero. Then on comparing the cases $\mathbf{r} = \mathbf{s}$ and $\mathbf{r} \neq \mathbf{s}$, one finds that their ratio depends simply on the statistics of the mode shapes $\langle u^4 \rangle / \langle u^2 \rangle^2$. This number is three if the modes are Gaussian random numbers. The effect has been called a "quantum echo" (Prigodin et al. 1994). It may be understood qualitatively by recognizing that a source at $\mathbf{s}$ preferentially excites those modes with antinodes at $\mathbf{s}$, and a receiver at $\mathbf{r}$ is preferentially sensitive to modes with antinodes at $\mathbf{r}$. If $\mathbf{r} = \mathbf{s}$, these preferences reinforce.

As described in Chapter 3 backscattering in a reverberant cavity is enhanced by a factor of two at early times (but not so early that single scattering dominates) in accord with the ray-based arguments given earlier. It is enhanced by a factor of three at at times greater than the Heisenberg time $T_H = 2\pi n(\omega)$, in accord with the above mode-based arguments. Weaver and Lobkis (2000b) confirmed this transition for laboratory ultrasonics.

8.2 Localization

In any structure for which the intuitive diffuse picture predicts energy transport on a timescale T_{TH} (the Thouless time, equal to L^2/D in a medium with a diffusivity D and size L) there is potential for localization. The key criterion is, as described by Thouless, whether T_{TH} is greater or less than the Heisenberg time T_{H}. The argument is simple: At times after T_{H}, the modes have been resolved; energy density is given by a incoherent sum of modes and has no time dependence. If this time is less than T_{TH}, then transport has not been completed, and yet no further transport is possible. The consequence is that, as Anderson put it, there is an "absence of transport," even if the system is allowed an infinite time. It has proved a challenge to make this argument precise and quantitative, but there are numerous demonstrations of its validity and many attempts to make it the basis for precise theoretical predictions of transport dynamics.

The literature on statistical energy analysis (SEA) and room acoustics appears to have largely neglected this effect, perhaps because for most practical structures dissipation limits transport to a degree such that additional limitation owing to localization is unimportant. Nevertheless, the literature does show systems for which localization is an issue (Hodges 1982, Pierre 1990, Weaver & Lobkis 2000a).

The idea is perhaps most transparent in a system consisting of two acoustic reverberation rooms coupled through a sufficiently small window. This is mathematically equivalent to the classic SEA system of two weakly coupled many-degrees-of-freedom vibrating substructures. As shown by Weaver and Lobkis (2000a) a transient source in one room leads, if coupling is sufficiently weak, to an asymptotic state in which that room has the majority of the energy, even after infinite time. There is no equipartition. Weaver and Lobkis (2000a) studied transport and its dynamics between two equal-size rooms, numerically, theoretically, and ultrasonically. Many would not call this Anderson localization, as they would save that term for infinitely extended structures. But the physics described by the Thouless argument applies to both.

That this absence of transport corresponds to localization of the modes may be seen by the following arguments: A field ψ due to a source at $\mathbf{s}$ is given by

$$\psi_r(t) = \sum_n \frac{u_r^{(n)} u_s^{(n)} \sin(\omega_n t)}{\omega_n}. \tag{8.4}$$

The square of the field at $\mathbf{r}$ is

$$\langle \psi_r(t)^2 \rangle = \sum_n \frac{u_r^{(n)2} u_s^{(n)2} \sin^2(\omega_n t)}{\omega_n^2} + \sum_{n \neq m} \frac{u_r^{(n)} u_s^{(n)} u_r^{(m)} u_s^{(m)} \sin(\omega_n t) \sin(\omega_m t)}{\omega_n \omega_m}. \tag{8.5}$$

On making a short-time average one sees that the second term vanishes on a time scale comparable to the inverse of the spacing $\omega_n - \omega_m$. In this case the mean square signal at $\mathbf{r}$ becomes the time-independent incoherent sum of modes:

$$\langle \psi_r(t)^2 \rangle \big|_{t->\infty} = \sum_n \frac{u_r^{(n)2} u_s^{(n)2}}{2\omega_n^2}. \tag{8.6}$$

If there is an absence of transport, then this is smaller for $\mathbf{r}$ far from $\mathbf{s}$ than it is when $\mathbf{r}$ is close to $\mathbf{s}$. Thus one concludes that the absence of transport corresponds to the localization of the normal modes; the same modes that have significant amplitude at $\mathbf{s}$ are not the modes with significant amplitude at $\mathbf{r}$.

The Thouless argument starts with the diffusion picture and supplements it with the constraint that energy evolution, being a function of interference of modes, has timescales no greater than the inverse of the spacing of the natural frequencies. Anderson localization pertains to the special case of an unbounded statistically homogeneous medium. In this case both the Thouless time and the Heisenberg time depend on the length scale of interest. Diffusion times scale with length squared, $T_{\mathrm{TH}} \sim L^2/D$. Heisenberg time is modal density; so it scales with volume in

d dimensions, $T_H \sim L^d$. In one dimension there is therefore a length scale beyond which transport is slower than mode resolution; this is the length scale beyond which transport must weaken; one concludes that one-dimensional random systems localize regardless of the (if nonzero) strength of the randomness. In two and three dimensions the arguments are more complex. Anderson showed, for his model at least, that if the randomness is sufficiently strong, the modes are localized in any number of dimensions. It is generally held that all random two-dimensional systems localize, although the length scale upon which they do so can be large, varying exponentially with the product of diffusivity and modal density. It is generally accepted that in three dimensions, one must have sufficiently strong disorder for the modes to be localized.

Examples of Anderson localization in one-dimensional acoustics and vibrations are well known (Pierre 1990) and are of some practical importance. It is also observed in two-dimensional systems (Weaver 1990a, Ye et al. 1992). It appears to be the case that the only unambiguous observation of Anderson localization to date in three dimensions is in ultrasonics (Hu et al. 2008). Actually, localization per se is not difficult to achieve. A system of decoupled or almost-decoupled oscillators with different natural frequencies (think of a loose collection of solid spheres of varying diameters) has modes that are localized on the individual oscillators. An excitation applied to one oscillator is never transferred to others. Thus there is no transport. It is not so much localization that is hard to observe, but rather the transition between Anderson localization and a more normal energy-conducting state. The challenge has been to construct a system for which varying a parameter such as frequency permits the character of the transport to rapidly change from localized to extended. In practice, if diffusivity is slow enough to cause localization, it is also so slow that even small levels of dissipation reduce signal level below noise levels before significant transport has transpired, even if that transport per se is governed by conventional diffusion, that is, even if the modes are extended. This underlines a recurring problem in studies of classical wave localization namely, dissipation masks, and mimics the effects.

8.3 Power-Law Dissipation

That diffuse fields dissipate exponentially is a canon of SEA and of room acoustics (the latter only if the dissipation is small on the scale of a transit time), $E \sim \exp(-\sigma t)$; rooms are typically first characterized by means of their reverberation time, the time required for a free field to decay by 60 dB. There are numerous sources for deviations from simple exponentials. One of the most interesting applies in the case in which there are only a small number of decay *channels*, as might be realized for example in a non-dissipative room with a small number of small windows to the outside world. This condition was first considered by Schroeder (1965) and later by Burkhardt (1997) and Lobkis et al. (2000, 2003).

We model a diffuse field as a superposition of independent real (i.e., standing wave) normal modes with Gaussian statistics. This is actually not correct. Lossy systems generically have complex eigenmodes and include propagating parts. However real Gaussian modal statistics does seem to be a good approximation. In this case each mode can be shown to have a decay rate (imaginary part of its eigenfrequency) given by first-order perturbation theory and equal to the sum of the squares of the mode's random Gaussian overlaps with each of the several decay channels. This leads to a χ-square-like distribution of modal decay rates and a net transient energy decay of the form

$$E(t) = E_0 \prod_{i=1}^{M} (1 + 2\sigma_i t)^{-1/2}, \tag{8.7}$$

where M is the number of channels and σ_i characterizes the strength of the ith channel. Interestingly, σ_i cannot be greater than an inverse Heisenberg time. Then (8.7) predicts a decay profile that begins like $\sim\exp(-\Sigma\sigma_i t)$ at early times and then slows. Decay profiles like this are observed in diffuse fields, and the behavior is successfully attributed to the stated conditions. The effect is readily understood by recognizing that those modes that overlap the channels well are the more strongly dissipated. After a short period only the less well overlapped, less strongly dissipated modes remain, and the rate of dissipation slows.

A further confounding of intuition is afforded by making corrections for the non-standing wave character of the modes. For the case of a small number of well-coupled decay channels, this character is strong, and deviations from (8.7) can be observed (Lobkis et al. 2003).

8.4 Other Mesoscopic Phenomena

Perhaps the most familiar of mesoscopic phenomena is the presence of normal modes. A picture of diffuse rays in a reverberant cavity is not hard to imagine. If it neglects phase coherences, though, it would not lend itself to inferring the detailed constructive and destructive interferences that give rise to the normal modes. Indeed, the construction of a modal picture from a ray picture is complex – as is discussed in Chapters 1, 2, 6, and 5. One hesitates to call modes mesoscopic, as any intuitions that might be confounded here would be unfamiliar to acousticians. We were all raised with the complementary (in Bohr's sense of the word) notions of ray propagation and modes; we no longer find the dual perspective to be remarkable. Nevertheless it is remarkable. What is even more remarkable is that these modes exhibit the statistics that they do, with level repulsion and spectral rigidity and with apparent consequences for signal level variances. These are all characteristics that follow only with difficulty from the diffuse ray picture.

The correspondence between a system Green's function and the field–field correlations of a diffuse field is another mesoscopic phenomenon. It has been confirmed in laboratory ultrasonics (Lobkis & Weaver 2001, Weaver & Lobkis 2001, 2003, Derode et al. 2003, Larose, Derode, Campillo & Fink 2004). It is being applied extensively by seismologists. The reader is referred to Chapter 12 by Campillo and Margerin.

Time-reversal acoustics (Chapter 10), in reverberant or multiply scattering or multiply refracting media, depends critically on the time invariance of the medium and thus on a lack of inelastic scattering. It also is a mesoscopic phenomenon.

8.5 Fidelity, Coda-Wave Interferometry, and Loschmidt Echo

The Loschmidt echo $M(t)$, or quantum fidelity, is a quantity used to measure the similarity of a wavefield or quantum state $\psi(\mathbf{r}, t)$ that has evolved from a specified initial condition in two slightly different systems. It is related to reversibility in quantum mechanics and is relevant to quantum computing. A slightly different concept is scattering fidelity, in which one compares the outgoing waveform $\psi(t)$ in a specified channel from two slightly different scatterers, each excited by the same incoming wave. Seismologists have introduced the term "coda wave interferometry," in which they compare multiply scattered responses in two slightly different systems to a specified input. In all cases one recognizes that the distributions of intensity are largely unchanged by perturbing the structure, but the phases can be very different. The quantity

$$M(t) = \int \psi(r,t)\psi'(r,t)\,\mathrm{d}r \left[\int \psi(r,t)^2 \mathrm{d}r \int \psi'^2(r,t)\,\mathrm{d}r\right]^{-1/2}, \tag{8.8}$$

where y and y' are the two fields, will reflect these changes; $M = 1$ if the waveforms are identical and $M < 1$ if different.

In practice it is difficult to compare entire wavefields, and so one instead constructs the scattering fidelity

$$F(T) = \int \psi(t)\psi'(t)W_T(t)\,\mathrm{d}t \left[\int \psi(t)^2 W_T(t)\,\mathrm{d}t \int \psi'(t)^2 W_T(t)\,\mathrm{d}t\right]^{-1/2}, \tag{8.9}$$

where W is a window function centered on time T and y and y' are the waveforms that result from the same excitation in two slightly different systems.

Seismologists have suggested that this quantity permits a measure of changing rock (Snieder 2002) and have successfully applied it to the rapidly changing environment near active volcanoes (Pandolfi et al. 2006). Engineers have suggested that it is a measure that might be useful for structural health monitoring. Multiply scattered waveforms or vibration responses in complex structures can be difficult to interpret, but changes are easily measured.

Lobkis and Weaver (2003) showed that ultrasonic scattering fidelity in isolated aluminum blocks (with perturbation due to temperature changes) was a measure of the mean lifetime of a ray against mode conversion among the various wave types. Gorin et al. (2006b) showed that the time rate of decay of fidelity could be predicted by random matrix theory arguments.

8.6 Summary

Mesoscopic acoustics is the study of the effects of residual phase coherence in a nominally diffuse field. These effects are often counterintuitive and surprising. They are furthermore often useful, suggesting measurements that are not obvious in more conventional acoustics.

9 Diagrammatic Methods in Multiple Scattering

Joseph A. Turner
Goutam Ghoshal
Department of Engineering Mechanics,
University of Nebraska–Lincoln, Lincoln, NE, USA

9.1 Introduction

The propagation of waves through heterogeneous media occurs in many forms, including acoustic, electromagnetic, and elastic. As these waves propagate, the wave front is altered because of spatial variations in properties. The result of the interaction with the medium is that the incident energy is dispersed in many directions – the input energy is said to be scattered. If the scattering is strong and one waits long enough, the signal received will become complex because of *multiple scattering* effects. Understanding this process is necessary for locating an object within a scattering medium and/or for quantifying the properties of the medium itself. The focus here is on the use of diagrams than can aid in analysis of the multiple scattering process.

Multiple scattering has been discussed by theorists since the time of Rayleigh (1892, 1945). Systems with distributions of discrete inclusions (scatterers) in a homogeneous background were studied by Foldy (1945), Lax (1951, 1952), Waterman and Truell (1961), and Twersky (1977) in terms of assumed exact descriptions of scattering by isolated inclusions. This approach may be contrasted with a model of the heterogeneous medium as having continuously varying properties. This approach entails stochastic operator theory and includes the work of Karal and Keller (1964), Frisch (1968), McCoy (1981), Stanke and Kino (1984), and Hirsekorn (1988). Both approaches seek the wave speed and attenuation of an ensemble average field, although the connection to measurements in a single sample is not always obvious. Interest in multiple scattering has also been examined with respect to energy transport. In this case, the multiple scattering problem is examined through an energy balance of the scattering process, leading to equations of radiative transfer (Schuster 1905, Ishimaru 1977). The general nature of this approach is evident in the wide array of problems that have been studied. These include not only acoustics but optics (Mishchenko et al. 2006) and elastic waves (Weaver 1990b, Turner & Weaver 1994), and seismology (Margerin et al. 2000, Weaver 2005). In much of this work, the degree of scattering that takes place is central to approximations made to the full multiple scattering problem. Relevant space and time restrictions of experiments can often be devised to confine the scattering to either a single scattering limit or a diffusion limit – two tractable approximations. However, the validity of such approximations, within the scope of a comprehensive multiple scattering formalism, is often not addressed.

Finally, one important application is noted. The ultrasonic biomedical imaging field (Insana & Hall 1990, Oleze & Zachary 2006) has a primary goal of quantifying ultrasonic images such that healthy and diseased tissues may be identified. This field has recently become greatly interested in improved quantitative models of the wave scattering, including the influence of multiple scattering. Biological tissue is, in fact, an excellent material that highlights the difficulty associated with understanding waves in heterogeneous media. On the one hand, it seems that tissue should be modeled from a continuum point of view with spatially varying properties. However, tissue also often contains local material discontinuities that are best represented using a discrete scatterer approach. Thus, for such materials, as with many others, the best choice for the modeling approach is not easily made. It is also clear that a comprehensive multiple scattering analysis must clearly define all possible interactions that may arise during the scattering process and provide a means to prioritize those interactions.

This type of goal was the motivation that led Feynman to develop a diagrammatic method for quantum field theory in which the numerous particle interactions that occur are catalogued and analyzed in a concise manner. The common expression "A picture is worth a thousand words" captures the value of diagrammatic methods for the complex integral equations that arise in these problems. As will be shown, it is the topology of the diagrams that can help us understand the approximations of interest. Unfortunately, a diagrammatic approach has been used by only a small number of researchers in the last several decades for acoustics problems despite the value of the method (Photiadis 1997, Weaver 1997, Derode et al. 2006). Thus, the goal of this chapter is to provide a basic introduction to diagrammatic methods. Some of the diagrammatic notation used here follows the tutorial article by Frisch (1968).

9.2 Wave Equations with Spatially Varying Coefficients

In the multiple scattering problem considered here, the material properties of the medium are taken to vary spatially but are assumed to be invariant in time. In addition, a scalar wave problem is the focus because such an example provides the best setting for the diagrams that follow without extraneous complications that can obscure the essential multiple scattering issues. In this case, consider the wave equation that governs the pressure in a fluid, $p(\mathbf{x}, t)$. It is a function of space and time:

$$\frac{1}{c^2(\mathbf{x})}\frac{\partial^2}{\partial t^2}p(\mathbf{x}, t) - \nabla^2 p(\mathbf{x}, t) + \frac{\nabla \rho_0(\mathbf{x})}{\rho_0(\mathbf{x})} \cdot \nabla p(\mathbf{x}, t) = f(\mathbf{x}, t), \tag{9.1}$$

where $c(\mathbf{x})$ is the local wave speed, $\rho_0(\mathbf{x})$ is the local material mass density, and $f(\mathbf{x}, t)$ is a source term. We shall henceforth assume that the material mass density ρ_0 is independent of $\mathbf{x}$. Thus the last term on the left-hand side is zero. On applying a temporal Fourier transform,

$$k(\mathbf{x})^2\tilde{p}(\mathbf{x},\omega) + \nabla^2\tilde{p}(\mathbf{x},\omega) = -\tilde{f}(\mathbf{x}), \tag{9.2}$$

where the wavenumber $k = \omega/c$ varies spatially. This equation has a Green's function described by

$$\left[\nabla^2 + k(\mathbf{x})^2\right] G(\mathbf{x},\mathbf{x}') = \delta^3(\mathbf{x}-\mathbf{x}'), \tag{9.3}$$

where G defines the response at position $\mathbf{x}$ owing to an excitation concentrated at position $\mathbf{x}'$. Our notation suppresses its dependence on frequency ω. Equation (9.3) serves as the ansatz for the subsequent derivations.

In statistically homogeneous media it is convenient to write the square of the wavenumber as a constant plus small fluctuations,

$$k(\mathbf{x})^2 = k_0^2 - \varepsilon(\mathbf{x}). \tag{9.4}$$

In the event that we had not assumed $\rho_0(\mathbf{x})$ to be constant, ε would be an operator that entailed spatial derivatives. Very often we will take the fluctuations to be centered; $\langle\varepsilon(\mathbf{x})\rangle = 0$, the angle brackets $\langle\rangle$ indicating an ensemble average. The statistics of $\varepsilon(\mathbf{x})$ are essential and are discussed next.

Substitution into Equation (9.3) allows the equation for the Green's function to be written,

$$\left(\nabla^2 + k_0^2\right) G(\mathbf{x},\mathbf{x}') = \delta(\mathbf{x}-\mathbf{x}') + \varepsilon(\mathbf{x})G(\mathbf{x},\mathbf{x}'). \tag{9.5}$$

The solution of Equation (9.5) when the fluctuations are not present (i.e., $\varepsilon(\mathbf{x}) = 0$) serves to define $G_0(\mathbf{x},\mathbf{x}')$, the non-stochastic, or bare, Green's function.

The differential form of the stochastic wave equation, given by Equation (9.5), is often recast into integral form, a form convenient for the present work. In this case, we multiply Equation (9.5) on the left by G_0 and integrate over all $\mathbf{x}$. The resulting equation governing the wave propagation is now of the form

$$G(\mathbf{x},\mathbf{y}) = G_0(\mathbf{x},\mathbf{y}) + \int G_0(\mathbf{x},\mathbf{x}')\varepsilon(\mathbf{x}')G(\mathbf{x}',\mathbf{y})\,\mathrm{d}\mathbf{x}'. \tag{9.6}$$

This equation can be written in a slightly more symmetric format if a different form of the material fluctuation is used. Thus, we define

$$\mu(\mathbf{a},\mathbf{b}) = \varepsilon(\mathbf{a})\delta(\mathbf{a}-\mathbf{b}), \tag{9.7}$$

which allows Equation (9.6) to be rewritten as

$$G(\mathbf{x},\mathbf{y}) = G_0(\mathbf{x},\mathbf{y}) + \iint G_0(\mathbf{x},\mathbf{w})\mu(\mathbf{w},\mathbf{z})G(\mathbf{z},\mathbf{y})\,\mathrm{d}\mathbf{w}\,\mathrm{d}\mathbf{z}. \tag{9.8}$$

Equation (9.8) is an integral equation that governs the response of the heterogeneous medium. At this stage, the equation is exact in the sense that very few assumptions have been made. However, such an equation is also very difficult (if not impossible) to solve in closed form. In addition, it should be noted that the quantity of interest $G(\mathbf{x},\mathbf{y})$ is stochastic. Thus, it is often of greater interest to determine the statistics of G rather than G itself. Equation (9.8) forms the basis for these calculations.

Shorthand notation for complex equations is often used to simplify further mathematical operations associated with integral equations, allowing the operator form of Equation (9.8) to be written,

$$G = G_0 + G_0 \mathscr{L} G, \tag{9.9}$$

where the operator $\mathscr{L}$ refers to the multiplication by μ and spatial convolution given in Equation (9.8). When needed for clarity, the complete integral forms of subsequent equations will be used. This operator form may also be written pictorially, using diagrams developed by Feynman for quantum field theory, in the following form:

(9.10)

where [thin line] denotes G_0, [thick line] denotes G, and ● denotes the operator $\mathscr{L}$.

One might attempt to solve such a problem for G using a perturbation approach, especially if the fluctuations of the medium are not large. Thus, we may attempt to write our solution as (sometimes called a Born series)

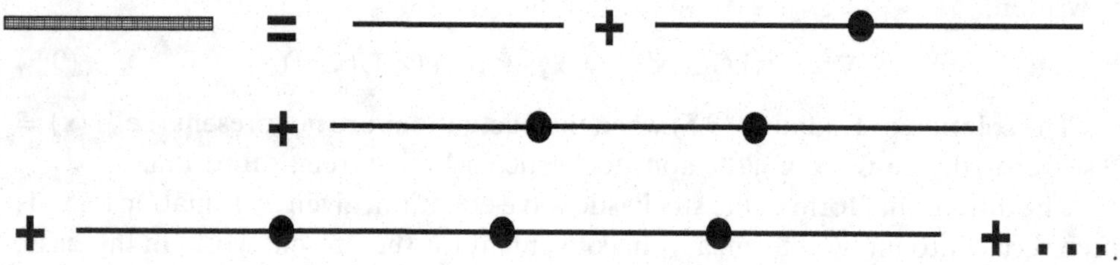

(9.11)

The fourth term in the expansion, given explicitly as

$$\int \cdots \int G_0(\mathbf{x}, \mathbf{p})\mu(\mathbf{p}, \mathbf{q})G_0(\mathbf{q}, \mathbf{r})\mu(\mathbf{r}, \mathbf{s})G_0(\mathbf{s}, \mathbf{t})\mu(\mathbf{t}, \mathbf{u})G_0(\mathbf{u}, \mathbf{y})\, \mathbf{dp}\, \mathbf{dq}\, \mathbf{dr}\, \mathbf{ds}\, \mathbf{dt}\, \mathbf{du}, \tag{9.12}$$

highlights the efficiency of using diagrams for this analysis. Although such a series will converge (assuming an infinitesimal amount of dissipation is present in the system), its convergence can be slow. As detailed by Frisch (1968), the validity of a truncated series is limited to short distances even when the fluctuations μ are small (an explicit example is presented in the next section). The presence of such "secular" terms in the expansion prevents the use of such an expansion. Thus, an alternative approach is needed, and this approach is focused on the statistics of the response.

9.3 Statistics of the Response

Inasmuch as the response will be a function of the stochastically fluctuating moduli, we realize that G itself is random and focus our attention on its statistics. Here we will discuss its ensemble average, or mean Green's function $\langle G \rangle$, and its mean square. Toward this end we need to first discuss the statistics of ε and the diagrams for its moments. The mean of the fluctuations $\langle \varepsilon(\mathbf{x}) \rangle = \varepsilon_0 = \langle \bullet \rangle$ will be zero if they are centered. The covariance $\langle \varepsilon(\mathbf{x})\varepsilon(\mathbf{y})\rangle = \varepsilon_0^2 + \Gamma_2(\mathbf{x},\mathbf{y}) =$  $^2 +$ 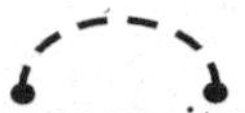 depends on the two-point function Γ_2. An assumption of statistical homogeneity implies that Γ_2 depends only on the separation $\mathbf{x}-\mathbf{y}$. Expressions for the higher-order statistics may be constructed as well; for example, $\langle \varepsilon(\mathbf{x})\varepsilon(\mathbf{y})\varepsilon(\mathbf{z})\rangle = \varepsilon_0^3 + \varepsilon_0\Gamma_2(\mathbf{x},\mathbf{y}) + \varepsilon_0\Gamma_2(\mathbf{y},\mathbf{z}) + \varepsilon_0\Gamma_2(\mathbf{x},\mathbf{z}) + \Gamma_3(\mathbf{x},\mathbf{y},\mathbf{z})$. In diagrammatic form, the third-order statistics are given by

$\langle \varepsilon(\mathbf{x})\varepsilon(\mathbf{y})\varepsilon(\mathbf{z})\rangle =$ + + + + .

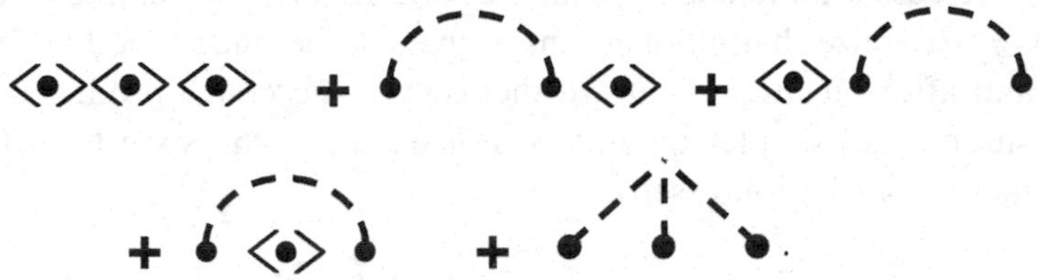

(9.13)

These concepts now permit the mean response $\langle G\rangle$ to be examined. Subsequently, the covariance of the response $\langle GG\rangle$, a quantity related to the mean energy of the wave field, will be examined as well.

9.3.1 Mean Response

A first attempt at calculating the mean response is made by taking the ensemble average of Equation (9.11). Thus, we write

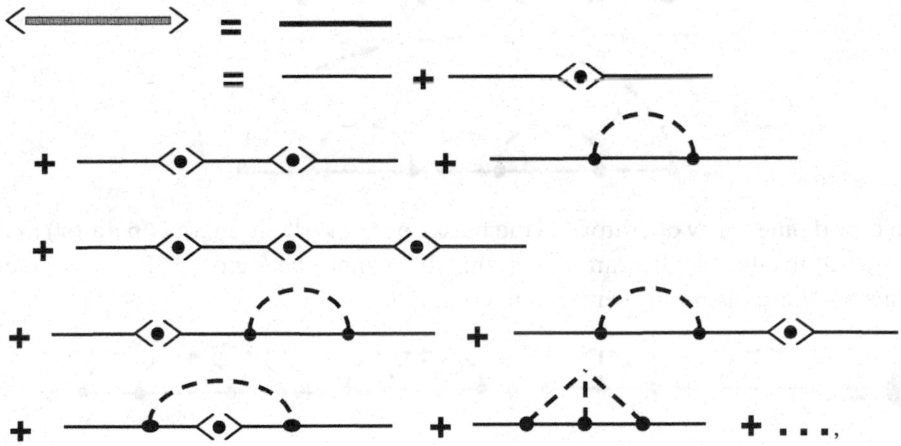

=

= +

+ +

+

+ +

+ + + . . . ,

(9.14)

where the mean response $\langle G\rangle$ is now denoted by a darker thick line . If the fluctuations are centered, only the first, fourth, and ninth diagrams shown above are nonzero. All diagrams with factors of $\langle \varepsilon(\mathbf{x})\rangle = \varepsilon_0 =$  vanish.

In the spirit of imagining that the fluctuations are small, it is natural to attempt to truncate a series like this after a finite number of terms. That this will be a disappointment is clear perhaps by considering doing so after two terms. In this case we would have

$$\langle G(\mathbf{x}, \mathbf{y}) \rangle = G_0(\mathbf{x}, \mathbf{y}) + \int G_0(\mathbf{x}, \mathbf{z}) \varepsilon_0 G_0(\mathbf{z}, \mathbf{y})\, d\mathbf{z}. \tag{9.15}$$

The integral can be done analytically and shown to grow linearly with distance $|\mathbf{x} - \mathbf{y}|$. The result is clearly nonsense at distances of order $1/\varepsilon_0$ or greater. Secular behavior like this is obtained at any finite truncation. Rather than truncating the series, we convert it to an approximate form that can be summed exactly.

We recognize that all diagrams above can be understood as factors of G_0 before and after quantities that are themselves successive products (actually spatial convolutions) of simpler quantities. For example, the sixth term (pre-factors and post-factors of G_0 suppressed)

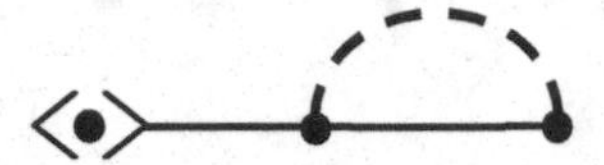

is the successive spatial convolution of three simpler parts

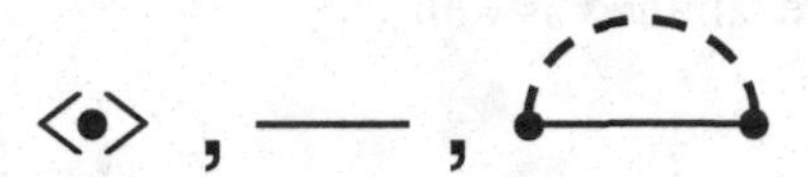

Some diagrams in (9.14) cannot be factored. For example,

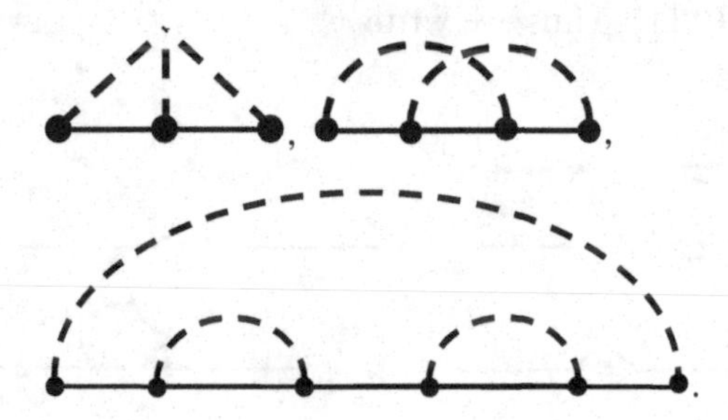

We now define a new operator M (the mass operator or self-energy operator) as the sum of all irreducible diagrams, diagrams that cannot be factored. The first several terms of M are (assuming centered fluctuations)

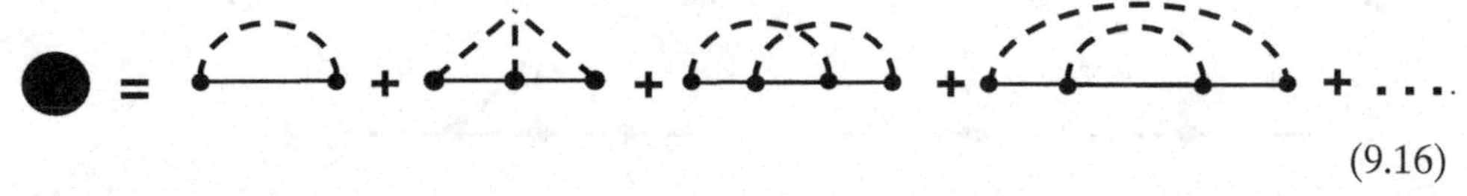

(9.16)

The series for M can be truncated without incurring secular terms. It is now not difficult to see that

$$\langle G \rangle = G_0 + G_0 M \langle G \rangle \tag{9.17}$$

or, diagrammatically

(9.18)

An expansion (9.18) in powers of M exactly reproduces Equation (9.14). This is the Dyson equation and precisely describes the mean Green's function. The only approximations are those in one's estimate for M.

9.3.2 Mean Response within the First-Order Smoothing Approximation

The mean response is governed by the Dyson Equation (9.18). It is readily solved in the presence of statistical homogeneity such that the Green's functions and mass operator are dependent only on a single variable, the difference between the two positions. In this case, the Dyson equation may be written as

$$\langle G(\mathbf{x}-\mathbf{y})\rangle = G_0(\mathbf{x}-\mathbf{y}) + \iint G_0(\mathbf{x}-\mathbf{w})M(\mathbf{w}-\mathbf{z})G(\mathbf{z}-\mathbf{y})\,\mathrm{d}\mathbf{w}\,\mathrm{d}\mathbf{z}. \tag{9.19}$$

This equation is solved by taking spatial Fourier transforms. In a three-dimensional domain, the necessary transform pair is defined for G_0 by

$$\begin{aligned} \tilde{G}_0(\mathbf{p}) &= \int G_0(\mathbf{r})\mathrm{e}^{\mathrm{i}\mathbf{p}\cdot\mathbf{r}}\,\mathrm{d}\mathbf{r}, \\ G_0(\mathbf{r}) &= \frac{1}{(2\pi)^3}\int \tilde{G}_0(\mathbf{p})\mathrm{e}^{-\mathrm{i}\mathbf{p}\cdot\mathbf{r}}\,\mathrm{d}\mathbf{p}, \end{aligned} \tag{9.20}$$

where the argument of G_0 is $\mathbf{r} = \mathbf{x} - \mathbf{y}$. Similar expressions can be written for $\langle G\rangle$ and M as well. These transforms allow the Dyson equation to be expressed algebraically,

$$\langle\tilde{G}(\mathbf{p})\rangle = \tilde{G}_0(\mathbf{p}) + \tilde{G}_0(\mathbf{p})\tilde{M}(\mathbf{p})\langle\tilde{G}(\mathbf{p})\rangle, \tag{9.21}$$

for which the solution is

$$\langle\tilde{G}(\mathbf{p})\rangle = \left[\tilde{G}_0(\mathbf{p})^{-1} - \tilde{M}(\mathbf{p})\right]^{-1}. \tag{9.22}$$

The spatial Fourier transform of the bare Green's function is given by $\tilde{G}_0(p) = (\omega^2 - c_0^2p^2)^{-1}$. The response of the bare system is governed by the zeros of the dispersion equation (i.e., $\omega^2 = c_0^2p^2$) because they determine the poles in transforms back to the time–space domain. For the mean response in the heterogeneous medium, we see a similar solution, with a slightly modified dispersion relation $\omega^2 = c_0^2p^2 + \tilde{M}(\mathbf{p})$. The solution given by Equation (9.22) is formally exact. However, in order to obtain a solution, M must be estimated. The simplest approximation for M is given by truncating the series defined in Equation (9.16) to a single term. Thus, we write the first-order smoothing approximation for M as

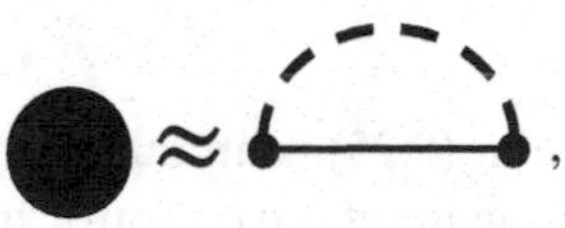
(9.23)

keeping the information related only to the two-point statistics of the medium, an approximation given explicitly by $M(\mathbf{x}-\mathbf{y}) = \Gamma_2(\mathbf{x}-\mathbf{y})G_0(\mathbf{x}-\mathbf{y})$. As the neglected terms in M are of higher order in ε, truncated mass operators are expected to work well as long as the fluctuations are not large. However, the validity of such truncations is also dependent on the strength and range of the correlations Γ_2.

In the spatial transform domain, the mass operator is written as

$$\tilde{M}(\mathbf{p}) = \int \Gamma_2(\mathbf{r}) G_0(\mathbf{r}) e^{i\mathbf{p}\cdot\mathbf{r}}\, d\mathbf{r}. \tag{9.24}$$

An alternative approximation is to replace the factor of G_0 in the first-order smoothing approximation expression for M with $\langle G \rangle$. This can be called the rainbow approximation for the shapes of the diagrams in its expansion. It is arguably more elegant than (9.23) but is, like (9.23), valid only to order ε^2.

The solution for mean response shows that the wavenumber is modified by the heterogeneities. The change in wavenumber is determined by solving for values of p that satisfy $\omega^2 - c_0^2 p^2 - \tilde{M}(\mathbf{p}) = 0$. It is particularly simple if p is only slightly modified by the heterogeneities, such that $\tilde{M}(\mathbf{p})$ can be approximated by $\tilde{M}(\omega/c_0)$. The modified dispersion equation is then $\omega^2 - c_0^2 p^2 - \tilde{M}(\omega/c_0) = 0$, which can be solved directly for $p = (1/c_0^2)\sqrt{\omega^2 - \tilde{M}(\omega/c_0)} \approx \omega/c_0 - i\Im\left\{\tilde{M}(\omega/c_0)\right\}/(2\omega c_0)$. This final expression shows the main impact of the heterogeneities – an imaginary component of the wavenumber arises, which introduces a spatial decay in the amplitude of the propagating wave. This decay, called attenuation, is the result of phase fluctuations across the wave front and is not a true energy loss. The energy that is lost from the main wave front is scattered into other directions but is conserved. Therefore, this type of decay is very different from losses in the medium due to true dissipation mechanisms (e.g., to conversion to heat). Next, we consider the evolution of the energy more specifically.

9.3.3 Covariance of the Response

In many applications, the statistics of the response of orders higher than the mean are of interest. For example, optics applications often involve energy-based quantities such as intensity. The most basic example is that of the covariance $\langle GG^* \rangle$, which is related to the energy of the wave (the superscript * denotes the complex conjugate of the quantity; here it also denotes G at a slightly shifted frequency, $\omega + \Omega$). In this case, the same diagrammatic approach used earlier can be exploited. Thus, we begin by writing an equation governing GG^* and then examine its ensemble average. Using the expansion given in Equation (9.11), we immediately write the diagrammatic expansion for the product $GG^* = G(\mathbf{x}, \mathbf{y}, \omega) G^*(\mathbf{x}', \mathbf{y}', \omega + \Omega)$ as

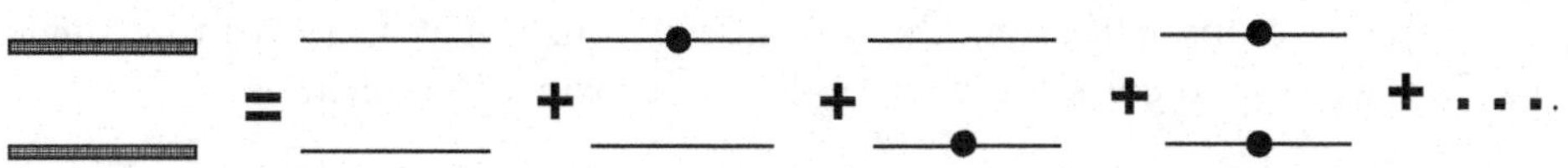

= + + + + ...

(9.25)

Each double diagram in Equation (9.25) is an outer product of the operator in the upper row and the complex conjugate of the operator in the lower row. The product is a function of four spatial variables. For example, the third and fourth terms are given, respectively, by (frequency dependence is now suppressed)

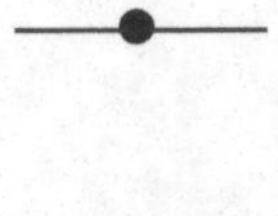

$$= \iint G_0(\mathbf{x}, \mathbf{y}) \left[G_0(\mathbf{x}', \mathbf{z}') \mu(\mathbf{z}', \mathbf{w}') G_0(\mathbf{w}', \mathbf{y}') \right]^* d\mathbf{z}'\, d\mathbf{w}', \tag{9.26}$$

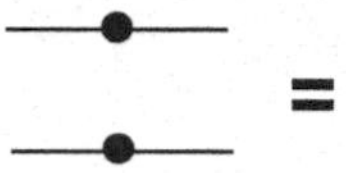

$$= \iiiint [G_0(\mathbf{x},\mathbf{z})\mu(\mathbf{z},\mathbf{w})G_0(\mathbf{w},\mathbf{y})]$$
$$\times [G_0(\mathbf{x}',\mathbf{z}')\mu(\mathbf{z}',\mathbf{w}')G_0(\mathbf{w}',\mathbf{y}')]^* \, d\mathbf{z}\, d\mathbf{w}\, d\mathbf{z}'\, d\mathbf{w}'. \quad (9.27)$$

We now take an ensemble average of Equation (9.25). We again assume centered fluctuations so that many of the diagrams are zero. Thus, we have

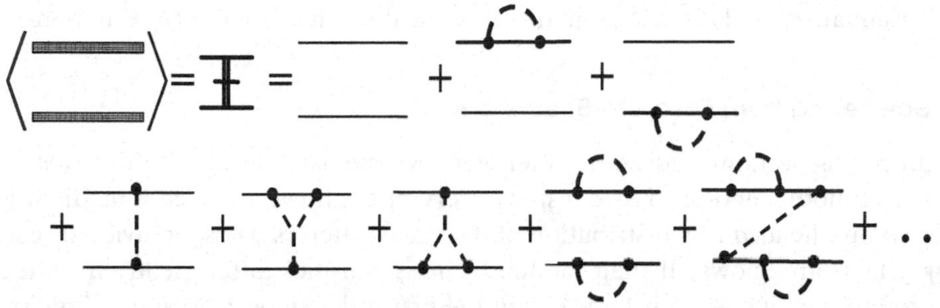

$$\quad (9.28)$$

where the diagram  denotes $\langle GG^* \rangle$. As an example, the fourth diagram in Equation (9.28) is given explicitly by

$$\iint G_0(\mathbf{x},\mathbf{p})G_0(\mathbf{p},\mathbf{y})G_0^*(\mathbf{x}',\mathbf{q})G_0^*(\mathbf{q},\mathbf{y}')\Gamma_2(\mathbf{p},\mathbf{q})\, d\mathbf{p}\, d\mathbf{q}. \quad (9.29)$$

Again we notice that each term of the preceding expansion can be factored. We again define an operator that is the sum of all diagrams that cannot be factored:

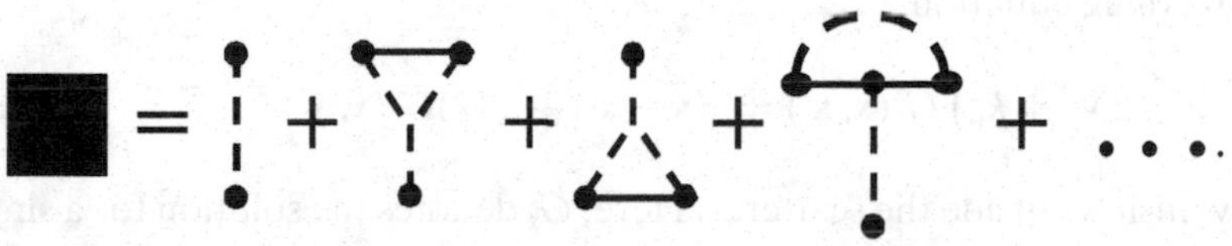

$$\quad (9.30)$$

This "intensity" operator allows the equation governing the covariance of the Green's function to be reduced to the form

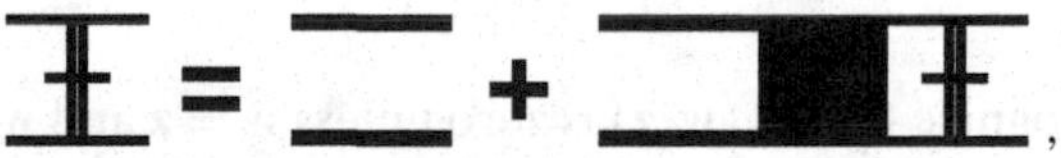

$$\quad (9.31)$$

known as the Bethe–Salpeter equation. This equation is again exact in the sense that all diagrams are included. On expanding it one can see that it is merely a re-ordering of (9.28). If approximations for the intensity operator are consistent with those for the self-energy, the Bethe–Salpeter equation can be shown to conserve energy.

The simplest choice for the intensity operator is to approximate it by its first term. Thus

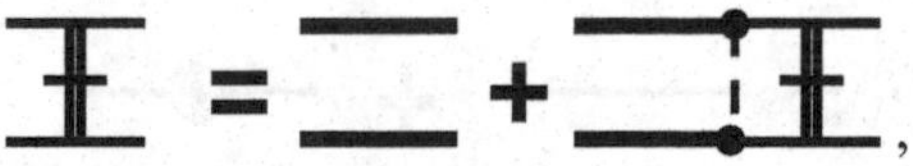

$$\quad (9.32)$$

which may be written explicitly as

$$\langle G(\mathbf{x},\mathbf{y})G^*(\mathbf{x}',\mathbf{y}')\rangle = \langle G(\mathbf{x},\mathbf{y})\rangle\langle G^*(\mathbf{x}',\mathbf{y}')\rangle + \iint \langle G(\mathbf{x},\mathbf{w})\rangle\langle G^*(\mathbf{x}',\mathbf{w}')\rangle\,\Gamma_2(\mathbf{w},\mathbf{w}')\,\langle G(\mathbf{w},\mathbf{y})G^*(\mathbf{w}',\mathbf{y}')\rangle\,\mathrm{d}\mathbf{w}\,\mathrm{d}\mathbf{w}'.$$

The Bethe–Salpeter equation is the starting point for derivations of radiative transfer equations and diffusion equations, depending on further approximations.

9.4 Scattering from Discrete Scatterers

The preceding sections addressed heterogeneous media in terms of their weakly fluctuation moduli and/or mass densities. Very often, however, such a medium is better comprehended as a distribution of discrete scatterers whose individual scattering effects are known. If their moduli or mass densities differ greatly from the background (i.e., if epsilon is large), then the first-order smoothing approximation is poor. A different formulation is needed. These types of scattering problems can be especially difficult to model, particularly if more than one type of scatterer is involved. In this section we illustrate the capability of diagrammatic methods to assist in construction of the full multiple scattering series as well as corresponding approximations.

The wave propagation problem described earlier must now be modified to account for the discrete nature of the scatterers and to exactly describe the effect of each single scatterer. We begin with the one-body problem of a single scatterer within an unbounded, homogeneous medium. For this problem we can again make use of the governing equation

$$\left(\nabla^2 + k_0^2\right) G_s(\mathbf{x},\mathbf{x}') = \delta(\mathbf{x}-\mathbf{x}') + \varepsilon_i(\mathbf{x})G_s(\mathbf{x},\mathbf{x}'), \qquad (9.33)$$

for which ε_i vanishes outside the scatterer. Here, G_s denotes the solution for a single, discrete scatterer. The integral form of this equation (Equation (9.8)) may be used again to write

$$G_s(\mathbf{x},\mathbf{y}) = G_0(\mathbf{x},\mathbf{y}) + \iint G_0(\mathbf{x},\mathbf{w})\mu_i(\mathbf{w},\mathbf{z})G_s(\mathbf{z},\mathbf{y})\,\mathrm{d}\mathbf{w}\,\mathrm{d}\mathbf{z}, \qquad (9.34)$$

where we again recognize that $\mu_i(\mathbf{w},\mathbf{z})$ is zero unless $\mathbf{w}=\mathbf{z}$ and $\mathbf{w}$ is within the scatterer volume. Here we now presume that the solution of this problem of a single scatterer is known. We first write Equation (9.34) in operator form for the ith scatterer located at position $\mathbf{x}_i$ as

$$G_s = G_0 + G_0\mathscr{L}_i(\mathbf{x})G_s. \qquad (9.35)$$

The solution may be written formally as

$$G_s = \text{———} + \text{—◐—}, \qquad (9.36)$$

where the scattering operator $t_i(x, y) =$  is the solution of an integral equation. A series expansion for it is

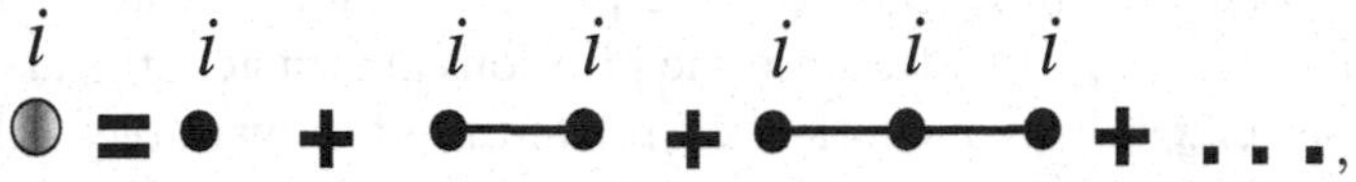

$$(9.37)$$

where the lines again denote G_0 and the dots with labels denote $\mathscr{L}_i(\mathbf{x})$. With these definitions, we can now consider the multiple scattering problem involving discrete scatterers.

We begin by writing the more general operator problem for N scatterers (all of the same type) as

$$G_N = G_0 + G_0 \sum_{i=1}^{N} \mathscr{L}_i(\mathbf{x}) G_N. \tag{9.38}$$

An expansion for G_N may be written diagrammatically, but we must be careful to track the various scatterers of the problem because the full expansion will involve sums over scatterers. Thus, we have

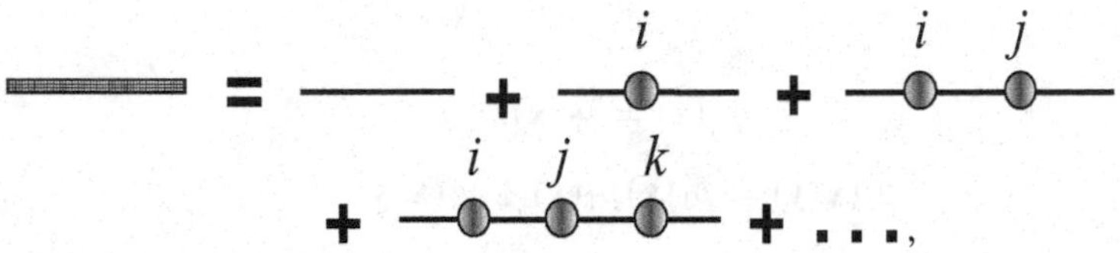

$$(9.39)$$

where the indicial labels i, j, and k denote different scatterers and sums from 1 to N are implicit. It is important to note that adjacent factors of the scattering operator cannot have the same index. Otherwise, some diagrams will be counted more than once. Since the indices are in fact dummy indices, their value is now primarily associated with tracking the order of operators with respect to each other. For example, the diagrams

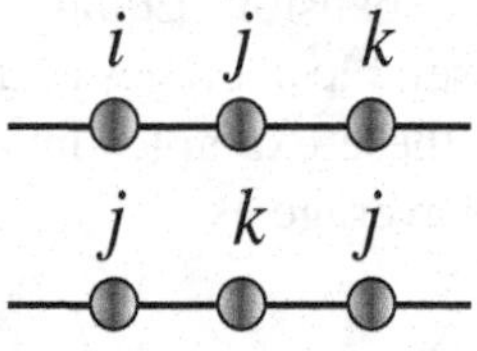

$$(9.40)$$

are identical if the indices are ignored. However, these diagrams are very different. Following Frisch, we use additional connections between the operators t to indicate whether or not they correspond to the same index. In this case, the indices may be dropped, and the diagrams from Equation (9.40) become

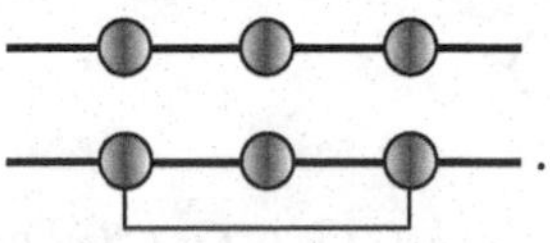

$$(9.41)$$

The rules prohibit connections between adjacent operators.

Before considering ensemble averages of the expansion of G_N, it is necessary to discuss the relevant statistics for discrete scatterers as was done earlier for the continuous medium. At this point, we assume that all scatterers are identical and of the same type and that each has a position (the particle center or other point) defined by $\mathbf{x}_i$, where $i = 1, 2, 3, \ldots, N$. Because the positions are, in general, assumed to be random, the set of $\mathbf{x}_i$ forms a set of N-correlated random variables. Thus, we define the N-particle distribution function $p_N(\mathbf{x}_1, \mathbf{x}_2, \mathbf{x}_3, \ldots, \mathbf{x}_N)$, which is normalized to the total volume of scatterers by

$$\int p_N(\mathbf{x}_1, \mathbf{x}_2, \mathbf{x}_3, \ldots, \mathbf{x}_N)\, \mathrm{d}^3x_1\, \mathrm{d}^3x_2 \ldots \mathrm{d}^3x_N = V^N. \tag{9.42}$$

This definition implies a joint probability density of $V^{-N}p_N$. By analogy, we also need s-particle distribution functions,

$$p_s(\mathbf{x}_1, \mathbf{x}_2, \mathbf{x}_3, \ldots, \mathbf{x}_s) = V^{s-N} \int p_N\, \mathrm{d}^3x_{s+1}\, \mathrm{d}^3x_{s+2} \ldots \mathrm{d}^3x_N, \tag{9.43}$$

defined for a subset s of the N particles.

Next, it is necessary to consider possible particle correlations using the scatterer correlation functions $\gamma_s(\mathbf{x}_1, \mathbf{x}_2, \mathbf{x}_3, \ldots, \mathbf{x}_s)$, which are analogous to Γ_s used earlier to denote the s-point statistics of the continuous medium. With these functions, we can write

$$p_1(\mathbf{x}) = \gamma_1(\mathbf{x}), \tag{9.44}$$

$$p_2(\mathbf{x}, \mathbf{y}) = p_1(\mathbf{x})p_1(\mathbf{y}) + \gamma_2(\mathbf{x}, \mathbf{y}), \tag{9.45}$$

$$\begin{aligned} p_3(\mathbf{x}, \mathbf{y}, \mathbf{z}) &= p_1(\mathbf{x})p_1(\mathbf{y})p_1(\mathbf{z}) + p_1(\mathbf{x})\gamma_2(\mathbf{y}, \mathbf{z}) \\ &\quad + p_1(\mathbf{y})\gamma_2(\mathbf{x}, \mathbf{z}) + p_1(\mathbf{z})\gamma_2(\mathbf{x}, \mathbf{y}) + \gamma_3(\mathbf{x}, \mathbf{y}, \mathbf{z}), \end{aligned} \tag{9.46}$$

and so on to the level of correlations desired. Note that if the positions of all scatterers are assumed to be uncorrelated, we then have $p_N(\mathbf{x}, \mathbf{y}, \mathbf{z}, \ldots, \mathbf{w}) = p_1(\mathbf{x})p_1(\mathbf{y})p_1(\mathbf{z}) \ldots p_1(\mathbf{w})$. However, such an idealization is most often not realistic except for cases involving a very low density of scatterers.

Before returning to the G_N expansion, Equation (9.39), it is also necessary to consider explicit outcomes of ensemble averages of diagrams such as those given by Equation (9.41). The second of these examples involves only two scatterers, such that we can write their ensemble average as

$j \quad k \quad j$

(9.47)

Using the definition of the two-point probability density, $V^{-2}p_2(\mathbf{x}, \mathbf{y})$, this term is given explicitly as

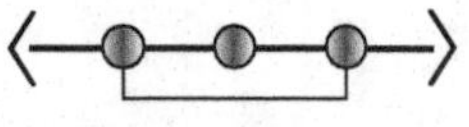

$$= \sum_{j=1, j\neq k}^{N} \sum_{k=1}^{N} V^{-2} \int G_0\, t_j\, G_0\, t_k\, G_0\, t_j\, G_0\, p_2(\mathbf{x}_j, \mathbf{y}_k)\, \mathrm{d}\mathbf{x}_j\, \mathrm{d}\mathbf{y}_k. \tag{9.48}$$

Since the function p_2 depends only on the position vectors of the specific scatterers, the summations may be evaluated giving

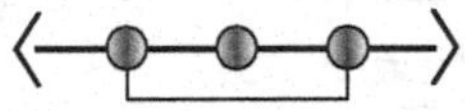

$$= N(N-1)V^{-2}\int G_0\, t_j\, G_0\, t_k\, G_0\, t_j\, G_0\, p_2(\mathbf{x}_j, \mathbf{y}_k)\, \mathrm{d}\mathbf{x}_j\, \mathrm{d}\mathbf{y}_k. \qquad (9.49)$$

The two-point probability density in the integral may then be expanded as appropriate to include any scatterer correlations. If correlations are included, these are indicated as in the continuum case by using a dotted line. Diagrams without a dotted line denote the uncorrelated statistics. Thus, we may write

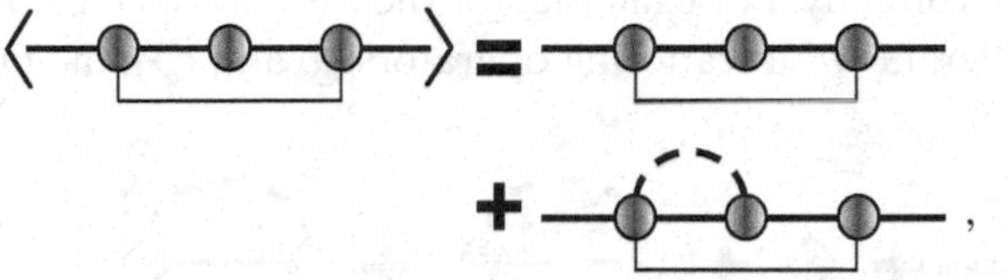

(9.50)

which is written explicitly as

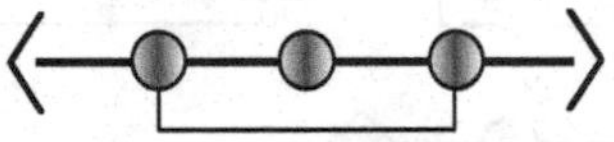

$$= N(N-1)V^{-2}\int G_0\, t_j\, G_0\, t_k\, G_0\, t_j\, G_0\, p_1(\mathbf{x}_j)p_1(\mathbf{y}_k)\, \mathrm{d}\mathbf{x}_j\, \mathrm{d}\mathbf{y}_k$$

$$+ N(N-1)V^{-2}\int G_0\, t_j\, G_0\, t_k\, G_0\, t_j\, G_0\, \gamma_2(\mathbf{x}_j, \mathbf{y}_k)\, \mathrm{d}\mathbf{x}_j\, \mathrm{d}\mathbf{y}_k. \qquad (9.51)$$

With this notation in mind, we now write $\langle G_N \rangle$, the ensemble average of Equation (9.39), as

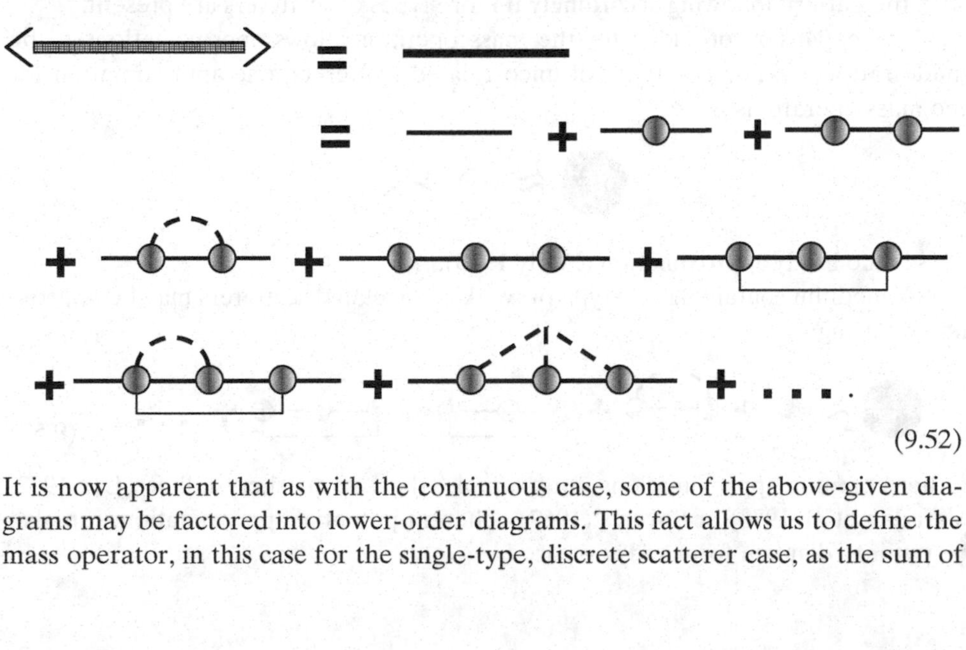

(9.52)

It is now apparent that as with the continuous case, some of the above-given diagrams may be factored into lower-order diagrams. This fact allows us to define the mass operator, in this case for the single-type, discrete scatterer case, as the sum of all diagrams that cannot be reduced further, as

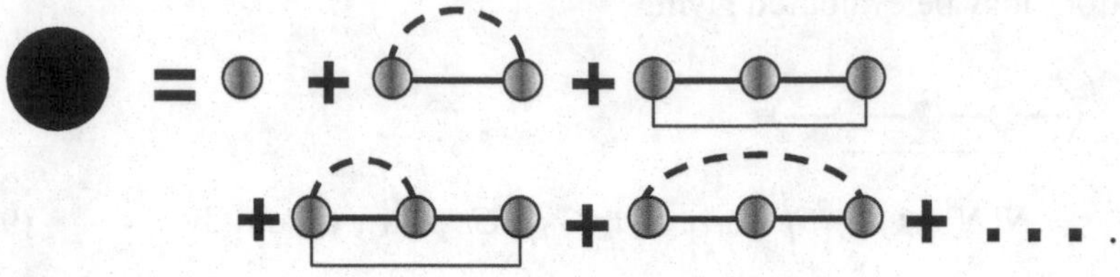

(9.53)

The first term is

$$M(\mathbf{x}, \mathbf{y}) = N\langle t_i(\mathbf{x}, \mathbf{y})\rangle = \frac{N}{V}\int t_i(\mathbf{x}, \mathbf{y})\,\mathrm{d}\mathbf{x}_i. \tag{9.54}$$

The Dyson Equation (9.18) follows as before. The extension to multiple scatterer types follows directly. For example, for the case in which the medium contains two types of scatterers (with scattering operators ◐ and ◇), the mass operator becomes

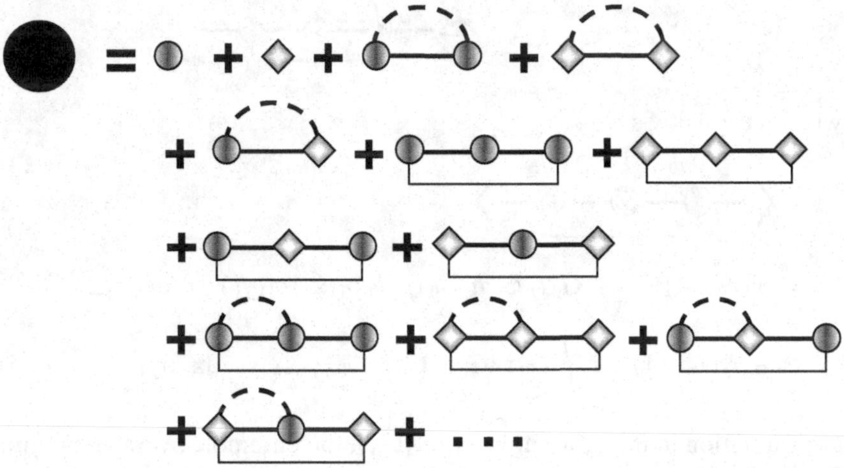

(9.55)

with the pattern following accordingly if more types of scatterers are present.

The expansion provided for the mass operator allows approximations to be made rationally. For two types of uncorrelated scatterers, one approximation for the mass operator is

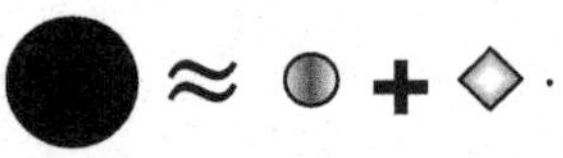

(9.56)

This is the Foldy approximation (Foldy 1945).

A medium containing one type of weakly correlated scatterers may be approximated using

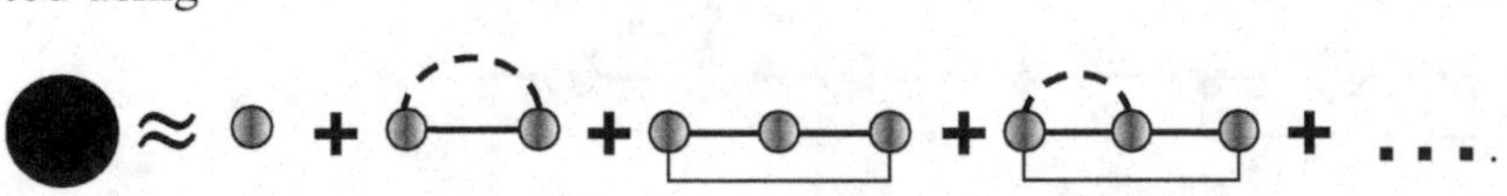

(9.57)

Other common approximations, for example, the Lax quasi-crystalline (Lax 1952, Varadan et al. 1978) or Soven's (1967) coherent potential approximation, can also be put into diagram form and compared.

We may also directly construct the relevant diagrams for the intensity operator for a medium with two scatterer types, the first few terms of which are given by

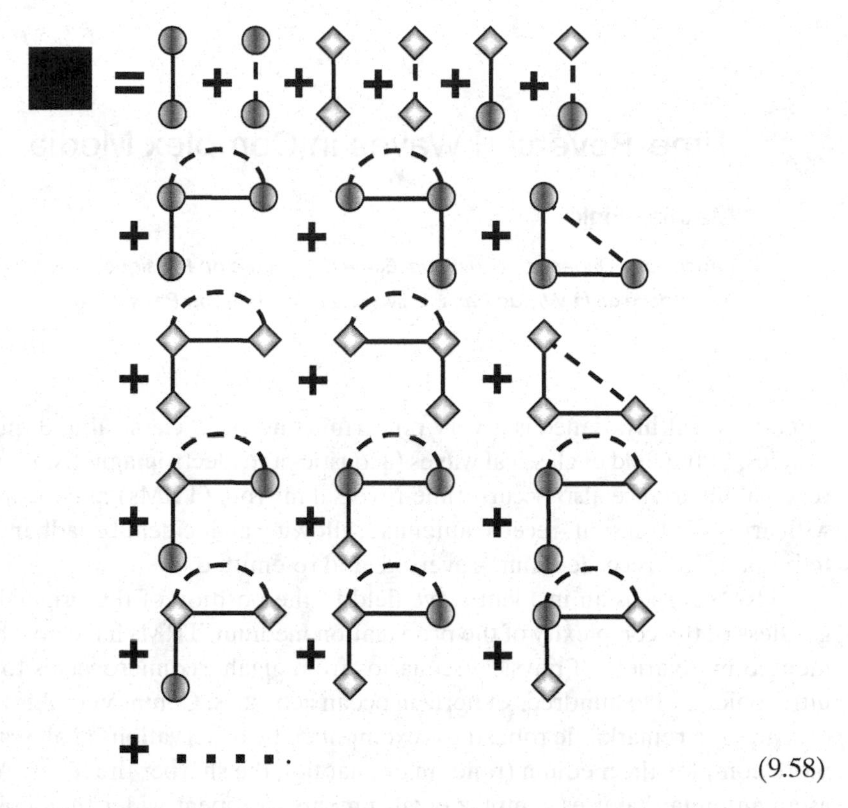

(9.58)

The simplest approximation to be used for the Bethe–Salpeter equation for a medium with two types of discrete scatterers would involve only the first row of the above-given diagrams.

9.5 Summary

In this chapter, the use of diagrams has been described for representing complex integral equations that arise from studies of waves in heterogeneous media. Diagrammatic methods allow the topology of the diagrams to help guide particular assumptions and/or simplifications that are often necessary – such approximations are often less clear when the equations are written explicitly. This approach may be used for continua with properties that are assumed to vary spatially as well as for examination of assemblages of discrete scatterers imbedded in a homogeneous medium. The diagrammatic approach allows the mean field, covariance of the field, and even higher-order statistics to be examined in a rational, logical manner. The simplicity of the diagrams allows complex wave interactions to be tracked easily. However, they also disguise the real complexity of the underlying terms in the resulting equation. Sometimes inclusion of only a few terms of an expansion results in an integral equation that is intractable.

10 Time-Reversed Waves in Complex Media

Mathias Fink

Laboratoire Ondes et Acoustique, Ecole Supérieure de Physique et de Chimie Industrielle de la Ville de Paris, Université Denis Diderot, Paris, France

Time-reversal invariance is a very powerful concept in classical and quantum mechanics. In the field of classical waves (acoustics and electromagnetism), where time-reversal invariance also occurs, time-reversal mirrors (TRMs) may be made simply with arrays of transmit–receive antennae, allowing an incident broadband wave field to be sampled, recorded, time-reversed, and re-emitted.

TRMs refocus an incident wave field to the position of the original source regardless of the complexity of the propagation medium. TRMs have now been implemented in a variety of physical scenarios from gigahertz microwaves to megahertz ultrasonics and to hundreds of hertz in ocean acoustics. Common to this broad range of scales is a remarkable robustness exemplified by observations at all scales that the more complex the medium (random or chaotic), the sharper the focus. A TRM acts as an antenna that uses complex environments to appear wider than it is, resulting, for a broadband pulse, in a refocusing quality that does not depend on the TRM aperture.

TRMs open the way to new methods for signal processing in imaging, detection, and telecommunications. TRMs have applications in ultrasonic therapy, medical imaging, non-destructive testing, telecommunications, underwater acoustics, seismology, sound control, and even home automation.

10.1 Introduction

The evolution of electronic components enables today the building of TRMs that make a wave relive the steps of its past life. These systems exploit the fact that in a majority of cases the propagation of acoustic and electromagnetic waves is a reversible process. Whatever the distortions (diffraction, multiple scattering, reverberation) suffered in a complex environment by a wave, there always exists, at least theoretically, a dual wave able to travel in the opposite direction all the complex travel paths and converge back to the initial source location, exactly as if a movie of the wave propagation had been played backward. The main interest of a TRM is to experimentally create this dual wave, thanks to an array of reversible transducers (able to work both in transmit and receive modes) driven using analog-to-digital and digital-to-analog converters and electronic memories. The TRM is thus able to focus the wave energy through very complex media. In acoustics, a TRM consists of a two-dimensional surface covered with piezoelectric transducers that successively play the role of hydrophones and loudspeakers. The ultrasonic wave emerging from a given source is detected by each of the microphones and is recorded in electronic memories. Then, in a second step (the time-reversal step), all memories are read in the reverse direction. More precisely, the chronology of the signals received by each hydrophone is reversed. The signals recorded at later times are read first. All hydrophones switch synchronously in a transmit mode (loudspeaker) and re-emit the "time-reversed" signals coming from the electronic memories. Thus, new initial conditions for the wave propagation are created, and thanks to reversibility, the diffracted wave has no other solution than living back step by step its past life in a reversed way. This kind of mirror is totally different from a classical mirror. A TRM builds the real image of the source at its location, whereas a classical mirror builds a virtual image of the source.

The great robustness of the time-reversal focusing ability has been verified in many scenarios ranging from ultrasonic propagation at millimeter wavelengths over several centimeters in the human body to ultrasonic propagation in the sea over several tens of kilometers at meter wavelength, and more recently in the propagation of electromagnetic waves over several hundreds of meters at centimeter wavelengths.

10.2 Time Reversal of Acoustic Waves: Basic Principles

Let us consider the propagation of an acoustic wave in a heterogeneous and non-dissipative medium, whose compressibility $\kappa(\mathbf{r})$ and density $\rho(\mathbf{r})$ vary in space. By introducing the sound speed $c(\mathbf{r}) = (\rho(\mathbf{r})\kappa(\mathbf{r}))^{-1/2}$, one can obtain the wave propagation equation for a given pressure field $p(\mathbf{r}, t)$:

$$\nabla \cdot (\frac{\nabla p}{\rho}) - \frac{1}{\rho c^2}\frac{\partial^2 p}{\partial t^2} = 0 \tag{10.1}$$

One can notice the particular behavior of this wave equation regarding the time variable t. Indeed, it only contains a second-order time-derivative operator. This property is the starting point of the time-reversal principle. A straightforward consequence of this property is that if $p(\mathbf{r}, t)$ is a solution of the wave equation, then $p(\mathbf{r}, -t)$ is also a solution of the problem. This property illustrates the invariance of the wave equation during a time-reversal operation, the so-called time-reversal invariance. However, this property is only valid in a non-dissipative medium. If wave propagation is affected by dissipation effects, odd-order time derivatives appear in the wave equation, and the time-reversal invariance is lost. Nevertheless, one should note that if the ultrasonic absorption coefficient is sufficiently small in the frequency bandwidth of the ultrasonic waves used for the experiments, the time-reversal invariance remains valid.

This symmetry has been illustrated by Stokes in the classical case of plane wave reflection and transmission through a plane interface between two layers of different wave speeds. Consider an incident plane wave of normalized amplitude 1 propagating from medium 1 to medium 2. It is possible to observe a reflected plane wave with amplitude R and a transmitted plane wave with amplitude T (Figure 10.1(a)).

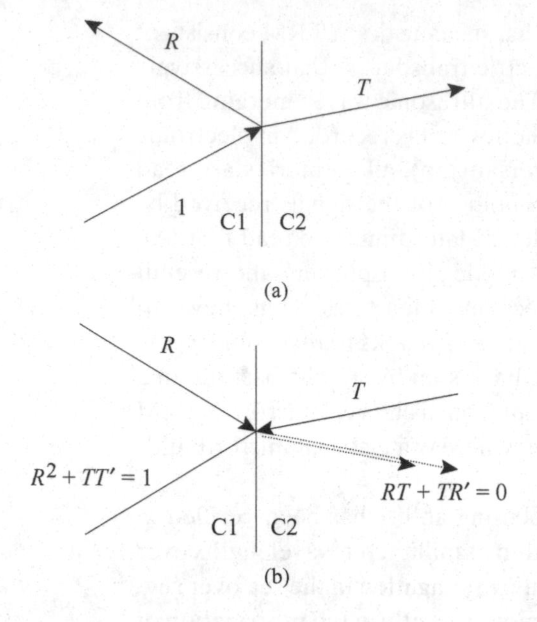

Figure 10.1. (a) Reflection and transmission of a plane wave at the interface between two media with different sound speed. (b) Time reversal of case (a). From Fink (1992). Copyright 1992, IEEE.

Starting from this simple configuration in which the total pressure field $p(\mathbf{r}, t)$ consists of three plane waves, Stokes tried to verify if this experiment could be "time reversed" or not. He used the fact that a time-reversal operation consists of the inversion of the wave vector direction for the simple case of a plane wave. Thus, the time-reversed field $p(\mathbf{r}, -t)$ can be described by a set of three plane waves: two incident waves of respective amplitude R and T going respectively from medium 1 to 2 and from medium 2 to 1, followed by a transmitted wave of amplitude 1 propagating in medium 1 (see Figure 10.1(b)). One can easily verify that this new theoretical wavefield is also a solution of the wave equation. Indeed, if we define the reflection and transmission coefficients R' and T' of an incident wave coming from medium 2, the superposition principle shows that the two incident waves generate four plane waves, two propagating in medium 1 with a resulting amplitude $R^2 + TT'$ and two other plane waves propagating in medium 2 with a resulting amplitude $RT + TR'$. An elementary calculation of the reflection and transmission coefficients R, T, R', and T' permits one to verify the equations

$$R^2 + TT' = 1, \tag{10.2}$$

$$R + R' = 0, \tag{10.3}$$

as is required if time-reversal invariance is to hold.

This example shows that the scene of Figure 10.1(b) can straightforwardly be interpreted as the time-reversed version of Figure 10.1(a).

10.3 Time-Reversal Mirrors

Taking advantage of this property the concept of a TRM has been developed and several devices have been built, which illustrated the efficiency of this concept (Fink 1992, 1997, Fink et al. 2000). In such a device an acoustic source, located inside a lossless medium, radiates a brief transient pulse that propagates and is potentially

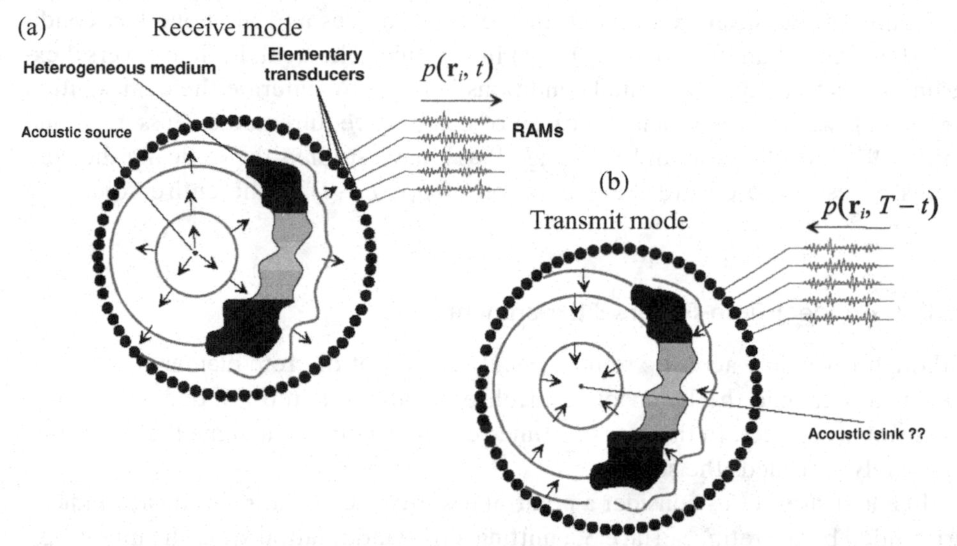

Figure 10.2. (a) Recording step: A closed surface is filled with transducer elements. A point-like source generates a wave front that is distorted by heterogeneities. The distorted pressure field is recorded on the cavity elements. (b) Time-reversed or reconstruction step: The recorded signals are time-reversed and re-emitted by the cavity elements. The time-reversed pressure field back-propagates and refocuses exactly on the initial source.

distorted by the medium. Time reversal as described earlier would entail the reversal, at some instant, of every particle velocity in the medium. As an alternative, the acoustic field could be measured on every point of an enclosing surface (acoustic retina), and retransmitted in time-reversed order; then the wave will travel back to its source (see Figure 10.2). Both time-reversal invariance and spatial reciprocity (Cassereau & Fink 1992) are required to reconstruct the time-reversed wave in the whole volume by means of a two-dimensional time-reversal operation. From an experimental point of view, a closed TRM consists of a two-dimensional piezoelectric, transducer array that samples the wavefield over a closed surface. An array pitch of the order of $\lambda/2$, where λ is the smallest wavelength of the pressure field, is needed to ensure the recording of all the information in the wavefield. Each transducer is connected to its own electronic circuitry that consists of a receiving amplifier, an analog-to-digital converter, a storage memory, and a programmable transmitter able to synthesize a time-reversed version of the stored signal. In practice, closed TRMs are difficult to realize and the time-reversal operation is usually performed on a limited angular area, thus apparently limiting focusing quality. A TRM consists typically of a small number of elements or time-reversal channels. The major interest of TRM, compared with classical focusing devices (lenses and beam forming) is certainly the relation between the medium complexity and the size of the focal spot. A TRM acts as an antenna that uses complex environments to appear wider than it is, resulting in a refocusing quality that does not depend on the TRM aperture.

The basic theory employs a scalar wave formulation $p(\mathbf{r}, t)$ and, hence, is strictly applicable to acoustic or ultrasound propagations in fluid. However, the basic ingredients and conclusions apply equally well to elastic waves in a solid and to electromagnetic fields.

In any propagation experiment, the acoustic sources and the boundary conditions determine a unique solution $p(\mathbf{r}, t)$ in the fluid. The goal, in time-reversal experiments, is to modify the initial conditions in order to generate the dual solution $p(\mathbf{r}, T - t)$, where T is a delay because of causality requirements. Cassereau and Fink (1992) and Jackson and Dowling (1991) have studied theoretically the conditions necessary to ensure the generation of $p(\mathbf{r}, T - t)$ in the entire volume of interest.

10.3.1 An Ideal Time-Reversal Experiment

Although reversible acoustic retinas usually consist of discrete elements, it is convenient to examine the behavior of idealized continuous retinas, defined by two-dimensional surfaces. In the case of a time-reversal cavity, we assume that the retina completely surrounds the source.

In a first step, let us consider a point-like source located at $\mathbf{r}_0$ inside a volume V surrounded by the retina surface S, emitting a time modulation $s(t)$. The inhomogeneous wave equation is given by

$$\nabla \cdot \left(\frac{\nabla p}{\rho}\right) - \frac{1}{\rho c^2}\frac{\partial^2 p}{\partial t^2} = -\delta(\mathbf{r} - \mathbf{r}_0)s(t). \tag{10.4}$$

Considering an impulsive source $s = \delta(t)$ at time 0, the causal solution to Equation (10.4) reduces to the retarded Green's function $G_{\text{ret}}(\mathbf{r}, \mathbf{r}_0; t)$ that takes into account the heterogeneities and the boundaries of the medium. Note that to respect causality in physics, only the causal Green's function (retarded) is selected, whereas the advanced Green's function (the anti-casual) is neglected. The goal of a time-reversed experiment is to generate this advanced Green's function by modifying the initial conditions. In the following we will use the notation $G(\mathbf{r}, \mathbf{r}_0; t)$ instead of $G_{\text{ret}}(\mathbf{r}, \mathbf{r}_0; t)$. During the second step of the time-reversal process, the initial source at $\mathbf{r}_0$ is removed, and we create on the surface of the cavity monopole and dipole sources that correspond to the time reversal of the fields p_s owing to s and measured during the first step.

Because of these secondary sources on S, a time-reversed pressure field $p_{\text{tr}}(\mathbf{r}, t)$ propagates inside the cavity. It can be calculated using a modified version of the Helmoltz–Kirchhoff integral:

$$p_{\text{tr}}(\mathbf{r}, t) = \int_{-\infty}^{\infty} \iint_S \left[G(\mathbf{r}, \mathbf{r}'; t - t')\frac{\partial p_s(\mathbf{r}', t')}{\partial n} - p_s(\mathbf{r}', t')\frac{\partial G(\mathbf{r}, \mathbf{r}'; t - t')}{\partial n} \right] \frac{d^2\mathbf{r}'}{\rho(\mathbf{r}')}\, dt'. \tag{10.5}$$

The Green's function definition, spatial reciprocity, and time-reversal invariance yield the following expression (Cassereau & Fink 1992, van Manen et al. 2005):

$$p_{\text{tr}}(\mathbf{r}, t) = G(\mathbf{r}, \mathbf{r}_0; T - t) - G(\mathbf{r}, \mathbf{r}_0; t - T). \tag{10.6}$$

This equation can be interpreted as the difference of advanced and retarded waves centered on the initial source position. The converging wave (advanced) collapses at the origin and is always followed by a diverging (retarded) wave. Thus the time-reversed field observed as a function of time shows two wavefronts of opposite sign. The wave re-emitted by the time reversal cavity looks like a convergent wavefield during a given period, but a wavefield *doesn't know how to stop.* When the converging wavefield reaches the location of the initial source location, it collapses and then continues its propagation as a diverging wavefield. Complete time reversal would require a time reversal of the original source as well.

For example, in the case of a *homogeneous* medium, assuming that the retina does not perturb the field propagation (free-space assumption), the free-space retarded Green's function G_0 reduces to a diverging spherical impulse wave that depends only on $\mathbf{r} - \mathbf{r}_0$ and propagates with a sound speed c. Thus, neglecting the causal time delay T, the time-reversed field can be written as

$$p_{\rm tr}(\mathbf{r}, t) \propto \frac{1}{4\pi r}\delta(t + r/c) - \frac{1}{4\pi r}\delta(t - r/c). \tag{10.7}$$

Although both converging and diverging spherical waves show a singularity at the origin, it is crucial to note that the time-reversed field remains finite for all time throughout the cavity. This result can be explained in the frame of the general theory for wave propagation. Indeed, during the reconstruction step the initial source is removed or remains passive. As a consequence there is no more spatial discontinuity, and the pressure field resulting from this back-propagation is not discontinuous inside V.

In the case of a narrow-band excitation (monochromatic excitation of pulsation ω), the interference between the converging and the diverging fields leads to the classical diffraction limits. Indeed by calculating the Fourier transform of Equation (10.7) over the time variable t, we obtain

$$\begin{aligned}\hat{P}_{\rm tr}(\mathbf{r}, \omega) &= \frac{\exp(-{\rm j}k\,|\mathbf{r} - \mathbf{r}_0|)}{4\pi\,|\mathbf{r} - \mathbf{r}_0|} - \frac{\exp({\rm j}k|\mathbf{r} - \mathbf{r}_0|)}{4\pi\,|\mathbf{r} - \mathbf{r}_0|} \\ &= -2{\rm j}\frac{\sin(k|\mathbf{r} - \mathbf{r}_0|)}{4\pi\,|\mathbf{r} - \mathbf{r}_0|} \\ &= -2{\rm j}\Im\left\{\hat{G}(\mathbf{r} - \mathbf{r}_0, \omega)\right\},\end{aligned} \tag{10.8}$$

where ${\rm j}^2 = -1$. The time-reversed field at the initial source position is finite because it is the difference between a converging and a diverging wave and not the sum (otherwise it would have a discontinuity there).

As a consequence, the time-reversed field is focused on the initial source position, with a focal spot size limited to one half-wavelength π/k that corresponds to the standard formulation for the complex field modulus where k is the wavenumber and $\hat{G}(\mathbf{r} - \mathbf{r}_0, \omega)$ is the monochromatic Green's function. The point spread function is proportional to the imaginary part of the monochromatic Green's function.

In the case of an inhomogeneous, and still lossless, medium a similar interpretation can be given, but the retarded Green's function G is now not dependent on $\mathbf{r} - \mathbf{r}_0$ but is

$$\begin{aligned}\hat{P}_{\rm tr}(\mathbf{r}, \omega) &= \iint_S \left[\frac{\partial \hat{G}^*(\mathbf{r}', \mathbf{r}_0; \omega)}{\partial n}\hat{G}(\mathbf{r}, \mathbf{r}'; \omega) - \hat{G}^*(\mathbf{r}', \mathbf{r}_0; \omega)\frac{\partial \hat{G}(\mathbf{r}, \mathbf{r}'; \omega)}{\partial n}\right]\frac{{\rm d}^2\mathbf{r}'}{\rho(\mathbf{r}')} \\ &= -2{\rm j}\Im\left\{\hat{G}(\mathbf{r}, \mathbf{r}_0; \omega)\right\},\end{aligned} \tag{10.9}$$

rather a function separately of both $\mathbf{r}$ and $\mathbf{r}_0$.

Note that the choice of the integration boundary S is arbitrary as long as it encloses $\mathbf{r}$ and $\mathbf{r}_0$. Equations (10.6) and (10.9) require that the TRM is composed of monopole and dipole sources. If, however, the TRM is located in the far field of source and observation points and heterogeneities, the expression can be simplified. In this case we may assume that

$$\frac{\partial}{\partial n}\hat{G}(\mathbf{r},\mathbf{r}';\omega) \approx \mathrm{j}k\hat{G}(\mathbf{r},\mathbf{r}';\omega). \tag{10.10}$$

Thus, the time-reversed field can be written as

$$\hat{P}_{\mathrm{tr}}(\mathbf{r},\omega) \approx 2\mathrm{j}\frac{\omega}{\rho c}\iint_S \hat{G}^*(\mathbf{r}',\mathbf{r}_0;\omega)\hat{G}(\mathbf{r},\mathbf{r}';\omega)\,\mathrm{d}^2\mathbf{r}'. \tag{10.11}$$

If we come back to the time domain, Equation (10.11) can be written as

$$p_{\mathrm{tr}}(\mathbf{r},t) \approx \frac{2}{\rho c}\frac{\partial}{\partial t}\iint_S G(\mathbf{r}',\mathbf{r}_0;-t)\otimes G(\mathbf{r},\mathbf{r}';t)\,\mathrm{d}^2\mathbf{r}'. \tag{10.12}$$

From an experimental point of view, it is not easy to measure and re-emit the field at any point of a surface S: experiments are carried out with transducer arrays that spatially sample the receiving and emitting surface. Assuming that the time-reversal retina consists of discrete elements located at position $\mathbf{r}_i$, this allows us to replace the integration over S in Equation (10.12) by a summation over N surface element positions:

$$p_{\mathrm{tr}}(\mathbf{r},t) = C\frac{\partial}{\partial t}\sum_{i=1}^{N} G(\mathbf{r}_i,\mathbf{r}_0;-t)\otimes G(\mathbf{r},r_i;t), \tag{10.13}$$

where C is a scaling factor. Taking into account spatial reciprocity this expression can be written as a summation of cross-correlation functions:

$$p_{\mathrm{tr}}(\mathbf{r},t) \approx C\frac{\partial}{\partial t}\sum_{i=1}^{N} G(\mathbf{r}_0,r_i;-t)\otimes G(\mathbf{r},r_i;t). \tag{10.14}$$

The spatial sampling of surface S by a set of elements may introduce grating lobes. These lobes can be avoided by using an array pitch smaller than $\lambda_{\min}/2$, where $\lambda_{\min}$ is the smallest wavelength of interest.

10.4 Time-Reversal Mirror in Complex Media

It is generally difficult to use acoustic arrays that completely surround the area of interest, and the closed cavity is usually replaced by a TRM of finite angular aperture. This yields an increase of the point spread function dimension that is related to the limited angular size of the mirror observed from the source. In the standard approach of diffraction in homogeneous free space, the point spread function is related to the angular spectrum of the aperture. For a closed TRM, the **k**-vectors of the radiated field span the whole 4π solid angle and the focal spot dimension is minimal ($\lambda/2$). When a TRM covers a limited solid angle, the spatial diversity of the **k**-vectors that interact with the TRM is reduced. Therefore the focal spot size is increased.

The main interest of focusing with TRM is that in media with complex structure the spatial diversity of the **k**-vectors captured by a small TRM can be significantly

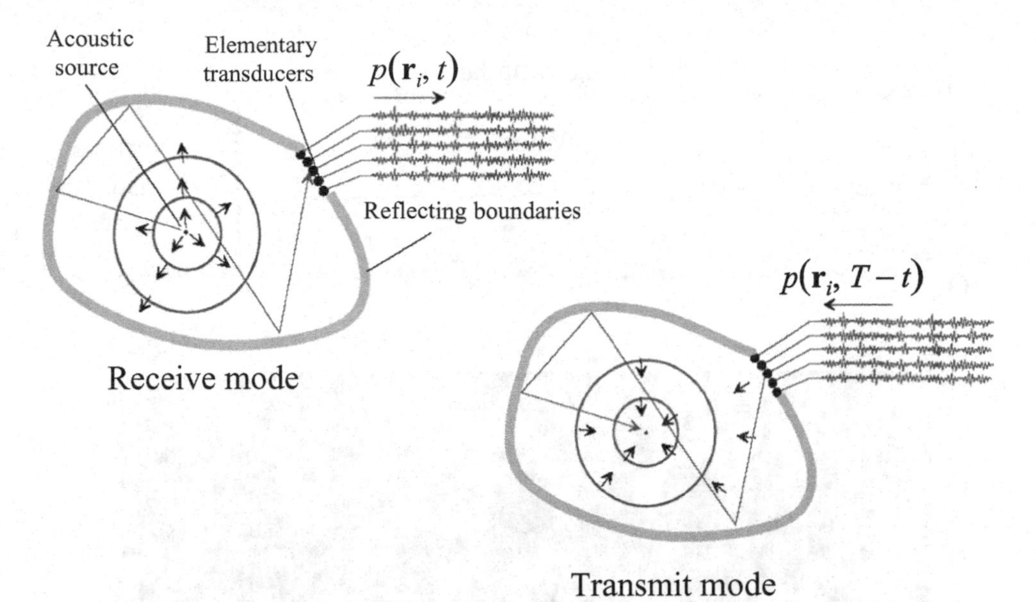

Figure 10.3. One part of the transducers is replaced by reflecting boundaries. In the first step (receive mode) the wave radiated by the source is recorded by a set of transducers through the reverberation inside the cavity. In the second step, the recorded signals are time-reversed and re-emitted by the transducers.

increased. Wave propagation in media with complex boundaries or random scattering can increase the apparent aperture of the TRM, resulting in a focal spot size smaller than that predicted by classical formula.

The basic idea is to replace part of the transducers needed to sample a closed time-reversal surface by reflecting boundaries that redirect parts of the incident wave toward the TRM aperture (see Figure 10.3). When a source radiates a wave field inside a closed cavity or in a waveguide, multiple reflections along the medium boundaries can significantly increase the apparent aperture of the TRM. Thus spatial information on the **k**-vectors that is usually lost with a finite-aperture TRM is converted into the time domain. The reversal quality then depends crucially on the duration of the time-reversal window, that is, the length of the recording that is reversed.

Such a concept is strongly related to a kaleidoscopic effect that appears thanks to the multiple reverberations on the waveguide boundaries. Waves emitted by each transducer are multiply reflected, creating at each reflection virtual transducers that can be observed from the desired focal, point. Thus we create a large virtual array from a limited number of transducers and a small number of transducers is multiplied to create a "kaleidoscopic" transducer array. Three different examples will be presented (a waveguide, a chaotic cavity, and a multiply scattering medium.)

10.4.1 Time Reversal in an Acoustic Waveguide

The simplest boundaries that can give rise to such a kaleidoscopic effect are plane boundaries as in rectangular waveguides or cavities. A first experiment conducted in the ultrasonic regime by Roux et al. (1997) and Roux and Fink (2000) showed

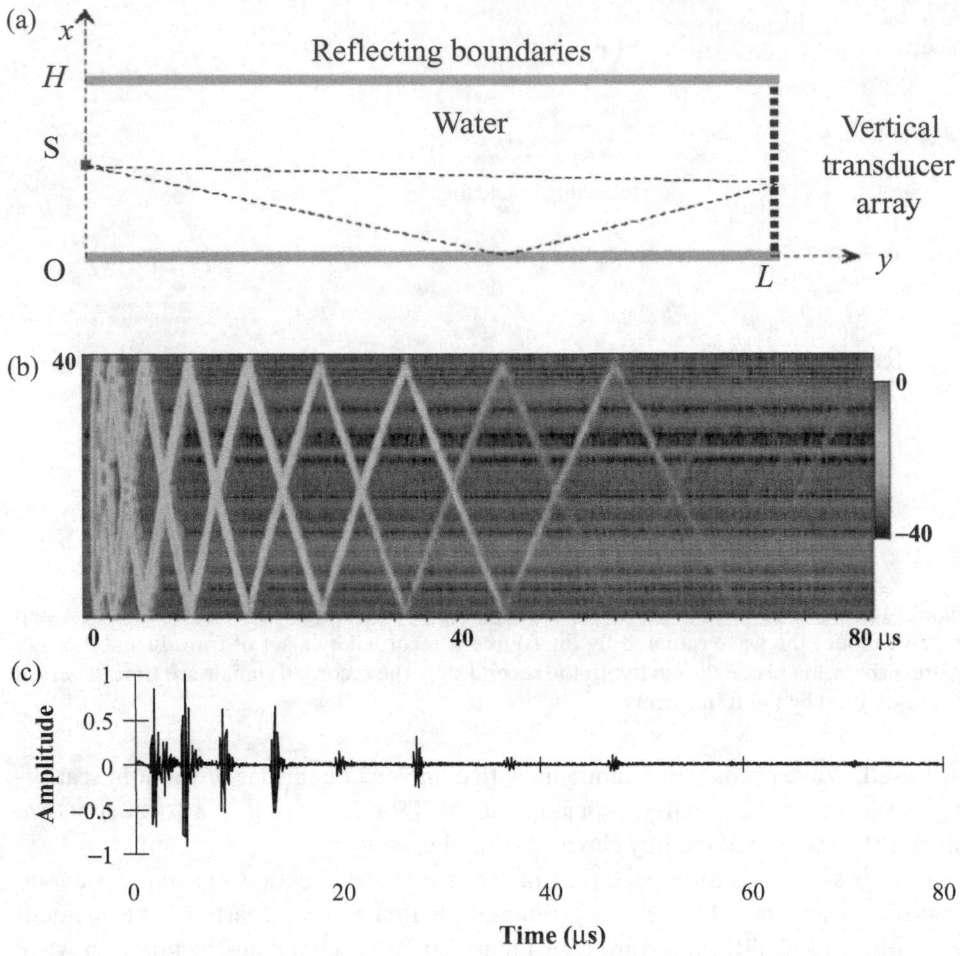

Figure 10.4. (a) Schematic of the acoustic waveguide: the guide length ranges from 40 cm to 80 cm and the water depth from 1 cm to 5 cm. The central acoustic wavelength (λ) is 0.5 mm. The array element spacing is 0.42 mm. The TRM is always centered at the middle of the water depth. (b) Spatial-temporal representation of the incident acoustic field received by the TRM; the amplitude of the field is in decibels. (c) Temporal evolution of the signal measured on one transducer of the array. Reprinted with permission from Fink (1997). Copyright 1997, American Institute of Physics.

clearly this effect with a TRM made of a one-dimensional transducer array located in a rectangular ultrasonic waveguide (see Figure 10.4(a)). For an observer located in the waveguide the TRM seems to be escorted by a periodic set of virtual images related to multipath propagation. An effective aperture 10 times larger than the real aperture was observed.

The experiment was conducted in a waveguide whose interfaces (water–air or water–steel) are plane and parallel. The length of the guide was $L \sim 800$ mm, on the order of 20 times the water depth of $H \sim 40$ mm. A sub-wavelength ultrasonic source was located at one end of the waveguide. On the other end, a one-dimensional TRM made of a 96 element array spanned the waveguide. The transducers had a center frequency of 3.5 MHz and 50% bandwidth. Because of experimental limitations the array pitch was greater than $\lambda/2$. A time-reversal experiment was then performed in the following way: (1) the point source emits a pulsed

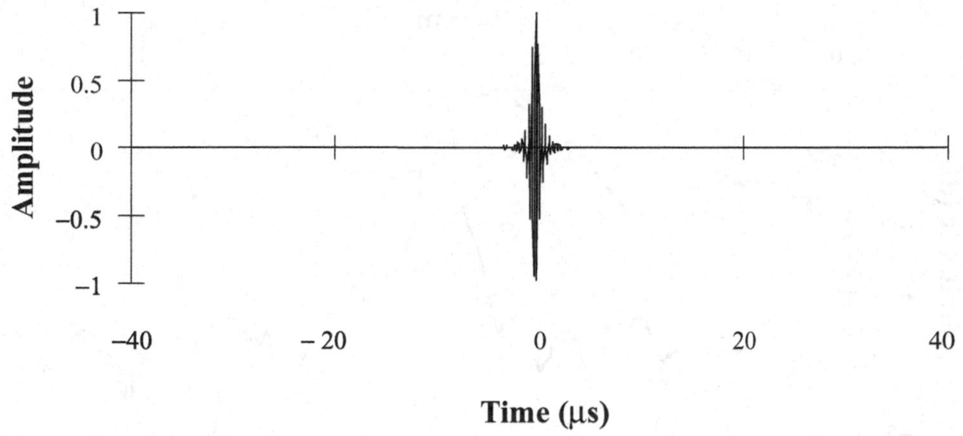

Figure 10.5. Time-reversed signal measured at the point source. Reprinted with permission from Roux et al. (1997). Copyright 1997, American Institute of Physics.

wave (1 μs duration); (2) the TRM receives, selects a time-reversal window, time-reverses, and re-transmits the field; (3) after back-propagation the time-reversed field is scanned in the plane of the source.

Figure 10.4(b) shows the incident field recorded by the array after forward propagation through the channel. After the arrival of the first wavefront corresponding to the direct path we observe a set of signals, owing to multiple reflections of the incident wave between the interfaces, that spread over 100 μs. Figure 10.4(c) represents the signal received on one transducer of the TRM.

After retransmission and propagation of the time-reversed signals recorded by the array during a window of 100 μs, we observe a remarkable temporal compression at the source location (see Figure 10.5). This means that multipath effects are fully compensated. Figure 10.5 shows that the time-reversed signal observed at the source is nearly identical to the one received in a time-reversed experiment conducted in free space. The peak signal exceeds its temporal side lobes by 45 dB.

The spatial focusing of the time-reversed field is also of interest. Figure 10.6 shows the directivity pattern of the time-reversed field observed in the source plane. The time-reversed field is focused on a spot that is much smaller than that obtained with the same TRM working in free space. In our experiment the −6 dB lateral resolution is improved by a factor of 9. This can be easily interpreted by the images theorem in a medium bounded by two mirrors. For an observer, located at the source point, the 40-mm TRM appears to be accompanied by a set of virtual images related to multipath reverberation.

The effective TRM is then a set of TRMs as shown in Figure 10.7. This is the kaleidoscope effect. When taking into account the first 10 arrivals, the theoretical effective aperture of the mirror array is 10 times larger than the real aperture. However in practice, as the replicas arrive later, their amplitudes decrease. Angular directivity of the transducers apodizes the effective aperture of the TRM, smoothing the focal spot and depressing its spatial side lobes. Figure 10.8 shows the effect of the time-reversed window duration ΔT on the width of the focal spot. The size of the focal spot decreases when the number of replicas selected by the window increases. This clearly shows that the effective aperture of the TRM is directly related

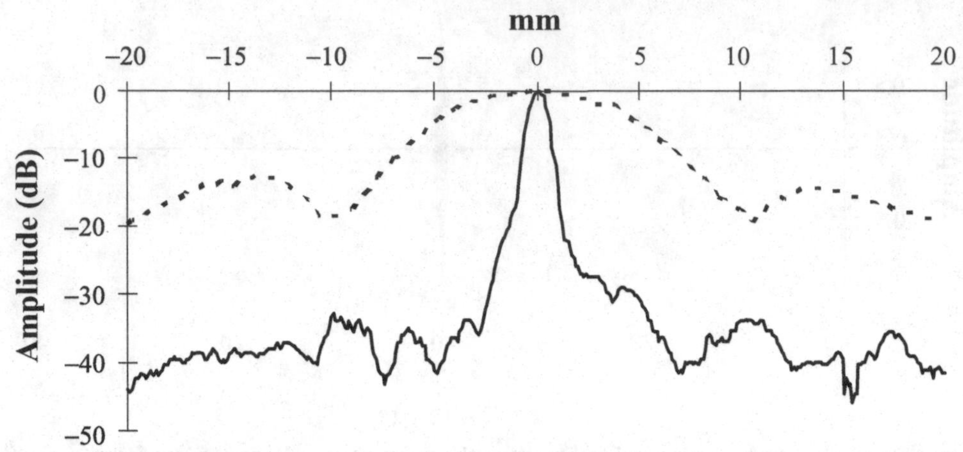

Figure 10.6. Directivity pattern of the time-reversed field in the plane of source: the dotted line corresponds to free space and the full line to the waveguide. Reprinted with permission from Roux et al. (1997). Copyright 1997, American Institute of Physics.

to the time-reversal window duration ΔT. It emphasizes the way the **k**-vector information is translated in the time domain through the interaction with the boundaries. In the set of replicas that are recorded in Figure 10.3(a), the last arrival corresponds to the highest transverse **k** values of the source angular spectrum. Thus, the −6 dB beamwidth of the focal spot is given by

$$\Delta \approx \frac{\lambda f}{D'}, \tag{10.15}$$

where D' is an effective aperture dimension of the TRM that depends on the duration ΔT. When all the replicas are selected and $\Delta T = 100$ μs, we measure $D' = 9D$, where D is the array aperture. The optimal duration ΔT of the time-reversal window is related to the waveguide dispersion and depends on the source–TRM distance.

10.4.1.1 Time Reversal: Matched Filter or Inverse Filter

The impressive time recompression can be interpreted in terms of matched filters. To do this we neglect the time derivatives that appear in Equations (10.12)–(10.14).

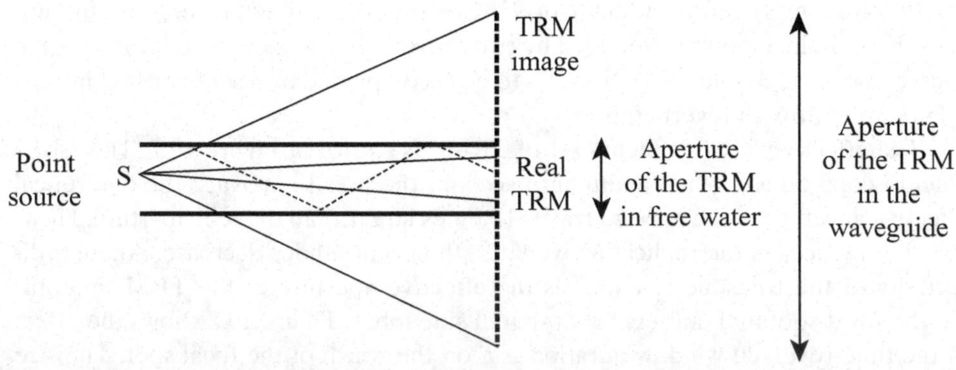

Figure 10.7. The principle of mirror images (kaleidoscope) applied to the waveguide.

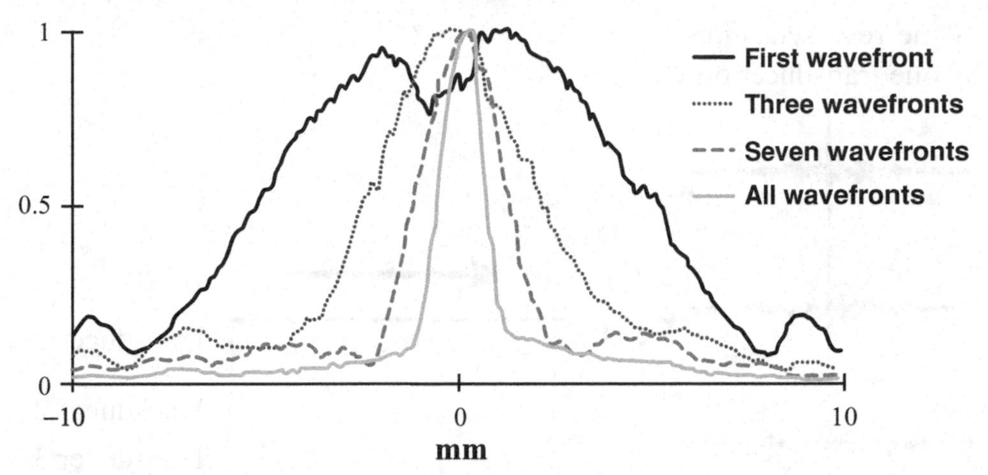

Figure 10.8. Directivity patterns of the time-reversed field versus the number of echoes selected in thc time-reversed window.

The experimental time-reversed field at the source location is then

$$p_{\text{tr}}(\mathbf{r} = \mathbf{r}_0, t) \propto \sum_{i=1}^{N} G(\mathbf{r}_0, \mathbf{r}_i; -t) \otimes G(\mathbf{r}_0, \mathbf{r}_i; t). \tag{10.16}$$

It is a sum of autocorrelation functions. Wave propagation through a waveguide may be described as a linear system with different impulse responses (Green's functions.) Each term in the summation (10.16) corresponds to a *matched filter*. Given a signal as input, a matched filter is a linear filter whose output after convolving with the input is optimal in some sense. Whatever the impulse response $h_i(t) = G(\mathbf{r}_0, \mathbf{r}_i; t)$, the convolution $h_i(-t) \otimes h_i(t)$ is maximum at time $t = 0$. This maximum is always positive and equals $\int h_i^2(t)\,\mathrm{d}t$, that is, the energy of the signal $h_i(t)$. This has an important consequence. Indeed, with an N-element array, the time-reversed signal recreated on the source depends on the sum

$$\sum_{i=1}^{N} h_i(-t) \otimes h_i(t). \tag{10.17}$$

Even if each of the $h_i(t)$ has different behavior, each term in this sum reaches its maximum at time $t = 0$. So all contributions add constructively around $t = 0$, whereas at earlier or later times uncorrelated contributions tend to destroy one another. Thus the re-creation of a sharp peak after time reversal on an N-element array can be viewed as an interference process between the N outputs of N matched filters (see Figure 10.9).

When the number N of channels is sufficient, that is to say when the TRM aperture fills exactly the whole waveguide aperture and when the sampling pitch is small enough, we observed that not only is the time-reversal process a spatio-temporal *matched* filter, but it is also a good approximation of an *inverse* filter of the propagation operator (Tanter et al. 2000, 2001). In the present case the 96 elements spanned the entire waveguide. A pitch of less than half a wavelength assured no grating lobes around the focus.

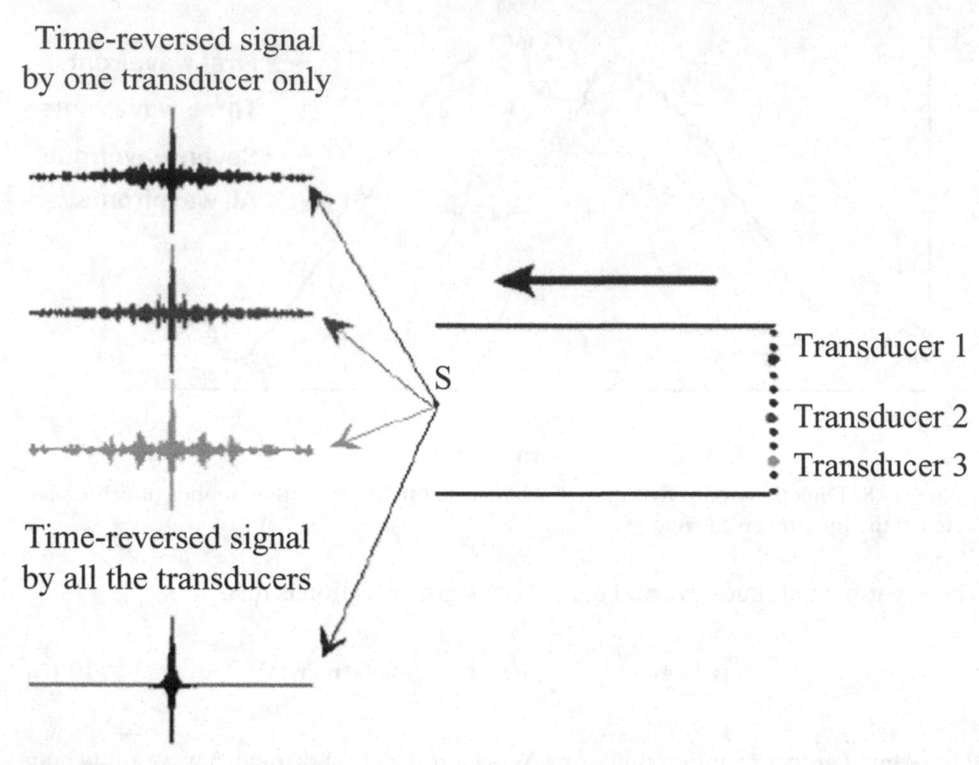

Figure 10.9. Each individual contribution of the time-reversed field is a symmetrical signal with a maximum at the same time T for each transducer. When all the transducers of the TRM work together the summation of all the signals give a perfect time recompression at the origin. Reprinted with permission from Roux and Fink (2000). Copyright 2000, Acoustical Society of America.

Acoustic waveguides are currently found in underwater acoustics, especially in shallow water, and TRMs can compensate for the multipath propagation there that limits the capacity of underwater communication systems. The problem arises because acoustic transmissions in shallow water reflect off the ocean surface and floor, so that a transmitted pulse gives rise to multiple copies of itself that arrive at the receiver. Underwater acoustic experiments have been conducted by W. Kuperman and his group from San Diego University in a sea water channel of 120-m depth, with a 24-element TRM working at 500 Hz and 3.5 kHz. They observed focusing with super resolution and multipath compensation at a distance up to 30 km (Kuperman et al. 1998). Such properties open the field of new discrete communications systems in underwater applications as was experimentally demonstrated by different groups (Rouseff et al. 2001, Edelmann et al. 2002).

In conclusion, two points must be emphasized. Boundaries help to obtain a better resolution by redirecting **k**-vectors of high transverse spatial frequency toward the TRM aperture. However a small-aperture TRM made of a reduced number of transducers does not work perfectly because regular sidewall reflections make the TRM into a periodic grating, one with significant sidelobes. There are two principal ways of reducing the grating lobes. One is to work with broadband signals. Thus for each spectral component, the grating lobe positions are shifted; so an averaging effect reduces their amplitude compared with the main lobe. An alternative is to work with boundaries that have less symmetry; this is the case of the so-called chaotic cavities that we will explore now.

10.4.2 Time Reversal in Chaotic Cavities

In this section, we are interested in another aspect of multiply reflected waves: waves confined in closed reflecting cavities with nonsymmetrical geometry. With closed boundary conditions, no information can escape from the system and a reverberant acoustic field is created. If, moreover, the geometry of the cavity shows ergodic and mixing properties, one may hope to collect all information at only one point. Ergodicity means that because of the boundary geometry, any acoustic ray radiated by a point source and multiply reflected would pass every location in the cavity. Therefore, all the information about the source can be redirected toward a single time reversal transducer (we will see later, using an eigenmode decomposition of the wavefield, that it is not exactly true). This is the regime of fully diffuse wave fields that can be also defined as in room acoustics as an uncorrelated and isotropic mix of plane waves of all propagation directions (Weaver 1982, Ebeling 1984; see also Chapter 4 of the present work). Draeger and Fink (1997, 1999) and Draeger et al. (1999) showed experimentally and theoretically that in this particular case a time reversal focusing with $\lambda/2$ spot can be obtained using only one time-reversal channel operating in a closed cavity.

The first experiments were made with elastic waves propagating in a two-dimensional cavity with negligible absorption. They were carried out using guided elastic waves in a monocrystalline D-shaped silicon wafer known to have chaotic ray trajectories (see Chapter 2). This property eliminates the effective regular gratings of the previous section. Silicon was selected also for its weak absorption. Elastic waves in such a plate are akin to Lamb waves.

An aluminum cone coupled with a longitudinal transducer generated waves at one point of the cavity. A second transducer was used as a receiver. The central frequency of the transducers was 1 MHz, and their bandwidth was 100%. At this frequency, only three propagating modes are possible (one flexural, one quasi-extensional, one quasi-shear). The source was considered point-like and isotropic because the cone tip is much smaller than the central wavelength. A heterodyne laser interferometer measured the displacement field as a function of time at different points on the cavity. Assuming that there is no mode conversion at the boundaries between the flexural mode and other modes, we have only to deal with one field, the flexural-scalar field.

The experiment is a "two-step process" as described earlier. In the first step, one of the transducers, located at point A, transmits a short omnidirectional signal of duration 0.5 μs. Another transducer, located at B, observes a long random-looking signal that results from multiple reflections from the boundaries of the cavity. It continues for more than 50 ms corresponding to some 100 reflections at the boundaries. Then, a portion ΔT of the signal is selected, time-reversed, and re-emitted by point B. As the time-reversed wave is a flexural wave that induces vertical displacements of the silicon surface, it can be observed using the optical interferometer that scans the surface around point A (see Figure 10.10).

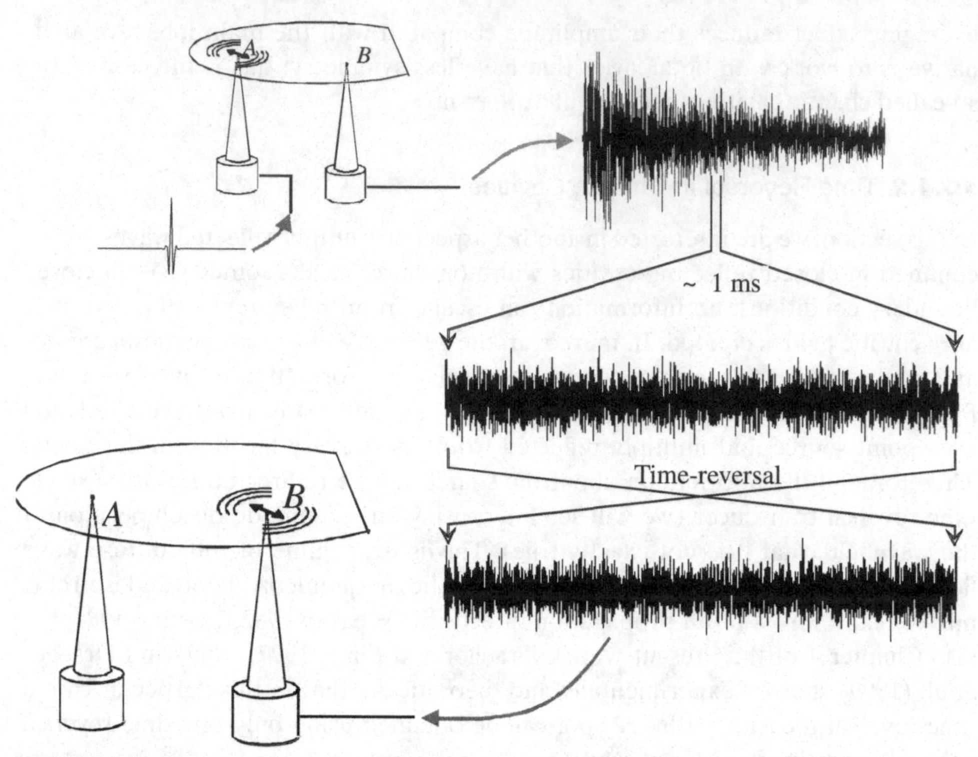

Figure 10.10. Time-reversal experiment conducted in a chaotic cavity with flexural waves. In a first step, a point transducer located at point A transmits a signal 1 μs long. The signal is recorded at point B by a second transducer. The signal spreads on more than 30 ms owing to reverberation. In the second step the experiment, a 1-ms portion of the recorded signal is time-reversed and retransmitted in the cavity.

For time-reversal windows of sufficiently long duration ΔT, one observes both an impressive time recompression at point A and a refocusing of the time-reversed wave around the origin (see Figures 10.11(a) and (b) for $\Delta T = 1$ ms), with a focal spot whose radial dimension is equal to half the wavelength of the flexural wave. Using reflections at the boundaries, the time-reversed wave field converges toward the origin from all directions and gives a circular spot, like the one that could be obtained with a closed time-reversal cavity covered with transducers. A complete study of the dependence of the spatiotemporal side lobes around the origin shows a major result (Draeger et al. 1999): a time duration ΔT of nearly 1 ms is enough to obtain a good focusing. For values of ΔT larger than 1 ms, the side lobes' shape and the signal-to-noise ratio (focal peak/sidelobes) does not improve further. There is a saturation regime. Once the saturation regime is reached, point B will receive redundant information. The saturation regime is reached after a time T_{H} called the Heisenberg time. It is the minimum time needed to resolve the eigenmodes in the cavity. It can also be interpreted as the time it takes for all a single ray to reach the vicinity of any point in the cavity within a distance $\lambda/2$. This guarantees enough interference between all the multiply reflected waves to build each of the eigenmodes in the cavity. The mean distance Δf between the eigenfrequencies is related to the Heisenberg time, $T_{\mathrm{H}} = 1/\Delta f$.

The success of this time-reversal experiment in a closed chaotic cavity is particularly interesting with respect to two aspects. First, it proves the feasibility of acoustic

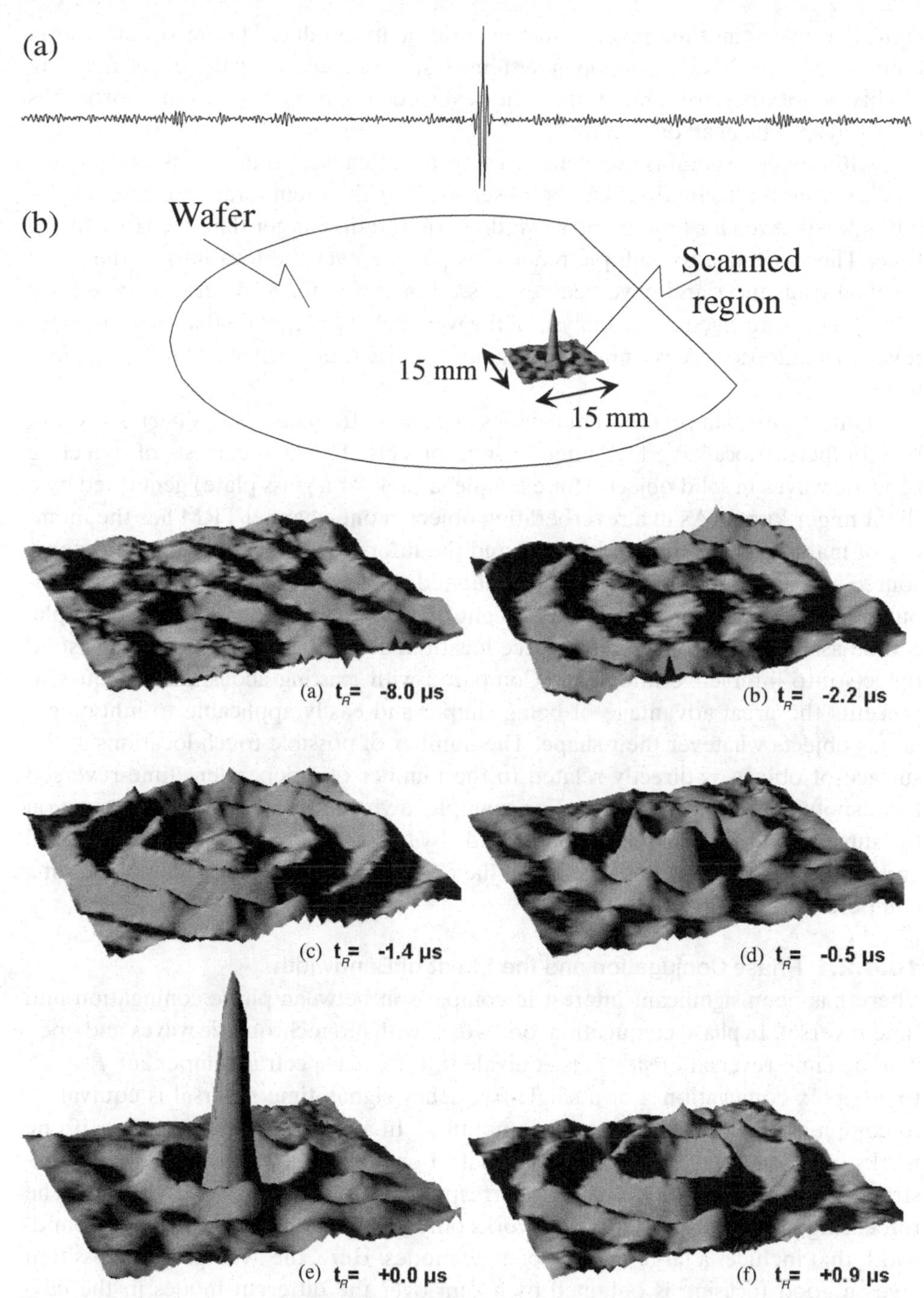

Figure 10.11. (a) time-reversed signal observed at point A. The observed signal is 210-μs long. (b) Time-reversed wavefield observed at different times around point A on a square of 15 mm × 15 mm.

time reversal in cavities of complex geometry that give rise to chaotic ray dynamics. Paradoxically, in the case of one-channel time reversal, chaotic dynamics is not only harmless but even useful, as it guarantees ergodicity and mixing. Second, using a source of vanishing aperture, there is an almost-perfect focusing quality. The procedure approaches the performance of a closed TRM, which has an aperture of 360°.

Hence, a one-point time reversal in a chaotic cavity produces better results than a limited-aperture TRM in an open system. Using reflections at the edge, focusing quality is not aperture limited; the time-reversed collapsing wavefront approaches the focal spot from all directions.

Although one obtains excellent focusing, a one-channel time reversal is not perfect, as residual fluctuations can be observed. Residual temporal and spatial sidelobes persist even for time-reversal windows of duration larger than the Heisenberg time. These are due to multiple reflections passing over the locations of the time-reversal transducer and have been expressed in closed form by Draeger and Fink (1999). Using an eigenmode analysis of the wavefield they explain that for long time-reversal windows there is a minimum signal-to-noise ratio even after the Heisenberg time.

Time reversal in reverberant cavities at audible frequencies has been shown to be an efficient localizing technique in solid objects. The idea consists of detecting acoustic waves in solid objects (for example, a table or a glass plate) generated by a slight finger knock. As in a reverberating object, a one-channel TRM has the memory of many distinct source locations, and the information location of an unknown source can then be extracted from a simulated time-reversal experiment in a computer. Different actions, turning on a light or a compact disk player, for example, can be associated with different source location. Thus, the system transforms solid objects into interactive interfaces. Compared with existing acoustic techniques, it presents the great advantage of being simple and easily applicable to inhomogeneous objects whatever their shape. The number of possible touch locations at the surface of objects is directly related to the number of independent time-reversed focal spots that can be obtained. For example, a virtual keyboard can be drawn on the surface of an object and the sound made by fingers when a text typed is captured and used to localize the impacts. Then, the corresponding letters are displayed on a computer screen (Ing et al. 2005).

10.4.2.1 Phase Conjugation and the Effect of Bandwidth

There has been significant interest in comparison between phase conjugation and time reversal. In phase conjugation, one works with monochromatic waves and open media. Time reversal of $p(\mathbf{r}, t)$ is equivalent, for each spectral component $\hat{P}(\mathbf{r}, \omega)$, to complex conjugation. For a single-frequency signal, time reversal is equivalent to complex conjugation of complex amplitude. In a closed cavity, as earlier, if one works only at a single frequency (say, that of one of the eigenmodes ω_i), one constructs only the eigenmode pattern corresponding to the selected frequency. The refocusing process discussed earlier works only with broadband pulses, over a bandwidth that includes a large number of eigenmodes. Here, the averaging process that gives a good focusing is obtained by a sum over the different modes in the cavity. By assuming that in a chaotic cavity we have a statistical decorrelation of the different eigenmodes, the time-reversed field can be computed by adding the various frequency components (each individual mode) and represented by a sum of Fresnel vectors (Figure 10.12). At the source position, all these phase-conjugated fields have a zero phase (this comes from the phase-conjugation operation that exactly compensates for the forward phase), and even if there is no amplitude focusing for each spectral contribution, there is a constructive interference between all these

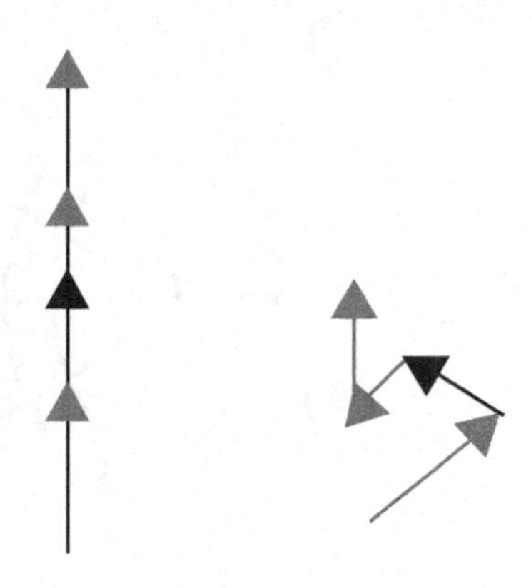

Figure 10.12. Schematic representation of a broadband time-reversal operation at the source (right) and off the source (left). Each arrow represents a different Fresnel vector corresponding to a frequency component. At the source position, all the phases are set back to 0; the amplitude of the resulting signal rises proportionally to the number of independent frequencies N. Outside the source, the different contributions are presumably decorrelated, and the standard deviation of their sum rises as $\sqrt{N}$.

fields at the focusing time as $\sum_i |H_{AB}(\omega_i)|^2$, where $H_{AB}(\omega)$ is the Fourier transform of $h_{AB}(t)$. Thus, the total field at the focusing time increases proportionally to the number I of modes (or arrows). Outside the source position, at point C, we observe $\sum_i H_{AB}(\omega_i)H^*_{CB}(\omega_i)$. The contributions of each individual mode are decorrelated because there is no more coherent phase compensation, and therefore the total length only increases as $\sqrt{I}$. The focusing peak emerges at the focusing time from the noise when the bandwidth is large enough to contain many different modes. Ideally, if we could indefinitely expand the bandwidth, the relative background level of the patterns should decrease as $1/\sqrt{I}$. As the number of eigenfrequencies available in the transducer bandwidth increases, the refocusing quality becomes better and the focal spot pattern becomes closed to the ideal Bessel function.

As a conclusion, it must be emphasized that in a closed cavity a one-channel TRM can be focused with $\lambda/2$ resolution if the duration of the time-reversal window is greater or equal to the cavity's Heisenberg time. Longer time windows do not improve the focusing quality. However, larger bandwidth $\Delta\omega$ reduces the sidelobe levels as $1/\sqrt{\Delta\omega}$.

10.4.3 Time Reversal in Open Systems: Random Medium

The ability to focus with a one-channel TRM is not only limited to experiments conducted inside closed cavities. Similar results have also been observed in time-reversal experiments conducted in open random media with multiple scattering (Derode et al. 1995, 2001, Blomgren et al. 2002). Derode et al. (1995) carried out the first experimental demonstration of the reversibility of an acoustic wave propagating through a random collection of scatterers with strong multiple scattering contributions. A multiple scattering sample is immersed between the source and a TRM array made of 128 elements. The scattering medium consisted of a set of 2,000 randomly distributed parallel steel rods (diameter 0.8 mm) arrayed over a region of thickness $L = 40$ mm, with average distance between rods 2.3 mm. The elastic mean free path in this sample was found to be 4 mm (see Figure 10.13). A source 30 cm away from the 128 element TRM transmitted a short (1 μs) ultrasonic pulse (three cycles at 3.5 MHz).

Figure 10.14(a) shows one part of the waveform received by one element of the TRM. It spread over more than 200 μs, that is, ~200 times the initial pulse duration. After the arrival of a first wavefront corresponding to the ballistic wave, a

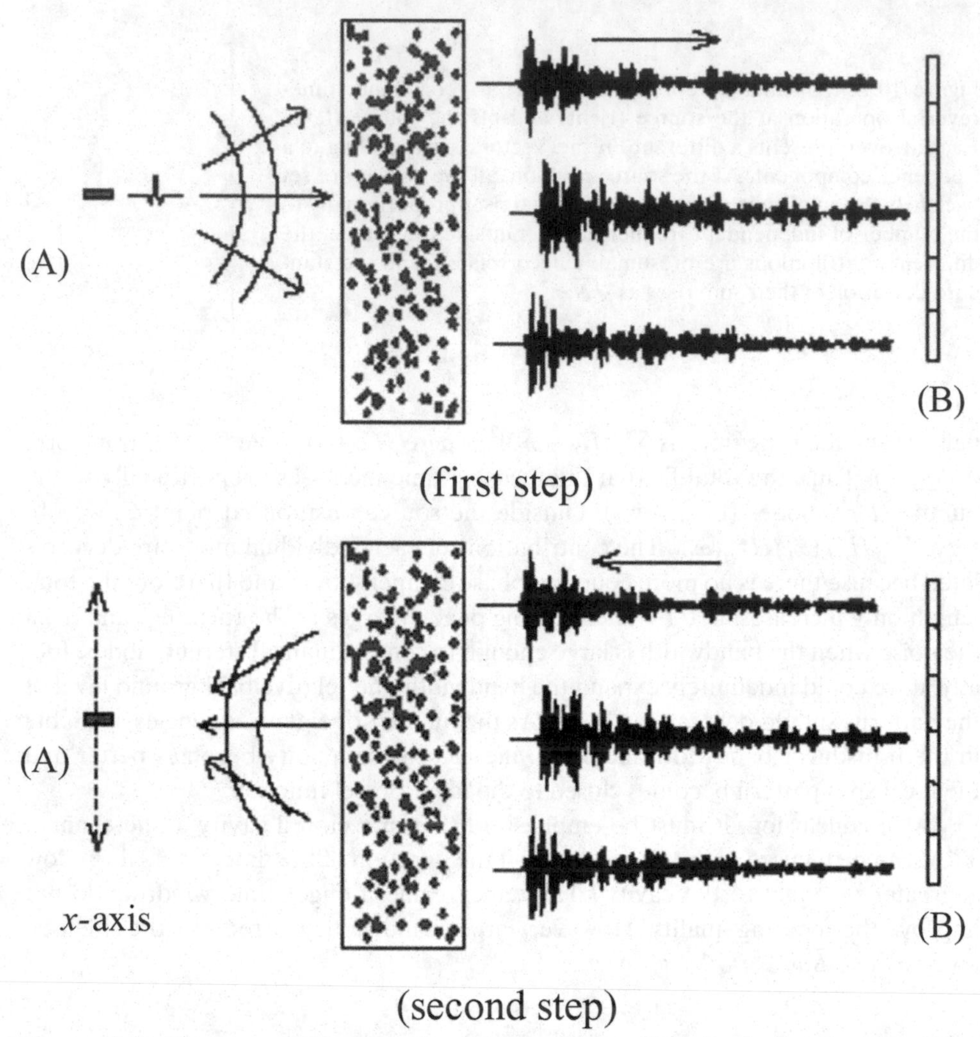

Figure 10.13. Time-reversal focusing through a random medium. In the first step the source (A) transmits a short pulse that propagates through the rods.The scattered waves are recorded on a 128-element array (B). In the second step, N elements of the array ($0 < N < 128$) retransmit the time-reversed signals through the rods. The piezoelectric element (A) is now used as a detector and measures the signal reconstructed at the source position. It can also be translated along the x-axis while the same time-reversed signals are transmitted by B, in order to measure the directivity pattern.

long diffuse wave is observed because of the multiple scattering. In the second step of the experiment, any number of signals (between 1 and 128) are time-reversed and transmitted, and a hydrophone measures the time-reversed wave in the vicinity of the source. For a TRM of 128 elements, with a time-reversal window of 300 μs, the time-reversed signal received on the source is represented in Figure 10.14(b): an impressive compression is observed because the received signal lasts about 1 μs, against over 300 μs for the scattered signals. The directivity pattern of the time-reversal field is also plotted on Figure 10.15. It shows that the resolution (i.e., the beam width around the source) is significantly finer than it is in the absence of scattering: the resolution is 30 times finer, and the background level is below −20 dB.

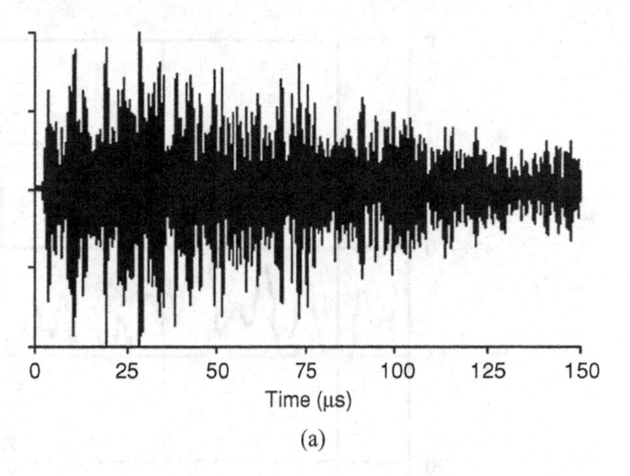

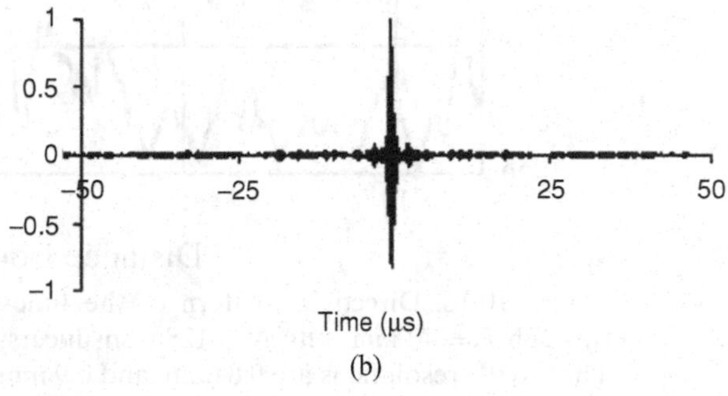

Figure 10.14. Experimental results: (a) signal transmitted through the sample ($L = 40$ mm) and recorded by the array element no. 64; (b) signal recreated at the source after time reversal.

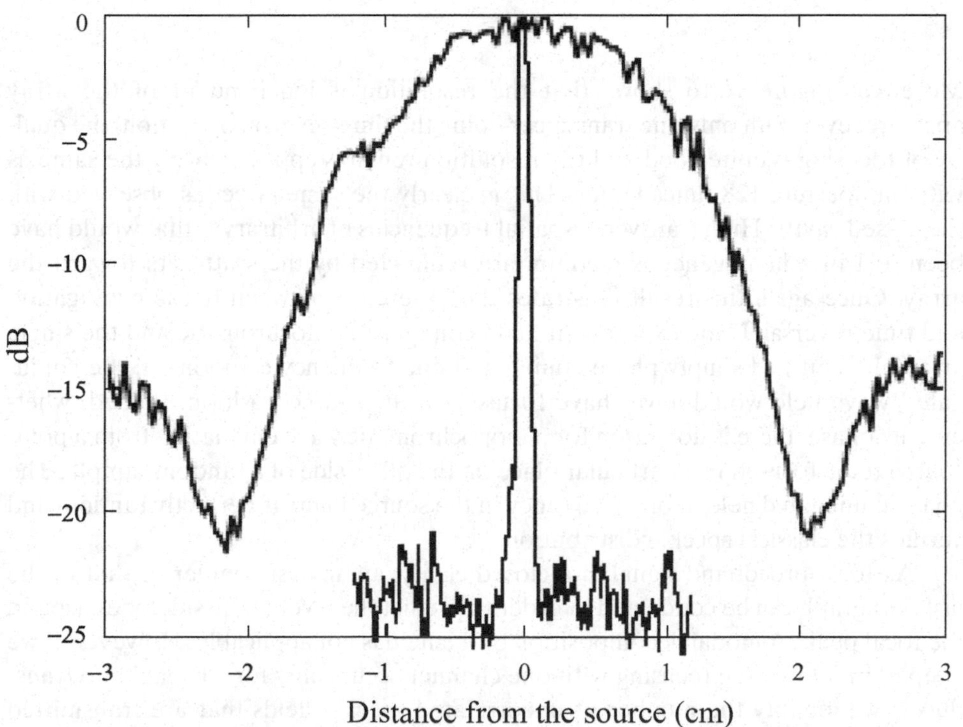

Figure 10.15. Directivity pattern of the time-reversed waves around the source position, in water (thick line) and through the rods (thin line), with a 16-element aperture. The sample thickness is $L = 40$ mm. The −6 dB widths are 0.8 mm and 22 mm, respectively.

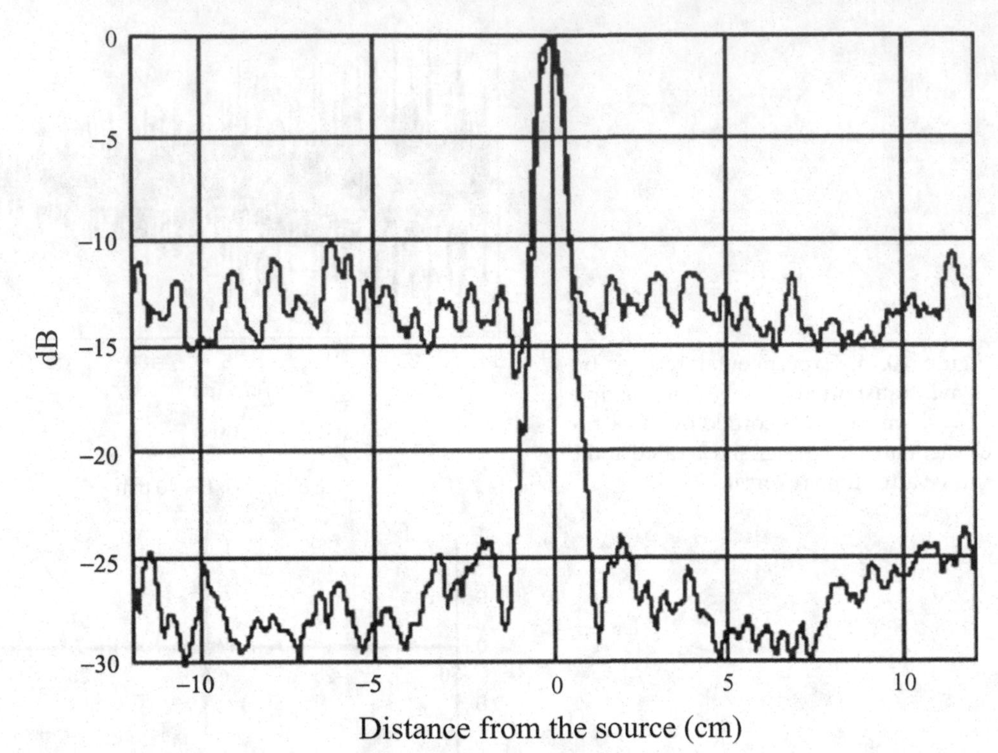

Figure 10.16. Directivity pattern of the time-reversed waves around the source position through $L = 40$ mm, with $N = 128$ transducers (thin line) and $N = 1$ transducer (thick line). The −6 dB resolutions are 0.84 mm and 0.9 mm respectively.

Moreover Figure 10.16 shows that the resolution is independent of the array aperture: even with only one transducer doing the time-reversal operation, the quality of focusing is quite good and the resolution remains approximately the same as with an aperture 128 times larger. This is clearly the same effect as observed with the closed cavity. High transverse spatial frequencies of arbitrary k that would have been lost in a homogeneous medium are redirected by the scatterers toward the array. Once again this result illustrates the difference between phase conjugation and time reversal. If the experiment had been quasi-monochromatic and the single array element had simply phase conjugated one frequency component, the conjugated wave field would never have focused on the source position. Indeed, whatever its phase, there is no reason for a monochromatic wave emanating from a point source to be focused in a particular place on the other side of a random sample. The phase-conjugated field at one frequency in the source plane is perfectly random and verifies the classical speckle distribution.

As for a broadband signal in a closed cavity, an analysis similar to that of the last paragraph can be conducted in order to predict the level of the sidelobes around the focal peak. A modal decomposition of the field is not applicable. However, if we keep in mind that the focusing with one channel occurs only for a broadband transducer, we identify the number of uncorrelated speckle fields that are transmitted in the transducer bandwidth $\Delta\omega$. For this, we have to define the spectral correlation length $\delta\omega$ of the scattered waves. Then there are $\Delta\omega/\delta\omega$ uncorrelated bits of spectral information in the frequency bandwidth, and the signal-to-noise is expected to vary like $\sqrt{\Delta\omega/\delta\omega}$.

10.5 Focusing below the Diffraction Limit

As discussed above, reversing an acoustic field using a TRM without reversing the source (i.e., without creating a local sink) does not correspond to an exact time reversal. A perfect time reversal would require that the original active source that injected energy into the system be replaced by a sink (the time reversal of a source). An acoustic sink is a device that absorbs all arriving energy without reflecting it. Taking into account the source term in the wave equation, reversing time leads to the transformation of source into a sink. For an initial point source transmitting a waveform $s(t)$, the wave field obeys Equation (10.4) (the inhomogeneous wave equation) but with a source term $s(t)$ replaced by $s(-t)$:

$$\nabla \cdot \left(\frac{\nabla p(\mathbf{r}, -t)}{\rho}\right) - \frac{1}{\rho c^2}\frac{\partial^2 p(\mathbf{r}, -t)}{\partial t^2} = \delta(\mathbf{r} - \mathbf{r}_0)s(-t). \tag{10.18}$$

To achieve a perfect time reversal the field on the surface of the cavity has to be time reversed, and the source has to be transformed into a sink. In this manner one may achieve time-reversed focusing below the diffraction limit. The role of the new source term $\delta(\mathbf{r} - \mathbf{r}_0)s(-t)$ is to transmit a diverging wave that exactly cancels the outgoing spherical wave.

In a monochromatic approach, taking into account the evanescent waves concept, the necessity of replacing a source by a sink in the complete time-reversed operation can be interpreted as follows: In the first step a point-like source much smaller than a wavelength radiates a field that can be described as a superposition of homogeneous plane waves propagating in the various directions $\mathbf{k}$ and of decaying, non-propagating, evanescent plane waves (Vesperinas & Wolf 1985). The evanescent waves contain information about fine-scale features of the source; they decay exponentially with distance and do not contribute in the far field. If the TRM is located in the far field of source, the time-reversed field retransmitted by the mirror does not contain these evanescent components. The role of the sink is to radiate exactly, with good timing, the evanescent waves that have been lost during the first step. The resulting field contains the evanescent part that is needed to focus below diffraction limits. Time reversal below the diffraction limit has been experimentally demonstrated in acoustics, using an acoustic sink placed at the focal point. Focal spots of size $\lambda/14$ have been observed (de Rosny & Fink 2002). One drawback is the need to use an active source at the focusing point to exactly cancel the usual diverging wave created during the focusing process.

An alternative procedure to enhance spatial resolution in time-reversed focal spots was explored in microwave experiments (Lerosey et al. 2007). The focusing point was surrounded by a microstructured medium with length scales well below the wavelength; strong scatterers were placed in the near field of the source. In this case, the microstructured medium modifies the spatial dependence of the imaginary part of the Green's function that now oscillates on scales smaller than the wavelength. The remarkable result was that two antennas separated by as little as 1/30 of a nominal wavelength could be separately focused upon. The diffraction limit was overcome although the TRM was in the far field of the source. Contrary to the acoustic sink experiment, the source remains passive, and high-spatial-frequency components of the field are created upon scattering in the disordered structure.

10.6 Conclusion

We have shown that in the presence of multiple reflections or multiple scattering, a small TRM manages to focus a pulse back to the source with a spatial resolution that beats the diffraction limit. The resolution is not dependent on the mirror aperture size, but it is only limited by the spatial correlation of the wave field. In these media, because of a sort of kaleidoscopic effect that creates virtual transducers, the TRM appears to have an effective aperture that is much larger that its physical size. Resolution can be improved in reverberating media using this concept. Time-reversal focusing also opens up completely new approaches to super-resolution. We have show that in a medium made of a random distribution of sub-wavelength scatterers, a time-reversed wave field interacts with the random medium to regenerate not only the propagating but also the evanescent waves required to refocus below the diffraction limit. Focal spots as small as $\lambda/30$ have been demonstrated with microwaves. This results in a large increase of the information transfer rate by time reversal in such disordered media.

11 Ocean Acoustics: A Novel Laboratory for Wave Chaos

Steven Tomsovic
Department of Physics and Astronomy, Washington State University, Pullman, WA, USA

Michael Brown
Rosenstiel School of Marine and Atmospheric Science, University of Miami, Coral Gables, FL, USA

One of the fascinating aspects of the field known colloquially as quantum chaos is the immense variety of physical contexts in which it appears. In the late 1980s it was recognized that ocean acoustics was one such context. It was discovered that the internal state of the ocean leads to multiple scattering of sound as it propagates and leads to an underlying ray dynamics which is predominantly unstable, that is, chaotic. This development helped motivate a resurgence of interest in extending dynamical systems theory suitably for applying ray theory in its full form to a "chaotic" wave mechanical propagation problem. A number of theoretical tools are indispensable, including semiclassical methods, action-angle variables, canonical perturbation theory, ray stability analysis and Lyapunov exponents, mode approximations, and various statistical methods. In the current work, we focus on these tools and how they enter into an analysis of the propagating sound.

11.1 Introduction

Acoustic wave propagation through the ocean became a topic of immense physical interest in the latter half of the twentieth century. Beyond the evident sonar applications, acoustic waves offer a means with which to probe the ocean itself. It is possible to monitor bulk mean ocean temperatures over time, which gives important information for studying global warming, and to obtain other information about the internal state of the ocean, that is, currents, eddies, internal waves, seafloor properties, and the like (Flatté et al. 1979, Munk et al. 1995). However, the present work owes its existence rather to the recognition, beginning in the late 1980s, that the ray dynamics underlying propagation of acoustic waves is chaotic (Palmer et al. 1988, 1991, Smith et al. 1992a, 1992b), and by inference, the subject can therefore be thought of as a unique problem in quantum chaos or, perhaps better stated, wave chaos (Beron-Vera et al. 2003, Brown et al. 2003). Our focus here is on the interconnections and synergies arising between research in wave chaos and ocean acoustics. The interested reader is encouraged to consult reference works such as Flatté et al.

(1979) and Munk et al. (1995) for more information about the motivations and foundations of ocean acoustics research.

Generally speaking, wave chaos studies require understanding theoretical techniques that can be loosely classified as random matrix (Brody et al. 1981, Mehta 1991) and certain field theories (Efetov 1999, Mirlin 2000), and semiclassical theories of dynamical systems (Gutzwiller 1990, Giannoni et al. 1991, Brack & Bhaduri 1997). We have chosen to focus on those features and techniques that are most directly related to semiclassical and dynamical system theory approaches. For the research practitioner, it is critical to become conversant with eikonal expansions, saddle point integration techniques, Wentgel–Kramers–Brillouin (WKB) theory, mode approximations, action-angle variables, canonical perturbation theory, ray stability analysis, and the geometry and measures of chaotic dynamics, among other skills. In order to simplify and focus the discussion, we cover these subjects as they apply to long-range ocean acoustics under conditions in which acoustic interactions with the sea surface and seafloor can be neglected.

11.2 Waves and Rays

The wave equation

$$\nabla^2 \Phi(\mathbf{r}, t) - \frac{1}{c^2(\mathbf{r}, t)} \frac{\partial^2}{\partial t^2} \Phi(\mathbf{r}, t) = 0 \tag{11.1}$$

accurately describes the acoustic waves of interest here, which are assumed to be sufficiently low frequency that dissipative losses can be neglected. The real part of $\Phi(\mathbf{r}, t)$ is the acoustic pressure and $c(\mathbf{r}, t)$ is the acoustic wave speed at location $\mathbf{r}$ and time t. At most midlatitude locations, the depth variations of $c(\mathbf{r}, t)$ lie between approximately 1480 $\mathrm{m\,s^{-1}}$ and 1540 $\mathrm{m\,s^{-1}}$. These variations are due mostly to the decrease in temperature with increasing depth and the nearly hydrostatic increase in pressure with depth; a 1°C increase leads to a 4 $\mathrm{m\,s^{-1}}$ increase in speed. Neglecting polar and coastal regions, the combined effects of temperatures and hydrostatic pressure lead to the formation of a sound speed minimum at approximately 1-km depth (in oceans that are typically approximately 5-km deep). The sound speed minimum is associated with a "sound channel" (Munk et al. 1995) that causes sound to refract away from the sea surface and seafloor, thereby allowing sound to travel great distances with only small absorption- and scattering-induced energy losses.

Superimposed on the background sound speed structure are fluctuations with length scales from millimeters to thousands of kilometers and timescales from seconds to millennia. The low-frequency sounds of interest for long-range propagation studies, with frequencies from tens to a few hundred hertz, are not sensitive to variability on scales shorter than approximately an acoustic wavelength (Hegewisch et al. 2001). It turns out that this roughly corresponds to the ocean's internal wave scale (Munk 1981), that is, a scale associated with bulk water motions because of buoyancy forces. In the upper ocean, these motions have a lower length scale of approximately 10 m. Owing to these waves and other forms of ocean variability, the possibility that the resulting sound speed variations have the requisite symmetry for separating variables in the wave equation is entirely negligible. Furthermore, the variations are not weak enough to be ignored. As a consequence, exact solutions

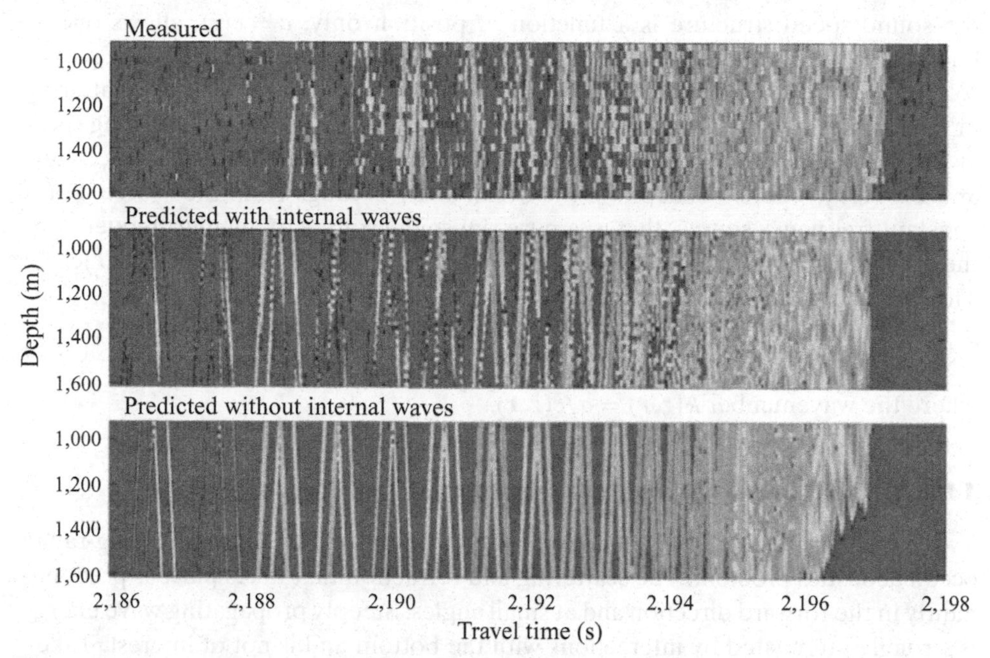

Figure 11.1. Measurements and simulations of acoustic intensity as a function of travel time and hydrophone depth in a long-range propagation experiment (Worcester et al. 1999). The top panel shows a single realization of the measured sound field. The middle panel shows a representative example of a parabolic equation simulation of the wavefield in a fairly realistic ocean environment that includes a simulated internal-wave-induced sound speed perturbation. The bottom panel shows a simulation of the wavefield in the same background environment but without the internal-wave-induced perturbation. In all three panels acoustic intensity relative to the peak intensity is color-coded in decibels. Reprinted with permission from Worcester et al. (1999). Copyright 1999, Acoustical Society of America. See http://www.cambridge.org/wrightandweaver for a colour version of this figure.

of realistic versions of Equation (11.1) cannot be compactly expressed in terms of some known standard functions. In one set of representative experiments in the eastern North Pacific Ocean, propagation over a distance of approximately 3000 km was studied (Colosi et al. 1999, Worcester et al. 1999). Figure 13 of Worcester et al. reproduced here in Figure 11.1, illustrates the form of the acoustic signal received downrange and compares it with numerical simulations both with and without internal wave effects. A number of wavefield properties, including the vertical extension of the acoustic arrivals, intensity fluctuations, time biasing, signal coherence, arrival time spread, and arrival time wander, were measured and remain only partially understood (Beron-Vera et al. 2003).

Those physical processes that lead to the time dependence of $c(\mathbf{r}, t)$ evolve on a much longer timescale than it takes for sound waves to pass through any given region. Internal waves, for example, are almost always treated as stochastic perturbations to the background. Midlatitude internal waves have a horizontal correlation distance of approximately 10 km and a correlation time of approximately 10 min in the upper ocean and much longer in the deep ocean. But it takes only about 6 s for sound to traverse 10 km. Locally therefore, $c(\mathbf{r}, t)$ can be taken to be time independent, "frozen" during the passage of sound waves. The assumption that the sound speed structure is a function of position only, $c = c(\mathbf{r})$, allows one to build up solutions to the full wave equation from a superposition of fixed-frequency solutions of a Helmholtz-like equation. Further, we shall assume that the scattering in the azimuthal direction is negligibly small, leading to cylindrical spreading (provided the range is not so large that Earth's curvature must be taken into account) and $\mathbf{r} = (z, r)$, where z is depth in the ocean and r is range from the source. For a constant frequency source, that is, a pure sinusoidal source of angular frequency σ, the wave field has a frequency response $\Phi_\sigma(z, r)$, where $\Phi(z, r, t) = \Phi_\sigma(z, r)\mathrm{e}^{-\mathrm{i}\sigma t}$. The Helmholtz-like equation in cylindrical coordinates is

$$\nabla^2\Phi_\sigma(z, r) + k^2(z, r)\Phi_\sigma(z, r) = 0, \tag{11.2}$$

where the wavenumber $k(z, r) = \sigma/c(z, r)$.

11.2.1 The Paraxial Optical Approximation

Two more approximations act together to further simplify the basic long-range ocean acoustics problem. The scattering and refraction that takes place is predominantly in the forward direction and at small angles. Steeply propagating wave energy is strongly attenuated by interactions with the bottom and is not of interest. Taken together, a Fresnel approximation (Tappert 1977) is valid, which gives the acoustic pressure as the product of an outgoing cylindrical wave and a slowly varying envelope function with a horizontal wavenumber $k_0 = \sigma/c_0$,

$$\Phi_\sigma(z, r) = \Psi_\sigma(z, r)\frac{\mathrm{e}^{\mathrm{i}k_0 r}}{\sqrt{r}}, \tag{11.3}$$

whose governing equation is to an excellent approximation

$$\frac{\mathrm{i}}{k_0}\frac{\partial}{\partial r}\Psi_\sigma(z, r) = -\frac{1}{2k_0^2}\frac{\partial^2}{\partial z^2}\Psi_\sigma(z, r) + V(z, r)\Psi_\sigma(z, r). \tag{11.4}$$

The constant c_0 is a reference sound speed. There is some flexibility in the choice of the potential function $V(z, r)$. The most common choice is $V(z, r) = [1 - (c_0/c(z, r))^2]/2$. An equally good choice of $V(z, r)$ is $(c(z, r) - c_0)/c_0$ (both choices are first-order accurate in $(c(z, r) - c_0)/c_0$). In most underwater acoustic applications a good choice of c_0 is 1.5 km s^{-1}. The potential can be decomposed into two contributions, $V(z, r) = V_0(z) + V_1(z, r)$. The first term $V_0(z)$ describes the large-scale wave guide effect mentioned earlier, which is either independent of range or very nearly so (adiabatic in the sense described below), and the second term $V_1(z, r)$ describes the remaining smaller-scale variations due to all the physical processes at work in the ocean. Flatté et al. (1979) contains details on all of the terms that have been dropped in arriving at Equation (11.4) and has order of magnitude estimates for the size of the various dropped contributions.

Interestingly, this so-called parabolic equation, Equation (11.4), maps without approximation onto the one-dimensional quantum mechanical Schrödinger equation through the substitutions $r \to t$ and $1/k_0 \to \hbar$. To the extent that the above-stated approximations are valid, this equivalence shows that there exists a perfect analogy between wave chaos and quantum chaos. The measurable quantities and specific questions asked are different in the acoustic and quantum domains, as is the specific nature of the potential, but the fundamental structure of the two problems is common.

It should also be noted, however, that the acoustic wave chaos/quantum chaos analogy is limited to fixed-frequency (monochromatic) acoustic wavefields. Equations (11.2)–(11.4) describe fixed-frequency acoustic wavefields. The acoustic response to a transient source $s(t)$, whose Fourier transform is $\bar{s}(\sigma)$, can be expressed as a Fourier integral,

$$\Phi(z,r,t) = \frac{1}{2\pi}\int_{\infty}^{\infty} \bar{s}(\sigma)\Phi_\sigma(z,r)\mathrm{e}^{-\mathrm{i}\sigma t}\,\mathrm{d}\sigma. \tag{11.5}$$

The quantity $\Phi(z,r,t)$ has no direct quantum mechanical counterpart.

11.2.2 Eikonal Approximations

The terminology and techniques of semiclassical theories, ray theories, eikonal approximations, WKB theory, and saddle point integration are so profoundly intertwined that often they can be applied interchangeably. Depending on one's starting point, any of these approaches can be used to generate the connection between ray, transport, and wave equations. Generally speaking, in semiclassical theory an asymptotic approximation to the exact wave field is represented by a superposition of simpler waves, each one possessing a well-defined phase with a slowly varying wavelength and envelope as a function of location. The resulting ray equations may have multiple solutions subject to the appropriate boundary conditions, and each one corresponds to one of the simpler waves and gives the phase. The transport equation describes the dynamical neighborhood of each ray and gives the slowly varying amplitude for the wave associated with that ray. A complete ray theory accounts for both the ray and transport equations. However, one often sees mention of ray theory in situations where it is clear that the authors are not considering the transport equations or even calculating the phases.

The theory can equally well be written down for non-stationary or stationary state/mode representations. Consider the non-stationary relations. Beginning with the eikonal ansatz

$$\Psi_\sigma(z,r) \approx \sum_j A_j(z,r)\exp\left(\mathrm{i}k_0 S_j(z,r)\right), \tag{11.6}$$

substitution into the parabolic equation and equating terms in like powers of k_0 generates the Hamilton–Jacobi and transport equations of classical mechanics (Goldstein 1980), respectively,

$$\begin{aligned} 0 &= \frac{\partial}{\partial r}S(z,r) + \frac{1}{2}\left[\frac{\partial}{\partial z}S(z,r)\right]^2 + V(z,r), \\ 0 &= \frac{\partial}{\partial r}A^2(z,r) + \frac{\partial}{\partial z}\left[A^2(z,r)\frac{\partial}{\partial z}S(z,r)\right]. \end{aligned} \tag{11.7}$$

The solution to the Hamilton–Jacobi equation for $S(z,r)$, which is referred to as Hamilton's principal function, can be constructed by integrating the equation

$$\frac{\mathrm{d}}{\mathrm{d}r}S(z,r) = p\frac{\mathrm{d}}{\mathrm{d}r} - \mathscr{H}(p,z;r) \tag{11.8}$$

together with the ray equations (Hamilton's equations)

$$\frac{\mathrm{d}z}{\mathrm{d}r} = \frac{\partial \mathscr{H}(p,z;r)}{\partial p}, \quad \frac{\mathrm{d}p}{\mathrm{d}r} = -\frac{\partial \mathscr{H}(p,z;r)}{\partial z}, \tag{11.9}$$

where

$$\mathscr{H}(p,z;r) = \frac{p^2}{2} + V(z,r) \tag{11.10}$$

is the Hamiltonian. Discrete values $A_j(z,r)$ and $S_j(z,r)$ of $A(z,r)$ and $S(z,r)$ appear in Equation (11.6). These discrete variables correspond to those rays, often referred to as "eigenrays," that connect fixed source and receiver locations. The eikonal function $S(z,r)$ in Equation (11.6) is a reduced travel time $S(z,r) = c_0\mathscr{T}(z,r) - r$, where $\mathscr{T}$ is ray travel time. Returning briefly to the analogy between quantum chaos and wave chaos, note that it follows from the relationship $S(z,r) = c_0\mathscr{T} - r$ that in the ocean acoustic context, the travel time is a measurable physical manifestation of Hamilton's principal function for a trajectory. In a quantum mechanical context, there is the association of $S(z,r)$ with a phase only.

The transport equation can be solved assuming a point source. For the jth eigenray,

$$A_j(z,r) = A_{0j}\, |q_{21}|_j^{-1/2} \mathrm{e}^{-\mathrm{i}\mu_j \frac{\pi}{2}}. \tag{11.11}$$

The matrix element q_{21}, defined next, describes the spreading of a ray bundle in the infinitesimal neighborhood of the jth ray. At any fixed r, one has

$$\begin{pmatrix} \delta p \\ \delta z \end{pmatrix} = Q_r(p_0, z_0) \begin{pmatrix} \delta p_0 \\ \delta z_0 \end{pmatrix}, \tag{11.12}$$

where the stability matrix (Tabor 1989)

$$Q_r(p_0, z_0) = \begin{pmatrix} q_{11} & q_{12} \\ q_{21} & q_{22} \end{pmatrix} = \begin{pmatrix} \left.\frac{\partial p}{\partial p_0}\right|_{z_0} & \left.\frac{\partial p}{\partial z_0}\right|_{p_0} \\ \left.\frac{\partial z}{\partial p_0}\right|_{z_0} & \left.\frac{\partial z}{\partial z_0}\right|_{p_0} \end{pmatrix}. \tag{11.13}$$

We usually drop the initial condition arguments for brevity, but clearly each initial condition leads to a unique range-dependent stability matrix. Elements of this matrix evolve according to

$$\frac{\mathrm{d}}{\mathrm{d}r} Q_r = K_r Q_r, \tag{11.14}$$

where $Q_{r=0}$ is the identity matrix and

$$K_r = \begin{pmatrix} -\frac{\partial^2 \mathscr{H}}{\partial z \partial p} & -\frac{\partial^2 \mathscr{H}}{\partial z^2} \\ \frac{\partial^2 \mathscr{H}}{\partial p^2} & \frac{\partial^2 \mathscr{H}}{\partial z \partial p} \end{pmatrix}. \tag{11.15}$$

At caustics q_{21} vanishes and the Maslov index μ advances by one unit. Jumps in the phase index actually preserve smooth phase variation for problems with initial wave packets. At these points diffractive corrections to Equation (11.11) must be applied. The normalization factor A_{0j} is chosen in such a way that close to the source it matches the Green's function for the parabolic equation.

11.3 The Separable Problem – Rays and Modes

In this section we focus on the special case in which the sound speed is a function of depth only, $c(z,r) = c(z)$ and $V(z,r) = V_0(z)$. For this special class of problems the parabolic wave equation admits separable solutions $\Psi_\sigma(z,r) = Z(z;\sigma)R(r;\sigma)$. One such solution, involving an expansion in normal modes, is discussed below. Solving the ray equations, Equation (11.9), is also greatly simplified in this case. A particularly useful technique to accomplish this goal is to make use of action-angle variables (I, θ); these variables are introduced in the next subsection. To make our discussion concrete, we focus on the most common class of deep-ocean acoustic waveguides in which $c(z)$ has a single minimum. For this class of problems there exists rays and modes with two internal turning depths that interact negligibly with the surface and the bottom.

As noted earlier, the class of problems treated here is too idealized to constitute a solution to Equation (11.4) under realistic conditions. Rather, the reason for discussing this problem is that almost all of the material introduced here turns out to be very useful as the basis of perturbation expansions that apply when a range-dependent environmental perturbation is introduced, as described in the sections that follow. In addition to this caveat, note that we do not consider here situations where tunneling and diffraction are of paramount interest. Caustic and turning point corrections, which can be accounted for by uniform approximations, are also not discussed here, as those subjects are beyond the scope of the present work and not the primary issue.

11.3.1 Rays and Action-Angle Variables

Action-angle variables are introduced by a canonical transformation from (p, z) to (I, θ) (Goldstein 1980). The same transformation replaces $\mathscr{H}(p, z)$ by $\mathscr{H}(I)$; in fact, $\mathscr{H}(p,z) = \mathscr{H}(I)$. The numerical values of $\mathscr{H}$ are unchanged by the canonical transformation, but the dependence of $\mathscr{H}$ on the independent variables is of course different in $\mathscr{H}(p, z)$ and $\mathscr{H}(I)$. Note also that if $c(z,r) = c(z)$, $\mathscr{H}$ is constant following rays. For the two internal turning point class of problems the action is defined by

$$I = \frac{1}{2\pi}\oint p\,\mathrm{d}z = \frac{1}{\pi}\int_{\check{z}(\mathscr{H})}^{\hat{z}(\mathscr{H})} \sqrt{2[\mathscr{H} - V_0(z)]}\,\mathrm{d}z, \tag{11.16}$$

where $V_0(\hat{z}(\mathscr{H})) = V_0(\check{z}(\mathscr{H})) = \mathscr{H}$. The transformation from (p, z) to (I, θ) makes use of the generating function

$$G(I, z) = \begin{cases} \int_{\check{z}(\mathscr{H})}^{z} \sqrt{2[\mathscr{H} - V_0(z')]}\,\mathrm{d}z', & p > 0, \\ \pi I + \int_{z}^{\hat{z}(\mathscr{H})} \sqrt{2[\mathscr{H} - V_0(z')]}\,\mathrm{d}z', & p < 0, \end{cases} \tag{11.17}$$

where $p = \partial G/\partial z$ and $\theta = \partial G/\partial I$. As z varies from one turning point to the next and back, $\check{z} \to \hat{z} \to \check{z}$, θ varies from $0 \to \pi \to 2\pi$, and for each complete ray cycle θ increases by 2π. The action-angle form of the ray equations are

$$\frac{\mathrm{d}I}{\mathrm{d}r} = -\frac{\partial \mathscr{H}}{\partial \theta} = 0, \qquad \frac{\mathrm{d}\theta}{\mathrm{d}r} = \frac{\partial \mathscr{H}}{\partial I} = \omega(I), \tag{11.18}$$

whose solution is $I(r) = I_0$, $\theta(r) = \theta_0 + \omega(I)r$. In this case, the total derivative of the Hamiltonian with respect to I is identical to the partial derivative and gives the angular frequency of motion. Similarly, and consistently, the transformed stability equations and solutions are

$$Q_r(I_0, \theta_0) = \begin{pmatrix} \left.\frac{\partial I}{\partial I_0}\right|_{\theta_0} & \left.\frac{\partial I}{\partial \theta_0}\right|_{I_0} \\ \left.\frac{\partial \theta}{\partial I_0}\right|_{\theta_0} & \left.\frac{\partial \theta}{\partial \theta_0}\right|_{I_0} \end{pmatrix} = \begin{pmatrix} 1 & 0 \\ \omega'(I_0)r & 1 \end{pmatrix}, \tag{11.19}$$

where the quantity $\omega'(I) = \mathrm{d}\omega(I)/\mathrm{d}I$ describes the rate of shearing in the dynamics, that is, how much the angular frequency of the motion shifts with a change in I.

The reduced travel time S (Hamilton's principal function) evolves according to

$$\frac{\mathrm{d}S}{\mathrm{d}r} = I\omega(I) - \mathscr{H}(I) + \frac{\mathrm{d}}{\mathrm{d}r}(G - I\theta), \tag{11.20}$$

which can also be integrated by inspection. The generating function G increases monotonically from 0 at $z = \check{z}$ to πI at $z = \hat{z}$ to $2\pi I$ at $z = \check{z}$. Because θ increases by 2π during each ray cycle, the endpoint correction term $G - I\theta$ in Equation (11.20) oscillates about zero. Except at short range, for most purposes this term gives a negligibly small contribution to S. In addition, because $I(r) = I_0$, that is, the action is a constant, it is possible to consider the depth z as depending on the angle variable $z_{I_0}(\theta)$, oscillating up-down one full cycle for θ increasing by 2π. The angular frequency $\omega(I) = 2\pi/R_\ell(I)$ where $R_\ell(I)$ is the spatial "period" in r of a ray. (The subscript ℓ denotes loop.)

Systems of the type just described are referred to as integrable, and their phase space ($\{p, z\}$) is filled (often said to be foliated) by a continuous family of manifolds, that is, continuous sets of phase space points called tori. In the class of systems discussed earlier, having a single degree of freedom, every ray is periodic and forms its own torus. In higher degrees of freedom, each ray lies on a particular torus, and if it is not periodic, then it will come arbitrarily close to every point on the torus eventually. If a ray is periodic, then a continuous family of like periodic rays combine to form the torus.

In anticipation of the material presented in Section 11.4.2, whose focus is wavefield properties in the presence of scattering by small-scale inhomogeneities, some brief remarks about wavefield structure in the absence of such scattering are in order. A typical experimental arrangement involves the production of a transient signal by a compact source at a fixed submerged location. The resulting transient wavefield is measured at a horizontal distance (range) r, typically with concurrent measurements at many depths. The challenge is to explain the distribution of acoustic energy in (z, t) as a function of range and central frequency.

From the action-angle description, a canonical transformation is needed to obtain the depth dependence of the wavefield (which follows the depth dependence of the rays); little more can be generally stated about this dependence. If one focuses on the temporal distribution of energy, without regard to the corresponding ray depth, a very simple picture emerges. This temporal dependence is described by the action-angle form of the ray equations and proves to be controlled by the quantity $\omega'(I)$. First, note that it follows from Equations (11.11) and (11.19) that geometric intensities (squared amplitudes) are proportional to the ray spreading factor $1/|q_{21}| \propto 1/|\omega'(I)r|$. Next, recall that the ray travel time $T = (S + r)/c_0$. It follows from this expression and Equation (11.20) that at fixed r, $\mathrm{d}T/\mathrm{d}I = \omega'(I)r/c_0$. Because I is a ray label, this simple equation shows that the dispersion or temporal spreading of the wavefield in the ray limit is controlled entirely by $\omega'(I)$. The foregoing comments reveal that when $c(z, r) = c(z)$, geometric amplitudes are controlled to a good approximation by $\omega'(I)$ and that, without approximation, geometric travel time dispersion is controlled by $\omega'(I)$.

11.3.2 Modes and Action Quantization[†]

For a point source at $(z, r) = (z_s, 0)$, the modal expansion of the solution to the parabolic wave equation (11.4) has the form

$$\Psi_\sigma(z, r) = \sum_m \varphi_m(z_s; \sigma)\varphi_m(z; \sigma)\mathrm{e}^{\mathrm{i}k_m r}, \qquad r \geq 0, \tag{11.21}$$

where the eigenfunctions $\varphi_m(z)$ and eigenvalues $k_m = k_0 \mathscr{H}_m$ satisfy

$$\frac{1}{2}\frac{\mathrm{d}^2\varphi_m}{\mathrm{d}z^2} + k_0^2(\mathscr{H}_m - V_0(z))\varphi_m = 0. \tag{11.22}$$

Together with appropriate boundary conditions, Equation (11.22) defines a Sturm–Liouville problem, guaranteeing that the functions $\varphi_m(z; \sigma)$ are complete and orthogonal,

$$\int \varphi_m(z; \sigma)\varphi_n(z; \sigma)\,\mathrm{d}z = \delta_{m,n}. \tag{11.23}$$

In the semiclassical (WKB) approximation, each mode can be expressed as a superposition of up- and down-going waves with reflection coefficients $\mathrm{e}^{\mathrm{i}\phi_u}$ and $\mathrm{e}^{\mathrm{i}\phi_l}$ at the upper and lower turning depths respectively. If these reflections take place without loss of energy, then ϕ_u and ϕ_l are real. Also, in the WKB approximation, $\mathscr{H}_m = \mathscr{H}(I_m)$ and

$$k_0 I_m = m - \frac{\phi_u + \phi_l}{2\pi}, \qquad m = 0, 1, 2, \ldots. \tag{11.24}$$

For modes with internal turning depths, $\phi_u = \phi_l = -\pi/2$, and the quantization condition (11.24) reduces to

$$k_0 I_m = m + \frac{1}{2}. \tag{11.25}$$

The quantization condition, Equation (11.24) or (11.25), is seen to be a statement that the action I is quantized. This relationship provides a simple and direct connection between the ray description and the asymptotic mode description of the underwater sound field.

[†] We have adapted the term quantization from quantum mechanics, but in this context, action quantization refers to the determination of the *discrete* set of rays and their classical actions, which directly correspond to the modes.

11.4 The Nonseparable Problem

The results presented so far apply to the idealized separable problem, $c(z, r) = c(z)$. The utility of those results stems in large part from the observation that those results provide a basis for two important classes of nonseparable problems. First is the adiabatic problem in which the range dependence is assumed to be slow $c(z, r) \to c(z, \varepsilon r)$. One can extend the action-angle formalism and modal expansion to incorporate this very slow range dependence into the definition of the action-angle variables themselves. This class of problems is fairly well understood (see, e.g., Virovlyansky et al. 2005) and will not be discussed further except to note that the principal results are that the action I is an adiabatic invariant, $\mathrm{d}I/\mathrm{d}r = O(\varepsilon^2)$ (compare with the first relation of Equation (11.18)), and mode coupling is negligible in a lower-frequency regime. The latter statement follows heuristically from Equation (11.24) or (11.25) and the smallness of $\mathrm{d}I/\mathrm{d}r$. The second class of nonseparable problems for which the action-angle description has proven to be very useful is the class for which $c(z, r) = c(z) + \varepsilon\delta c(z, r)$. This class of problems is critical to understanding ocean acoustics and gives rise to wave chaos in that context.

11.4.1 Stability Analysis in Range-Dependent Systems: Numerical Approach

In principle, it is by way of the stability analysis, all of whose information is embedded in Q_r, that one assesses how regular (stable) or chaotic (unstable) the ray dynamics are. A certain amount of analytical mathematics has been developed in this domain, which can be loosely termed KAM theory (Arnol'd 1978) after Kolmogorov, Arnol'd, and Moser, which is discussed in the next subsection. Before entering that discussion however, consider that Equations (11.12)–(11.14) are quite suitable for numerical investigations because they present the same level of difficulty as Hamilton's equations for determining the rays themselves. In fact, the same numerical techniques are invoked for both.

From a numerical analysis perspective, an important property of chaotic rays is that they exhibit extreme sensitivity to initial conditions in their local neighborhood, which expresses itself as an exponential growth in the separation of nearby rays. For finite range r, the growth can be approximately characterized by a positive, finite-range, stability exponent ν_r (Wolfson & Tomsovic 2001),

$$\nu_r = \frac{\log|\operatorname{tr}(Q_r)|}{r}, \tag{11.26}$$

In contrast with the Lyapunov exponent, which can roughly be thought of as the infinite-range limit

$$\nu_L = \lim_{r\to\infty} \nu_r, \tag{11.27}$$

ν_r depends on the particular ray, varies with range, and thus fluctuates. It turns out that Equation (11.26) is not sophisticated enough to answer questions of instability in a range-dependent environment. What follows illustrates the main difficulty and resolution.

Although regular rays do not separate exponentially, they can exhibit algebraic range dependence (Grassberger et al. 1988), and this leads to some ambiguity as to whether a ray is actually regular or chaotic (stable or unstable). For example, consider a one-degree-of-freedom system for which $\mathscr{H}(p, z; r)$ is periodic in range. If at range r a ray corresponds to a periodic orbit (i.e., $z_0 = z_r$ and $p_0 = p_r$), then the trace of its stability matrix indicates (Ott 1993)

$$|\mathrm{tr}\, Q_r| \begin{cases} < 2, & \text{orbit is stable and local dynamics is rotational (regular),} \\ = 2, & \text{orbit is marginally stable and local dynamics is shearing (regular),} \\ > 2 & \text{orbit is unstable and local dynamics is hyperbolic (chaotic),} \end{cases} \tag{11.28}$$

and there is no ambiguity. It turns out that for such systems there is a simpler approach to getting at this information, the construction of Poincaré surfaces of a section. By considering a section of the available phase space that the rays pass through, recording where rays intersect, the section gives an image of the stability throughout the phase space. This is because the recorded points of a stable or marginally stable ray must lie on a lower-dimensional set than a chaotic ray. The advantage is that without calculating Q_r one can see at a glance in low-dimensional systems which parts of the phase space contain regular rays and which parts contain chaotic rays. For nonperiodic dynamical systems, this construction of a Poincaré surface of section does not work. How can we still get at least an approximate sense for which parts of phase space are regular and which parts are chaotic?

There is no shortcut; one must consider Equations (11.12) and (11.14). It is worth showing a concrete example. A simple periodic map is given by the standard map (Chirikov 1979),

$$p_{j+1} = p_j - \frac{K}{2\pi} \sin 2\pi z_j \quad (\text{mod } 1), \tag{11.29}$$

$$z_{j+1} = z_j + p_{j+1} \quad (\text{mod } 1). \tag{11.30}$$

For certain values of K, the phase space contains both regular and chaotic rays. See the upper left panel of Figure 11.2. Each solid line indicates a single ray possessing regular motion, whereas the randomly looking filled in zones indicate the region of phase space belonging to chaotic rays.

The relations in Equation (11.28) for $|\,\mathrm{tr}(Q_r)|$ are not true even for other-than-full-period periodic orbits and are certainly not true for nonperiodic orbits or range-dependent dynamical systems. In other words, if one applies Equation (11.28) to a large ensemble of rays with initial conditions uniformly covering the available phase space, one would construct something like a surface of section that has the appearance of the upper right panel in Figure 11.2. It possesses some similarities to the proper surface of section but is obviously incorrect because the vast majority of stable orbits mistakenly show up as unstable.

There is, nevertheless, a means by which to separate approximately stable and unstable regions of phase space as a function of propagation range; that is, the proportion of rays mimicking stable behavior up to some range will decrease with increasing range of propagation. It is easy to recognize the origin of the difficulty with a trivial example. Imagine an orbit exists, which is periodic after two iterations of its

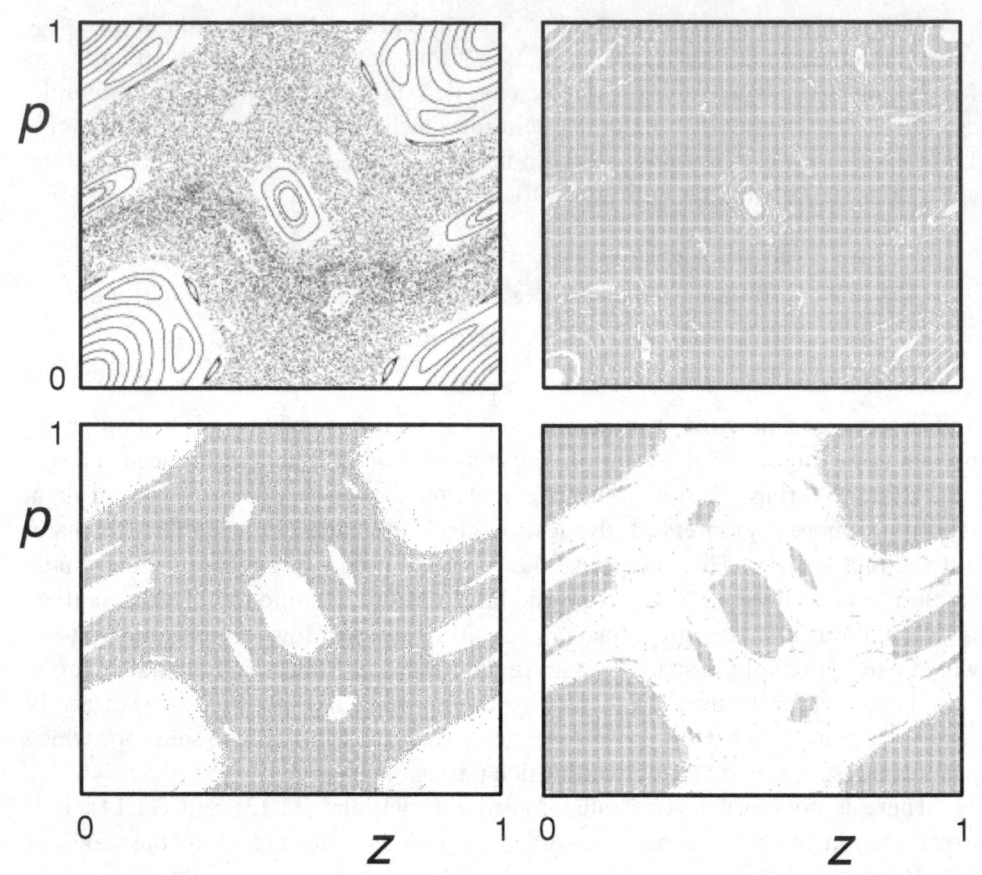

Figure 11.2. An approximate surface of section. The upper left panel shows the true surface of section for the standard map for a value of $K = 1.365$ found by running a small number of rays for thousands of iterations of the map. The points falling on lines are in regular regions, and the seemingly randomly placed points fill the chaotic zone. The upper right panel shows those initial conditions taken from a large uniform ensemble for which $|\mathrm{tr}\, Q_r| > 2$ after 100 iterations of the map. By comparison with the upper left panel, it demonstrates that the majority of the stable orbits are mistakenly determined to be unstable if one blindly follows the criterion of Equation (11.28) for nonperiodic orbits. The bottom left panel shows the initial conditions for which a fit as described in the text gives a positive finite-range Lyapunov exponent. The lower left panel shows those initial conditions for which the same fit gives a vanishing Lyapunov exponent to within the precision allowed (determined by the range of propagation considered). By comparing with the information contained in the upper left panel, it demonstrates that this method approximately separates correctly the unstable and stable rays, respectively.

map, and that its stability matrix for a single period is given by $Q = Q_2 Q_1$. In this example, suppose

$$Q_2 = \begin{pmatrix} 1 & -1 \\ \frac{3}{2} & -\frac{1}{2} \end{pmatrix}, \qquad Q_1 = \begin{pmatrix} \frac{1}{2} & 1 \\ -\frac{1}{2} & 1 \end{pmatrix}. \tag{11.31}$$

For a full period and n retracings, the orbit would be seen to be marginally stable:

$$Q = \begin{pmatrix} 1 & 0 \\ 1 & 1 \end{pmatrix} \quad \text{and} \quad Q^n = \begin{pmatrix} 1 & 0 \\ n & 1 \end{pmatrix}. \tag{11.32}$$

However at odd integer iterations,

$$Q_1 Q^n = \begin{pmatrix} \frac{1}{2}+n & 1 \\ -\frac{1}{2}+n & 1 \end{pmatrix}, \tag{11.33}$$

giving $\mathrm{tr}(Q_1 Q^n) = 3/2 + n$, which grows without bound as n increases. It is incorrect to identify chaotic nonperiodic rays with $|\,\mathrm{tr}(Q_r)| > 2$ and regular ones with $|\,\mathrm{tr}(Q_r)| \leq 2$.

Rather it is necessary to distinguish between exponential growth and algebraic growth (often found with oscillatory behavior as given earlier). Using the relation (not valid for ranges that are very short) (Grassberger et al. 1988)

$$|\,\mathrm{tr}(Q_r)| \sim r^{\mu} \mathrm{e}^{\alpha + \nu r}, \tag{11.34}$$

a least squares method can be generated to determine whether exponential growth is occuring. Minimizing the function

$$F(r) = \int_{r_{\min}}^{r} \left(\log\left|\mathrm{tr}(Q_s)\right| - \alpha - \mu \log s - \nu s\right)^2 \, \mathrm{d}s \tag{11.35}$$

with respect to $\{\alpha, \mu, \nu\}$ leads to an inhomogeneous set of three linear equations that can be solved using a standard method involving a ratio of determinants; only the solution for ν is of interest. The coarse graining is evident here. Roughly speaking, if up to the maximum propagation range, $\nu r \lesssim 1$ or 2, it is not possible to distinguish algebraic growth from exponential. Thus, as a practical matter, we consider rays for which $\nu \leq 2/r$ as regular. The bottom two panels of Figure 11.2 show the improvement that results from incorporating the least squares analysis. The separation of regular and chaotic trajectories now approximates the true dynamics as seen in the Poincaré surface of section. This approach does not depend upon whether a system is nonperiodic or not.

11.4.2 Chaos: Resonances, KAM Theory, and Extreme Sensitivity

Turning now to analytic mathematical theory associated with the problem for which $c = c(z) + \delta c(z, r)$, we shall assume that the perturbation term δc is an order-ϵ multiperiodic function of N spatial frequencies Ω_i, $1 = 1, 2, \ldots N$,

$$\delta c = \delta c(z, \Omega_1 r, \Omega_2 r, \ldots, \Omega_N r). \tag{11.36}$$

The only restriction on N is that it is finite. This class of perturbations is sufficiently large such that it encompasses commonly used models of ocean sound speed perturbations – due, for example, to internal waves (Colosi & Brown 1998).

Consistent with Equation (11.36), the action-angle form of the Hamiltonian is

$$\mathscr{H}(I, \theta, r) = \mathscr{H}_0(I) + \varepsilon \mathscr{H}_1(I, \theta, \Omega_1 r, \Omega_2 r, \ldots, \Omega_N r), \tag{11.37}$$

and the corresponding ray equations are

$$\frac{\mathrm{d}I}{\mathrm{d}r} = -\varepsilon \frac{\partial \mathscr{H}_1}{\partial \theta}, \qquad \frac{\mathrm{d}\theta}{\mathrm{d}r} = \omega(I) + \varepsilon \frac{\partial \mathscr{H}_1}{\partial I}. \tag{11.38}$$

Here the action-angle variables are defined in the background environment; that is, Equations (11.16) and (11.17) apply except $\mathscr{H}$ is replaced by $\mathscr{H}_0$. Also, consistent with the preceding approximations, $\varepsilon\mathscr{H}_1 = \delta c/c_0$. Similarly, Equations (11.12)–(11.14) have an action-angle form by the substitution $(p \to I, z \to \theta)$.

The stability of trajectories in a Hamiltonian system under a small perturbation to an integrable system is addressed by the previously mentioned KAM theory (Arnol'd 1978). The principal result is the KAM theorem, of which there are many variants. According to each such variant of the theorem, most of the tori of the unperturbed system, on which trajectories lie, survive under a sufficiently small perturbation provided certain conditions are met. Proofs of KAM theorems for systems of the form of Equations (11.37) and (11.38) can be found in Jorba and Simo (1996) and Sevryuk (2007). Both of these references assume that $\mathscr{H}_1$ is a quasiperiodic function of the timelike variable r. With no loss of generality, commensurable (rationally related) frequencies Ω_i in Equation (11.37) may be eliminated, reducing that system to a quasiperiodic system.

In realistic ocean environments one generally does not encounter situations in which the perturbation is so small that unperturbed tori survive in the perturbed environment. One might then question why KAM theory should be emphasized. The relevance of KAM theory to systems of this type stems in large part from the insight that KAM theory provides into the mechanism that leads to torus destruction. That mechanism is the excitation and overlapping of resonances when the unperturbed frequency of motion $\omega(I)$ is rationally related to the vector of forcing frequencies $\mathbf{\Omega}$; the resonance condition is $k\omega(I) = \mathbf{j} \cdot \mathbf{\Omega}$, where k is an integer and $\mathbf{j}$ is a vector of integers. Low-order resonances, for which both k and all of the elements of $\mathbf{j}$ are small are most strongly excited. Each resonance has a characteristic width (in I or ω), which scales like

$$\Delta I \propto \sqrt{\varepsilon / |\omega'(I)|}, \qquad \Delta\omega \propto \sqrt{\varepsilon |\omega'(I)|}, \tag{11.39}$$

where $\omega'(I)$ is evaluated on the resonant torus; note that the resonance width also falls off rapidly with increasing k in a characteristic way. Refinements to the estimates Equation (11.39) on degenerate tori where $\omega'(I) = 0$ are presented in Rypina et al. (2007). When resonances overlap, intervening tori are broken; this leads to Chirikov's definition of chaos (Chirikov 1959).

A second point to emphasize is that among the class of systems of the form Equation (11.37), the choice $N = 1$ is special only insofar as it allows the construction of Poincaré sections (as discussed in the previous subsection) – by viewing trajectories stroboscopically at r-values that are integer multiples of the period of the perturbation $2\pi/\Omega_1$, modulo an arbitrary constant. Although this is an important simple visualization tool for the $N = 1$ special case, it should be emphasized that independent of N, (1) KAM theory holds; (2) torus destruction is caused by the excitation and overlapping of resonances whose widths scale like Equation (11.39); and (3) surviving tori or secondary islands that are formed when tori are broken serve as impenetrable barriers to other trajectories (for an explanation of why this is so, see Rypina et al. 2007). Taken together, these properties reveal that the essential physics of torus destruction and ray chaos are independent of N. Also, the foregoing comments strongly suggest that the most important environmental property controlling the physics of torus destruction and ray chaos is $\omega'(I)$, a property of the unperturbed environment. We will return to this point below. It is worth connecting heuristically the preeminent role of $\omega'(I)$ with Equation (11.14). The matrix K_r constructed along a ray determines Q_r and ultimately everything about a ray's stability. From the action-angle form of Equation (11.15) and the action–angle form of Hamilton's Equations (11.38), only the lower left element of K_r has a contribution that is not $O(\varepsilon)$; its leading behavior is given by $\omega'(I) = \mathrm{d}^2\mathscr{H}_0/\mathrm{d}I^2$ (see Equation (11.19)).

Although the essential physics of ray chaos is independent of N, it should be noted that for small N (and for $N = 1$, in particular) torus destruction is dominated by the excitation of a small number of low-order resonances; for small ε most of the tori that are not entrained into one of the low-order resonances survive under perturbation. In contrast, when N is large and a broad range of Ω_i's are present, many low-order resonances are excited and, typically, all of the original tori are destroyed. Numerical evidence strongly suggests that this is the case in typical deep-ocean underwater acoustic environments, where for internal-wave-induced sound speed perturbations $\varepsilon = O(\delta c/c_0) = O(10^{-3})$. Under such conditions it is natural to adopt a stochastic framework to describe ray motion and underwater acoustic wavefields approximately.

11.4.3 Stochastic Approximations: Action Diffusion

In the presence of a weak but highly structured sound speed perturbation field, N large in Equation (11.36), a very useful model of sound scattering involves the diffusion of rays in action I, which implies spreading of acoustic energy. Conceptually, this model follows from the assumption that at each of a sequence of discrete scattering events, rays are scattered from one action surface to another. It is easiest to illustrate the basic assumptions and the utility of the model by discussing an example.

Consider the scattering-induced perturbation to the range of a ray, Δr, as a result of undergoing a sequence of n scattering events. In typical deep-ocean environments, steep rays are most strongly scattered near their upper turning depths, leading to the so-called apex approximation in which all scattering takes place at these locations. After n such scattering events,

$$\Delta r_n = \frac{\mathrm{d}R_\ell(I_0)}{\mathrm{d}I} \sum_{i=1}^{n} (I_i - I_0), \tag{11.40}$$

where $R_\ell(I)$ is a ray cycle (or loop) distance; partial ray loops have been neglected; I_0 is the unperturbed ray action; and I_i is the action after the ith scattering event. Let δI_j denote the jump in I at the jth scattering event. Then

$$I_i - I_0 = \sum_{j=1}^{i} \delta I_j. \tag{11.41}$$

To a good approximation, scattering events can be assumed to be independent and δI_j can be assumed to be a delta-correlated zero-mean random variable,

$$\langle \delta I_j \rangle = 0, \qquad \langle \delta I_j \delta I_k \rangle = \langle (\delta I)^2 \rangle \delta_{jk}, \tag{11.42}$$

where $\langle \dots \rangle$ denotes ensemble averaging. It follows from Equations (11.40)–(11.42) that $\langle (\Delta I_n)^2 \rangle = \langle (I_n - I_0)^2 \rangle = \langle (\delta I)^2 \rangle n \simeq \langle (\delta I)^2 \rangle r / R_\ell$ and that for large n,

$$\langle (\Delta r)^2 \rangle = \left(\frac{\mathrm{d}R_\ell(I_0)}{\mathrm{d}I} \right)^2 \langle (\delta I)^2 \rangle \frac{1}{3} \left(\frac{r}{R_\ell} \right)^3 . \tag{11.43}$$

But $\omega(I) = 2\pi / R_\ell(I)$; so $\omega'(I) = -2\pi R_\ell'(I) / (R_\ell(I))^2$. We may replace $\langle (\delta I)^2 \rangle / R_\ell$ by a general action diffusivity D, where $\langle (I(r) - I_0)^2 \rangle = Dr$. With these substitutions

$$\langle (\Delta r)^2 \rangle = \left(\frac{\omega'(I_0)}{\omega(I_0)} \right)^2 D \frac{r^3}{3} . \tag{11.44}$$

In spite of the seemingly strong assumptions on which Equation (11.44) is based, this expression turns out to provide a very good approximation to distributions of range spreads of rays in realistic deep-ocean environments. Two features of this expression are noteworthy. First, the scattering process is completely parameterized by the action diffusivity D (which is typically a slowly varying function of I_0). Second, the observable $(\Delta r)_{\text{rms}}$ is proportional to $\omega'(I_0)$. In other words, the scattering is parameterized entirely by D, but the scattering is amplified by an amount proportional to $\omega'(I_0)$ in the observable $(\Delta r)_{\text{rms}}$.

The argument just presented focused on a ray-based calculation. It should be emphasized, however, that the central ingredient of the scattering argument presented – the diffusion of action – is equally relevant to the modal description of the sound field. The reason is that it follows from the quantization condition (11.25) that scattering of mode energy from mode m to $m \pm 1$ is accompanied by a jump in action by the amount $1/\sigma$. Thus as energy diffuses in action it also diffuses in mode number, $(\Delta m(r))_{\text{rms}} = \sigma \sqrt{Dr}$. This simple approximation provides a direct connection between ray- and mode-based descriptions of wavefields in the presence of weak scattering.

With this as background we are now in a position to state, without derivation, the dependence of various wavefield properties on $\omega'(I)$ and the action diffusivity D. Most of the properties enumerated here are measurable. All of the derivations are straightforward (most are very simple) but are not presented here. Several different measures of time spreads that are constrained in different ways, corresponding to different experimental scenarios, have been shown to be proportional to $\omega'(I)$ (Beron-Vera & Brown 2003, 2004). Both the spatial and temporal spreading of directionally narrow beams have been shown to be proportional to $|\omega'(I)|D$ (Beron-Vera & Brown 2008). The diffractive contribution to the effective width of a ray (the Fresnel zone width) has been shown to be approximately proportional to $\sqrt{|\omega'(I)|}$ (Rypina & Brown 2007). The scattering-induced contribution to the effective width of a ray has been shown to be proportional to $|\omega'(I)|D$ (Rypina & Brown 2007). In a transient wavefield, a modal group arrival can be defined as the contribution to the wavefield associated with a fixed mode number, but containing frequencies across the entire excited frequency band. The deterministic dispersive contribution to the modal group time spread has been shown to be proportional to $|\omega'(I)|$ (Udovydchenkov & Brown 2008), and the scattering-induced (associated with mode coupling) contribution to the modal group time spread has been shown to be proportional to $|\omega'(I)|D$ (Udovydchenkov & Brown 2008). Also, numerical

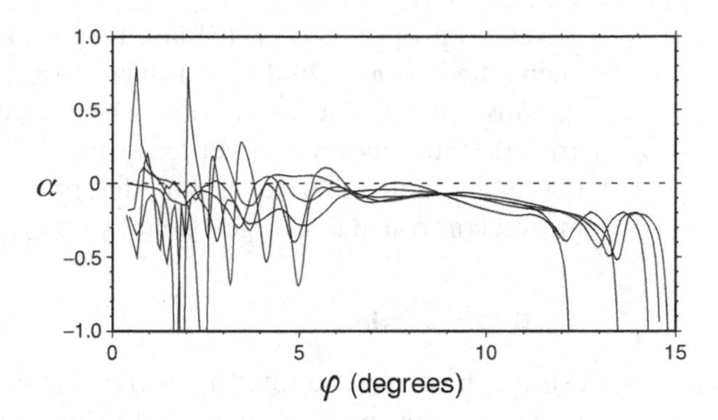

Figure 11.3. Stability parameter α versus axial ray angle φ in the environment corresponding to the wavefields shown in Figure 11.1. Each of the five $\alpha(\varphi(z_{\text{axis}}))$ curves was computed from a 650-km range average of environmental measurements (Worcester et al. 1999) along the 3250-km transmission path. Reprinted with permission from Beron-Vera et al. (2003). Copyright 2003, Acoustical Society of America.

simulations reveal that average Lyapunov exponents scale like $|\omega'(I)|$ (Beron-Vera & Brown 2003).

The scattered wavefield properties just described have two common properties. First, the scattering process itself is described by the diffusion of rays (and hence also acoustic energy) in I. This diffusion process is parameterized by the diffusivity D. Second, observable wavefield properties are proportional to $\omega'(I)D$. In other words, the scattering itself, as embodied in D, is amplified by a factor that is proportional to $\omega'(I)$. Thus, a large scattering-induced time spread, for instance, could be the result of a large $\omega'(I)$ together with a modest valuc of D. The important role of $\omega'(I)$ in controlling observable scattering-induced wavefield properties (e.g., time spreads) is not well appreciated by underwater acousticians. Rather, it is widely assumed, incorrectly, by most practitioners in the field that scattering-induced contributions to wavefield properties are controlled entirely by the sound speed perturbation field $\delta c(z, r)$. Finally, we note that the simple arguments above leading to Equation (11.44) have been put on a more rigorous basis by making use of results from the study of stochastic differential equations (Virovlyansky et al. 2007). As earlier, the critical conceptual notion is diffusion of rays in I. Part of the utility of that approach is that it leads to explicit ray action probability density functions, which are both insightful and useful. Also, that approach gives a proper treatment of scattering near the sound channel axis. (The treatment above – Equation (11.44), for example – does not properly treat near-axial scattering, as it does not account for the fact that I is nonnegative, with $I = 0$ on the sound channel axis.)

Figure 11.3 shows a quantity closely related to $\omega'(I)$ for experimental conditions corresponding to the wavefields shown in Figure 11.1. The quantity plotted in Figure 11.3, $\alpha(I) = I\omega'(I)/\omega(I)$, is a nondimensional measure of $\omega'(I)$. Under conditions appropriate for this figure, axial ray angle $\varphi(z_{\text{axis}})$ is a monotonically increasing function of action I. The relationship between ray angle $\varphi(z)$ and action I follows from $\mathrm{d}z/\mathrm{d}r = p = \sqrt{2[\mathcal{H}(I) - V(z)]} = \tan\varphi(z)$.) This figure shows that there are large variations in $\omega'(I)$ for shallow angle (less than about 5°) rays, corresponding to late-arriving energy, whereas $|\omega'(I)|$ is small for the steep (between approximately 6° and 12°) rays, corresponding to early-arriving energy.[†] This dependence is consistent with the aforementioned dependence of scattered

[†] The fact that shallow rays arrive later than steep rays may be counterintuitive because the shallow rays travel a shorter distance, however in the ocean's sound channel the path length decrease turns out to be more than offset by the slower mean velocity.

wavefield properties on $\omega'(I)$ and Figure 11.1: The early steep ray arrivals (which encounter strong sound speed fluctuations) are relatively stable with small time spreads; the late flat ray arrivals (which encounter weak sound speed fluctuations) are less stable and have larger time spreads. These observations suggest strongly, albeit in a nonquantitative fashion, that properties of finite-frequency wavefields are indeed controlled to a large extent by $\omega'(I)$.

11.5 Discussion

Although we have adopted a tutorial tone in this review, it should be clear from the discussion that many fundamental issues remain unsolved. The first such set of issues relate to the usual problems associated with wave chaos. Realistic deep-ocean environments have range dependence as well as depth dependence, and in such environments ray trajectories exhibit extreme sensitivity associated with chaotic motion. In spite of this, finite-frequency wavefields, both measured and simulated, show remarkable stability. The wave chaos challenge is to reconcile ray instability with wavefield stability. Surprisingly, wavefield stability can be partially explained using ray arguments: Although individual rays are predominantly chaotic, continuous collections of rays (corresponding to a point source with a broad angular aperture, for instance) are surprisingly stable. This is linked to a phenomenon that has been referred to as "manifold stability" (Cerruti & Tomsovic 2002). Finite-frequency effects, of course, also serve to mitigate against extreme sensitivity (Hegewisch et al. 2001), but this introduces a new level of difficulty and more questions. The relationship between ray amplitude statistics (Wolfson & Tomsovic 2001) and wavefield intensity statistics (Colosi et al. 1999) is not well understood; the proliferation of eigenray contributions and the necessity of applying caustic corrections to the ray amplitudes constitute significant complications.

In view of these difficulties one might ask whether one should bother at all with the ray description. There are at least two strong reasons to continue to use ray methods and to try to better understand the wave chaos enigma. First, the ray-like character of underwater acoustic wavefield measurements strongly suggests that the ray description is both useful and approximately valid. Second, the ray description is physically insightful, providing critically important insight into the underlying propagation physics that is difficult to obtain by any other means.

In this chapter we have focused on the class of problems for which $c(z,r) = c(z) + \delta c(z,r)$, as this assumption is applicable in a wide variety of underwater acoustic environments. Provided $\delta c(z,r)$ can be accurately approximated as a multiperiodic function of r, a fairly complete understanding of the ray dynamics follows in the form of a KAM theorem and associated theory. Also, for this class of problems a stochastic description of ray motion is approximately valid and has proven to be useful. Interestingly, however, the connection between the dynamical-systems-based results and the stochastic description is not well understood. For instance, because the mechanism leading to torus destruction and the loss of ray stability is the excitation and overlapping of low-order resonances, and because resonance widths are controlled by $|\omega'(I)|$, one expects that ray diffusion in action (I) is controlled in part by $|\omega'(I)|$, but no such connection has been established. At least in the study of the class of problems on which we have focused, there is reason for optimism. Namely, the action–angle formalism ties together all of the subjects that we have discussed: integrable systems, modes and mode coupling, deterministic chaos and resonances, and stochastic approximations. Thus, it is reasonable to expect that an action–angle-based description will provide valuable insight into many of the questions that we have raised and tie together seemingly unrelated issues.

12 Mesoscopic Seismic Waves

Michel Campillo
Laboratoire de Géophysique Interne et Tectonophysique, Observatoire de Grenoble, Université Joseph Fourier, Grenoble, France; (NRS, Paris, France)

Ludovic Margerin
Centre Européen de Recherche et d'Enseignement de Géosciences de l'Environnement, Université Paul Cézanne, Aix-en-Provence, France; CNRS, Paris, France

This chapter presents recent developments of multiple scattering and mesoscopic concepts in seismology. After a brief review of classical elastic wave propagation in the Earth, we focus on the scattered waves that form the tail or coda of the signal. The stabilization of vertical to horizontal kinetic energy ratios in the coda is illustrated with data from small crustal earthquakes recorded at a temporary network in California. Using a priori geological data, we show that the measurements agree very well with the equipartition principle applied to a layered elastic medium. This confirms that the formation of the coda results from the multiple scattering of elastic waves by Earth's heterogeneities. The concepts of equipartition, diffusion, and energy stabilization are carefully discussed and distinguished with the aid of numerical Monte Carlo simulations. We underline that energy stabilization occurs much earlier than equipartition because the latter concept asks for an isotropic energy flux distribution. Having established the importance of multiple scattering in the Earth we explore the role of interference in random seismic wavefields. A brief review of recent experimental and theoretical works on the weak localization of seismic waves is presented. We further explore mesoscopic concepts by demonstrating the close relations between long-time correlation of random seismic waves and the Green's function. Using data from a permanent seismic network in the Alps, we give strong experimental evidence that the correlation of seismic noise contains not only ballistic but also multiply scattered seismic waves. The power of Green's function reconstruction from random signals is illustrated with applications to seismic tomography and passive monitoring of volcanic activity.

12.1 Introduction: Seismic Waves and Data

Our present knowledge of Earth's dynamics, including plate tectonics, mantle convection, the origin of the magnetic field, and the like, is based on the progress of seismological models of the inner structure. Piecewise homogeneous reference Earth models have been a fundamental step forward in the understanding of the deep

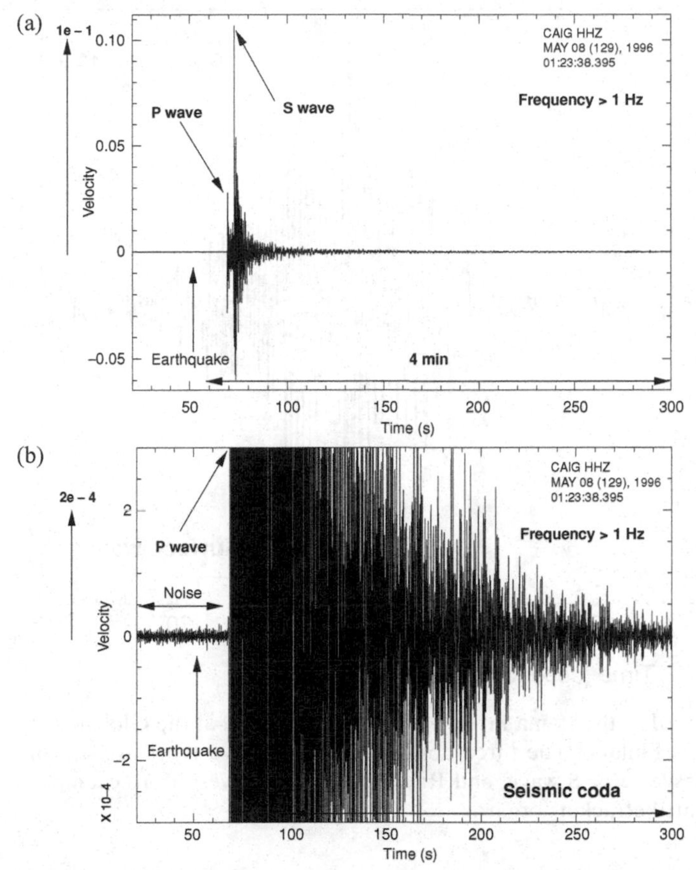

Figure 12.1. Example of crustal seismogram records. (a) The full range of data is shown. The most prominent arrivals are the ballistic P and S waves. (b) The vertical scale has been magnified by a factor of 500. The late arrivals form the seismic coda.

interior processes. It has been recognized that those smooth models cannot account for some major observations. One of them is the strong apparent attenuation of ballistic waves, and another is its counterpart, the long duration of seismograms with energy arrivals at times that greatly exceed the travel times of the direct waves (see Figure 12.1). This leads to the consideration of the scattering of seismic waves by the inhomogeneity of the material, an inhomogeneity that is widely attested by geological surface observations. In recent years, the attention of seismologists has turned to multiple scattering of elastic waves in the Earth, with a renewal of the investigation of the physics of seismic wave propagation in complex media and of possible mesoscopic effects.

The analysis of the records of ground motion at the surface of the Earth has allowed imaging of the interior of the planet at different scales and the characterization of earthquakes. The interpretation of seismograms is made in the framework of the classical theory of elastodynamics. Seismic data consist of time series of the three-dimensional motions recorded at the surface of the Earth. Seismic signals currently cover a frequency band between 0.01 Hz and 10 Hz. Nowadays, most of the recording stations are operated continuously. The sources in most cases produce pulses that are very short with respect to the travel times of the waves. This is not

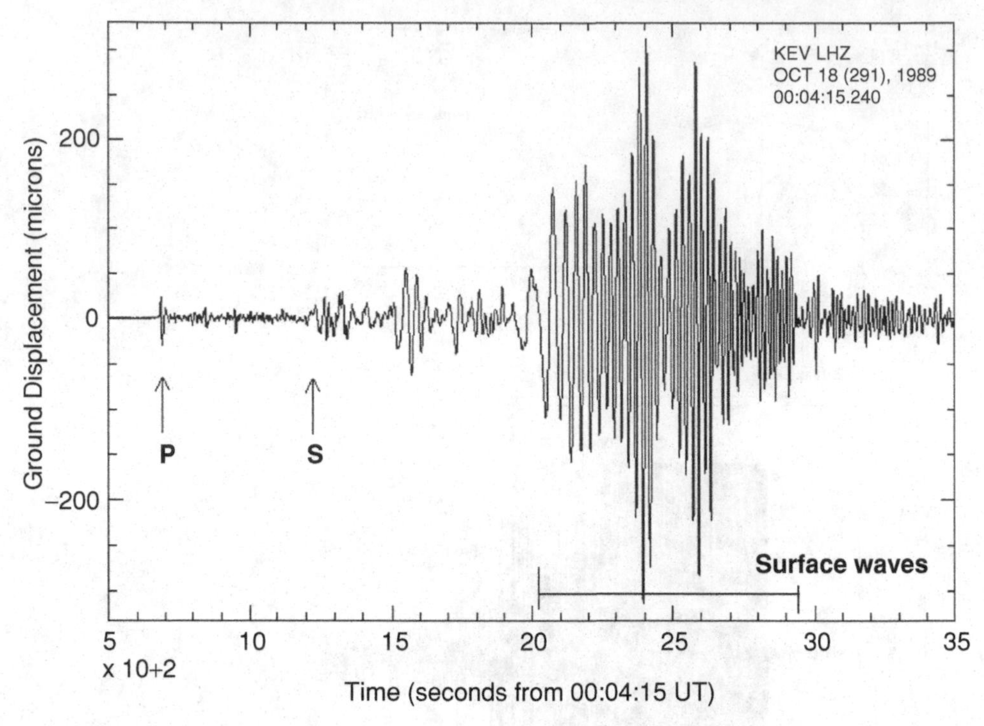

Figure 12.2. Broadband record of the Loma Prieta earthquake recorded at the Global Seismic Network station KEV in Finland. The three dominant seismic wave types are shown: longitudinal or P wave, transverse or S wave, and Rayleigh surface waves. More examples are available at http://www.quaketrackers.org.nz.

the case with large earthquakes, but that is an issue out of the scope of the present work. Ballistic waves are usually clearly identified. Their travel times, and often amplitudes, can be described in the framework of the classical ray theory and can be used for imaging purposes. Whatever the imaging technique – body wave tomography or surface wave tomography – this approach leads to the determination of the local properties of an effective medium (seismic speeds) that can be related to a particular constituent or geological unit. Nevertheless, the ballistic arrivals represent only a small part of the records (see Figure 12.1), and it was realized early that smooth velocity models cannot account for some of the characteristics of seismic signals. The numerous arrivals following the ballistic waves are interpreted as resulting from scattering in a heterogeneous Earth. These late arrivals are called the seismic coda. For long lapse times, it is characterized by a smooth, exponential-like decay of the energy. It appears that scattering is strong at short periods and tends to be almost negligible for period larger than 30 s. The long coda of short-period (around 1 s) seismograms is the most spectacular and direct expression of multiple scattering in the Earth (Figure 12.1).

When considering wave propagation, the solid Earth can be considered as an elastic body. The reader can find a detailed treatment of elasticity in the seismological context in Aki and Richards (2002), for instance. A sample seismogram is shown in Figure 12.2, where three ballistic arrivals can be identified. In an isotropic medium, two types of propagating waves exist simultaneously. The first mode is the

P wave, a longitudinal wave associated with velocity $\alpha = \sqrt{(\lambda + 2\mu)/\rho}$, where λ and μ denote the two Lamé coefficients and ρ is the medium density. A second type of wave is the transversely polarized S wave with velocity $\beta = \sqrt{\mu/\rho}$. In the solid materials of the Earth, we often observe that $\alpha \approx \sqrt{3}\beta$. These two modes are coupled and cross-converted at each reflection, transmission, or scattering. Because of the general trend of increase of velocity with depth, these waves can be trapped in upper layers and give rise to modes of guided waves. A specific coupling between P and S waves is associated with the Rayleigh surface wave. This is the dominant arrival of a broadband seismogram for moderate to large shallow earthquakes as illustrated in Figure 12.2. This perturbation propagates at the surface of an elastic body, even in the case of a homogeneous half-space. It is a combination of inhomogeneous P and S waves (with exponential decay of amplitude with depth) that complies with the zero-stress condition at the surface. It is a specificity of elastodynamics.

The long duration of seismic signals for local earthquakes recorded in a frequency range between 0.1 and 10 Hz is a clear evidence of scattering in the Earth. This duration greatly exceeds the travel time of direct paths, by a factor of 100 in some examples. These late, apparently incoherent, arrivals build up the seismic coda. In Figure 12.1, a typical crustal seismogram illustrates the very large duration of the coda compared with the travel time of ballistic waves. It is also interesting to consider the very large dynamics of seismic records. The coda waves are hardly visible when the full range of the data is shown. Aki and Chouet (1975) showed that the decay with time of the coda envelope is a regional characteristic, independent of the precise location of the station or of the earthquake, or of the earthquake magnitude. According to Rautian and Khalturin (1978), this stationarity of the decay is observed for lapse times larger than twice the travel time of direct shear waves. This rule of thumb has been widely used in coda studies, but it cannot be applied to the case in which the source–receiver distance is smaller than the characteristic scale of the scattering process, that is, the mean free path. The pertinent time scale in coda analysis is the mean free time, that is, the ratio between the mean free path and the typical wave velocity.

It is difficult to assess the nature of coda waves and their regime of propagation. By using a small aperture array of seismometers, it is possible to estimate the distribution of energy on a set of incoming plane waves. Although the first arrivals are concentrated in particular directions, the coda is produced by waves arriving from various azimuths and incidence angles, resulting in a speckle pattern that fluctuates with lapse time. This indicates that these waves are produced by scattering on inhomogeneities distributed in the Earth. Whether single scattering, multiple scattering, or diffusion should be used to model coda waves has been debated by seismologists since Aki and Chouet (1975).

12.2 Equipartition of Seismic Waves

It is difficult to assess the regime of propagation of the waves in a medium with absorption. With elastic waves we can rely on the concept of equipartition that leads eventually to a measurable marker of the diffusive regime, namely, the ratio of P to S wave energies (Weaver 1982, Ryzhik et al. 1996). In the seismological context, the predominance of S waves in the coda was discussed in Aki (1992). He showed that the scattering from P to S is much stronger than from S to P. Using an elegant reciprocity argument, he established the following relation between the scattering coefficients:

$$\frac{g_{PS}}{g_{SP}} = \frac{\alpha^4}{\beta^4}, \tag{12.1}$$

where α and β are the velocity of P and S waves, respectively. We therefore expect a predominance of S waves in the coda. Dainty and Toksöz (1990) applied array analysis to NORSAR data to show that the coda is dominated by waves with apparent velocities less than 4 km/s, that is, by S waves.

More generally, in the diffusive regime that emerges after several scatterings, the wavefield is expected to consist of contributions of all possible modes of propagation. For example in a full space, modes of propagation are pure P and S plane waves. In this case, equipartition means that the wavefield consists of waves propagating in all directions, with all possible polarizations and equal weights on average. This implies that the relative contribution of P and S waves to the local energy tends to a constant ratio, whereas the energy density itself continuously decays because of the spatial expansion and the anelastic dissipation. A simple mode counting argument allows to compute the ratio of P to S energy at equipartition in a full space (Weaver 1982, Ryzhik et al. 1996):

$$\frac{E_S}{E_P} = 2\frac{\alpha^3}{\beta^3}. \tag{12.2}$$

In the case of a Poisson solid ($\alpha/\beta = \sqrt{3}$), the full-space energy ratio equals approximately 10.4. When there is preferential absorption of one of the modes (P or S), a different stabilization of the energy ratio occurs in the multiple scattering regime (Margerin, van Tiggelen & Campillo 2001). The effect of absorption is to shift the ratio in favor of the mode that is less absorbed. With realistic values of dissipation in rocks, this effect is not expected to strongly affect the observations.

It is useful to further clarify the concept of equipartition and the definition of the underlying set of modes. The set of modes we consider is the one of a reference model on which the disorder is added. The disorder is assumed to be large enough to provoke the coupling between the modes but not to change drastically the structure of the spectrum and eigenfunctions. In other words, the modes we consider are not formally the ones of the actual Earth with all its complexity but the ones of a fictitious model close enough to essentially share the same spectrum, that is, essentially the same propagation properties. For a first-order computation of the energy ratio, it could be a simple stratified model. Furthermore, we deal here with "local" propagating modes of the lithosphere, that is, fundamental and higher-modes surface waves, and body waves leaking into the mantle. It is important to notice that equipartition is expected in the phase space so that it does not mean that the local distribution of energy is the same everywhere in the real space. Locally the ratio is governed by the eigenfunctions of the modes. For example, in the equipartition regime, the contribution of surface waves at the free surface will be larger than at depth because of the decay of the fundamental mode eigenfunction (see Hennino et al. 2001).

Energy stabilization in the seismic coda is illustrated in Figure 12.3. Margerin et al. (2009) have analyzed the vertical to horizontal kinetic energy ratio for 10 small earthquakes recorded at a temporary seismic network in Pinyon Flats, California. For a diffuse elastic wavefield in an unbounded medium, this ratio would simply equal 1/2. In the vicinity of a free surface, the Rayleigh wave slightly modifies the theoretical ratio to a value close to 0.56, in excellent agreement with results of Hennino et al. (2001). In Figure 12.3, we remark that the vertical-to-horizontal kinetic energy ratio in the coda is independent of earthquake location or mechanism in the coda window, whereas this ratio shows large and rapid fluctuations in the noise and in the direct arrivals. In the example shown in Figure 12.3, the central frequency of the signal is around 8 Hz, and the stabilization ratio lies between 0.1 and 0.2, which is three to four times smaller than the value reported by Hennino et al. (2001). To understand the difference between this measurement and previous theoretical results obtained for a homogeneous half-space, we have analyzed the frequency dependence of the kinetic energy ratio in the coda in the frequency band of 5–25 Hz. The lower frequency is imposed by the sensor sensitivity, whereas the choice of upper bound is dictated by the quality of the signal-to-noise ratio. The results are shown in Figure 12.4, where a strong frequency dependence is observed. In the frequency band of 5–10 Hz, the vertical to horizontal kinetic energy ratio is lower than expected (0.2 instead of 0.56) and shows a sharp peak above 1 around 18 Hz. Because the network was installed on a weathered granite zone, one may expect large differences between the true velocity profile and a simple homogeneous half-space model. Fortunately, Fletcher et al. (1990) have carefully studied the velocity structure under the array. Using borehole measurements they provided a detailed model of the subsurface at Pinyon Flats. Applying the equipartition principle to the complete set of modes of a layered elastic half-space including surface and body waves, we were able to calculate the vertical-to-horizontal kinetic energy ratio for a realistic velocity model at Pinyon Flats. As shown in Figure 12.4, the agreement between observation and theory is satisfactory. This confirms the validity of the equipartition principle and also demonstrates that the coda energy partition contains information on the local geological structure.

What is the significance of the stabilization of energy ratios with respect to the validity of the diffusion approximation and the equipartition itself? Is there an unambiguous relation between stabilization of energy ratio and equipartition? To answer these questions, Figure 12.5 shows numerical solutions of the elastic radiative transfer equation computed with the Monte Carlo approach of Margerin et al. (2000) compared with solutions of the diffusion equation (Paul et al. 2005). The calculations are performed for monodisperse spherical inclusions, with slightly different (5%) density and velocities with respect to a homogeneous matrix. The elastic and scattering properties are representative of the Earth's crust: the shear velocity is of the order of 4 km/s; the ratio between P and S velocities equals $\sqrt{3}$; and the ratio between the size of the scatterers and the shear wavelength approximately equals 1. In this regime, scattering occurs preferentially in the forward direction for both P and S waves. Note that because of the elastic isotropy of the medium, mode conversions (P to S and S to P) do not occur in the exact forward and backward directions. The source is assumed to release shear waves isotropically, which is a rough but reasonable approximation for earthquakes. The source station distance and the shear

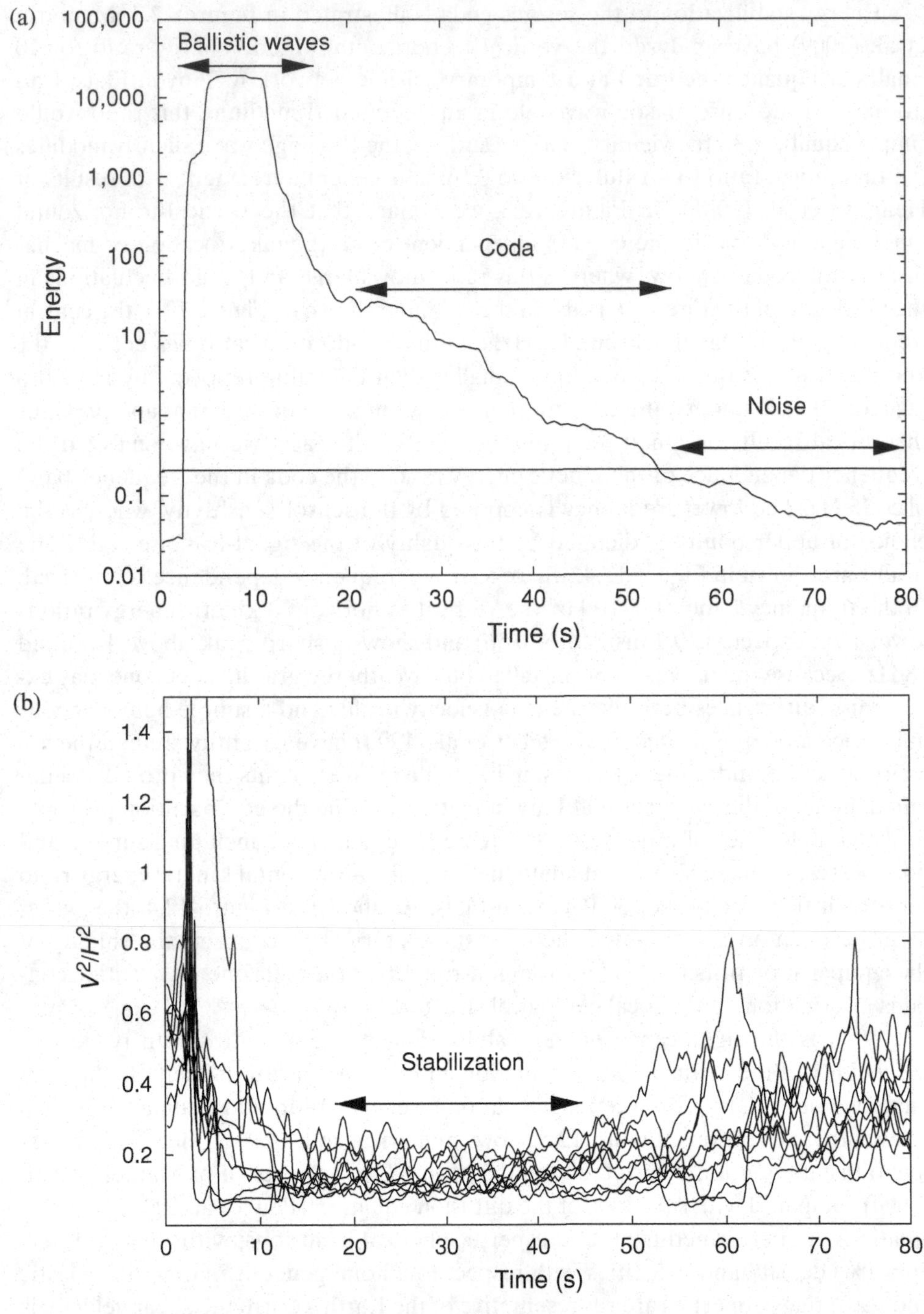

Figure 12.3. Stabilization of energy ratios in the seismic coda of small earthquakes recorded in California. (a) Typical seismogram energy envelope. The horizontal line gives the upper bound for the noise level. (b) Stabilization of the vertical to horizontal kinetic energy ratio in the coda. Notice the large fluctuations of the energy ratio in direct waves and in the noise.

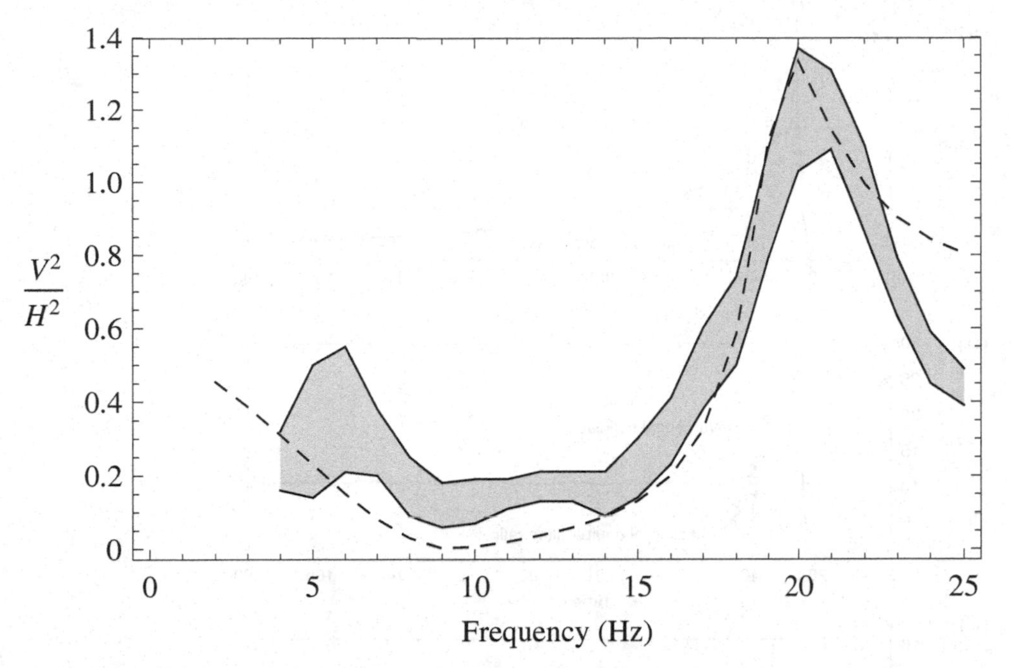

Figure 12.4. Frequency-dependent vertical to horizontal kinetic energy ratio measured at Pinyon Flats (California). The gray area delimits the measurements with uncertainties. The dashed line is the theoretical equipartition prediction from a detailed seismological model of the site. From Margerin et al. (2009). Copyright 2008 Wiley-Blackwell.

mean free path approximately equal 100 and 30 km, respectively. The value of the mean free path is representative of the crust for short-period waves (frequency > 1 Hz) in tectonically active regions. The numerical simulations indicate that the stabilization of the ratio between P and S energies occurs when the total energy is well approximated by the solution of the diffusion equation (Figures 12.5 (a) and (b)). On the contrary, when considering the angular dependence of the energy flow (Figure 12.5 (c)), it is clear that the stabilization occurs very early in the evolution of the wavefield toward isotropy. At a time when the stabilization is reached (60 s in Figure 12.5), the anisotropy of the field remains strong with a ratio larger than 4 between energies propagating in the forward and backward directions. The diffusion solution itself includes a flow of energy from the source and therefore an anisotropy of the field. Figure 12.5 also indicates that when using the diffusion approximation, the anisotropy is underestimated with respect to the radiative transfer equation. We must conclude that the stabilization of S-to-P energy ratio is a good indication that the field is entering a regime in which the total energy is described by the diffusion equation and therefore will evolve towards equipartition and isotropy. Note that although the computations we just presented are performed in a full space, we argue that the conclusion would be much the same in a stratified medium like the Earth.

12.3 Weak Localization

As shown in the previous section, the propagation of multiply scattered waves is often described by considering the transport of the energy. The energy transport

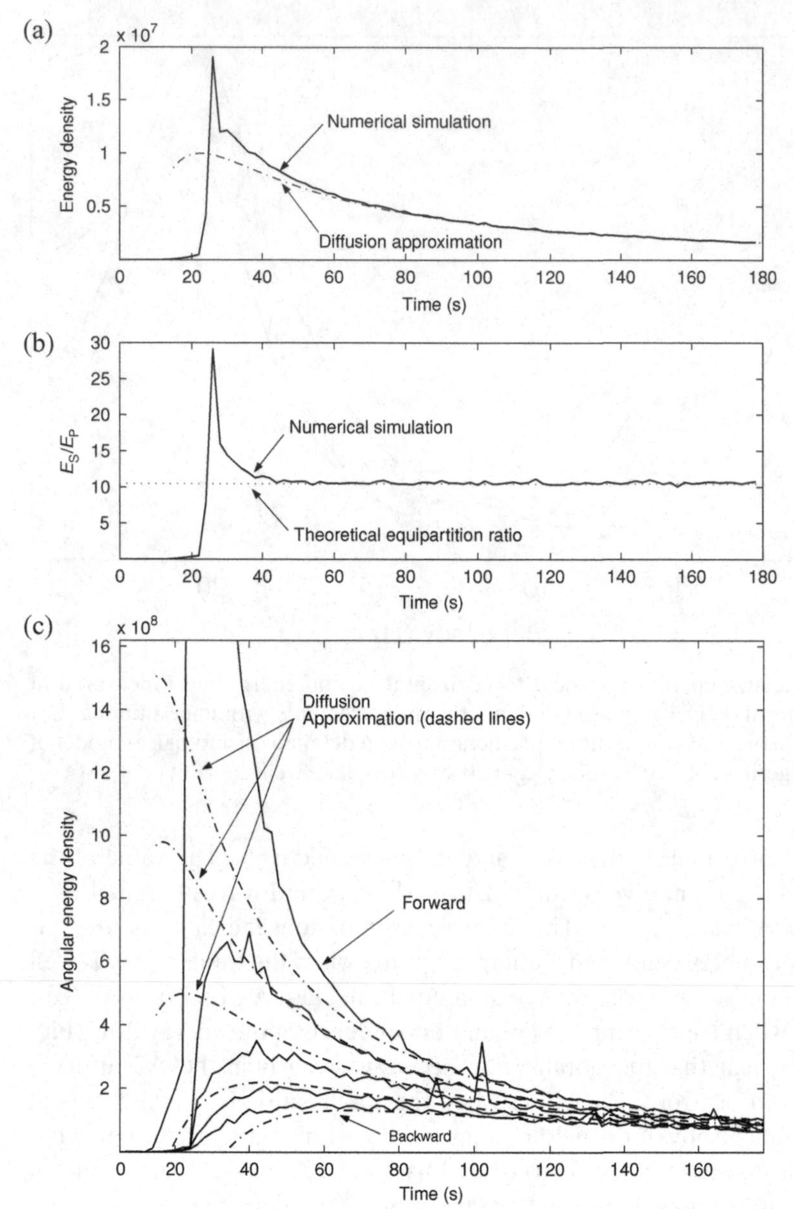

Figure 12.5. Comparison between numerical (Monte Carlo) solutions of the radiative transfer equation, and analytical solutions of the diffusion equation. (a) Energy density, (b) P-to-S energy ratio, and (c) angular distribution of elastic energy flux. The dashed and solid lines show the results of the diffusion approximation and radiative transfer equation respectively. The energy flux decreases monotonically from $\theta = 0$ (forward direction) to $\theta = \pi$ (backward direction), where θ denotes the angle between the propagation direction and the source–observer vector. The results for $\theta = \pi/4$, $\pi/2$, and $3\pi/4$ are also plotted. The source station distance and the shear mean free path equal 100 and 30 km, respectively. The shear wavespeed is 4 km/s. From Paul et al. (2005).

approach has been developed by astrophysicists at the beginning of the twentieth century and has given birth to the theory of radiative transfer (Chandrasekhar 1960, Apresyan & Kravstov 1996). Phenomenologically, the transfer equation for acoustic, electromagnetic, and elastic waves can be derived from a detailed local balance of energy that neglects the possible interference between wave packets. This important assumption is justified by the fact that the phase of the wave is randomized by the scattering events. Thus at a given point, the field can be written as a sum of random phasors, and on average, intensities can be added, rather than amplitudes.

The above-mentioned reasoning misses an important aspect of the wave propagation in multiply scattering media that is closely related to the reciprocity principle. Let us consider a simple but more detailed argument, where the role of reciprocal scattering paths is emphasized. We represent a scalar partial wave as a complex number $\psi = Ae^{i\phi}$, where A and ϕ are real numbers denoting the amplitude and phase, respectively. Each partial wave follows an arbitrarily complicated scattering path from source to receiver in the medium. At a given point, the measured field u is a superposition of a large number of partial waves that have propagated along different scattering paths: $u = \sum_j A_j e^{i\phi_j}$, where the A_j and ϕ_j are random and uncorrelated because of the multiple scattering events and j can be understood as a "label" for the different paths. Typical examples of scattering paths are shown in Figure 12.6.

Let us now pair direct and reciprocal scattering paths to obtain the following: $u = \sum_{j'}(\psi^d_{j'} + \psi^r_{j'})$, where the ψ denote the complex partial waves; the superscripts d and r stand for "direct" and "reciprocal"; and a new label j' has been introduced to emphasize the new representation of the field. The intensity I is proportional to $|u|^2$ and reads

$$I = \sum_{j',k'}(\psi^d_{j'} + \psi^r_{j'})(\psi^d_{k'} + \psi^r_{k'})^*, \tag{12.3}$$

where the asterisk denotes complex conjugation. In Equation (12.3), it is reasonable to assume that the waves visiting *different* scatterers will have random phase differences and after averaging over scatterer positions will have no contribution. Thus, we can restrict the summation to the case $j' = k'$ to obtain

$$I = \sum_{j'} |\psi^d_{j'}|^2 + |\psi^r_{j'}|^2 + \sum_{j'}(\psi^d_{j'}\psi^{r\,*}_{j'} + \psi^{d\,*}_{j'}\psi^r_{j'}). \tag{12.4}$$

The first term on the right-hand side of Equation (12.4) represents the usual incoherent contribution to the measured intensity, which can be calculated with radiative transfer theory. The second term can be interpreted as the *interference* between the direct and reciprocal paths in the scattering medium. In a reciprocal medium, the amplitude and phase of the direct and reciprocal wave paths are exactly the same, that is, $A^d_j = A^r_j$ and $\phi^d_j = \phi^r_j$, provided, that source and receiver are located at the same place. Therefore the total intensity, which includes the interference term, is exactly double of the classical incoherent term. This is the interference term that is at the origin of the coherent backscattering or weak localization effect.

In seismic experiments performed at the surface, weak localization appears as an enhancement of seismic energy in the vicinity of a source for long lapse times.

Configuration 1

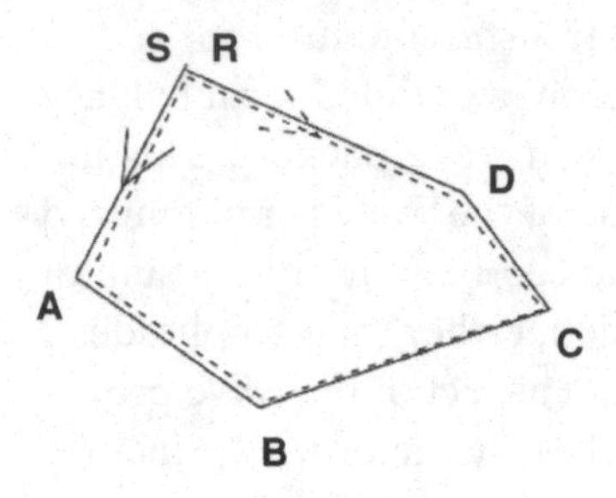

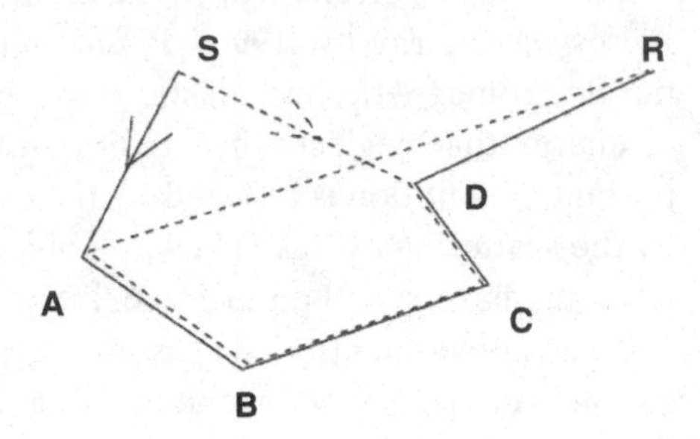

Configuration 2

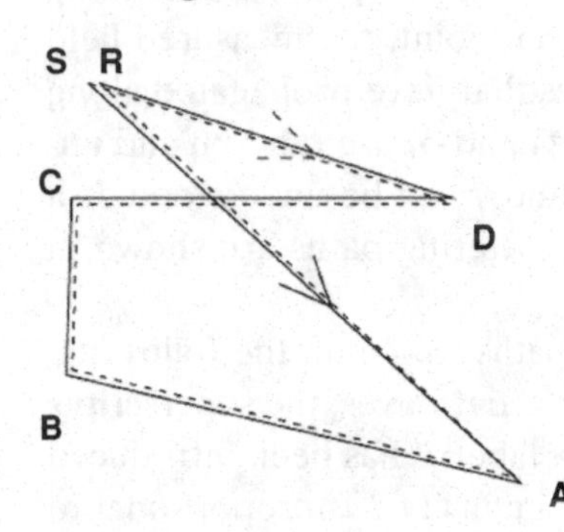

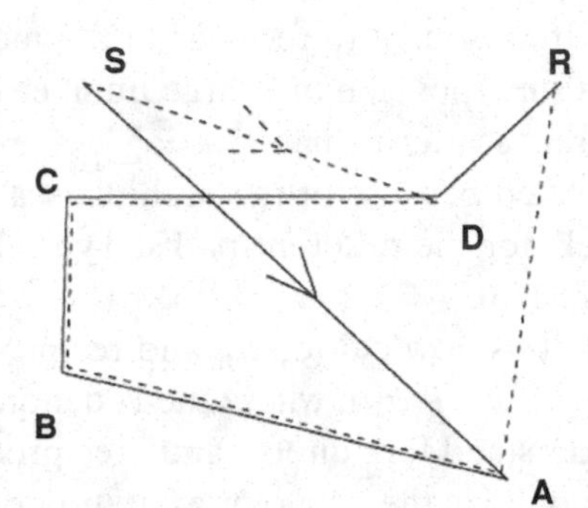

Figure 12.6. Examples of multiple scattering paths from source S to receiver R. Scattering events are labeled with letters A, B, C, and D. The solid and dashed lines represent direct and reciprocal paths respectively. The two configurations differ by the position of the scattering events. Source and receiver coincide (left). Source and receiver are typically a few wavelengths apart (right). From (Margerin, Campillo & van Tiggelen 2001). Copyright 2001 Wiley-Blackwell.

Margerin, Campillo & van Tiggelen (2001) studied the scalar case in the configuration of a seismic experiment. Note that special care must be given to rules of reciprocity with polarized waves. The elastic case was treated by van Tiggelen et al. (2001). A field experiment was performed in a volcanic environment using a sledgehammer as a source recorded along a profile (Larose, Margerin, van Tiggelen & Campillo 2004). The energy of the signal is computed in different time windows of 0.4-s duration. In each window, the energy is normalized by the maximum over the array and is then averaged over 12 different configurations. Figure 12.7 shows the average energy enhancement along the profile measured at different lapse times after the passing of the direct waves. As expected from the theory, the enhancement shows up progressively for the longer lapse times.

This simple experiment shows a nonintuitive mesoscopic effect that demonstrates that the phase cannot be neglected for seismic waves in the multiply reflected regime. The characteristic time for the onset of the enhancement spot (approximately 0.7 s in the case of Figure 12.7) is the scattering mean free time, which is a

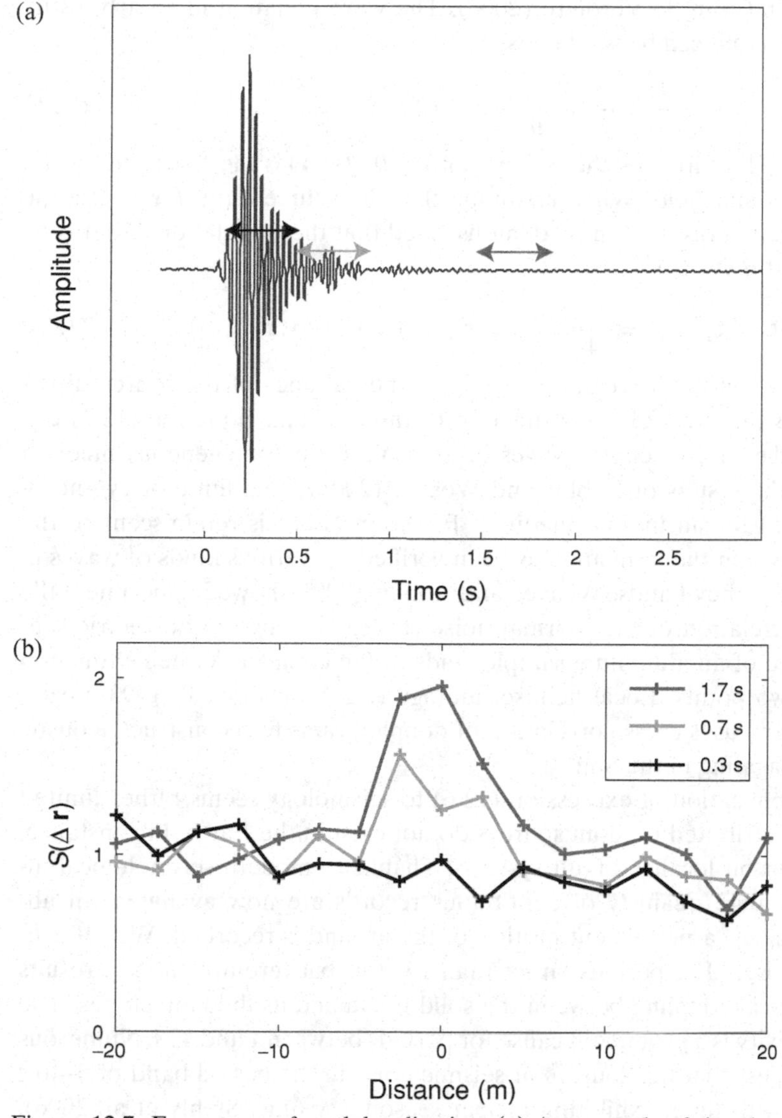

Figure 12.7. Emergence of weak localization. (a) An example of signal produced by a hammer strike. (b) The normalized average energy in the different time windows is plotted as a function of the distance from the source. Note the absence of enhancement spot for the early coda and the progressive onset of the weak localization. From (Larose, Margerin, van Tiggelen & Campillo 2004). Copyright 2004 by the American Physical Society.

measurement of the heterogeneity of the medium. This type of experiment is a way to measure the mean free time independent of dissipation.

12.4 Field Correlations of Seismic Waves and Green's Function

Weak localization is an illustration that in complex, apparently random fields, the wave phases are preserved. In the following, we show that the Green's function between two points can be retrieved from observations of random fields, in absence of a deterministic source at one of the points. Let us recall a mathematical result that is demonstrated in Colin de Verdière (2006). The wave equation in an arbitrarily heterogeneous medium can be written as

$$\frac{\partial^2 u}{\partial t^2} + 2\,a\,\frac{\partial u}{\partial t} - c^2(\mathbf{r})\,\Delta\,u = f. \tag{12.5}$$

The attenuation is described by the parameter $a > 0$, $f(t, \mathbf{r})$ is the source field, and $u(t, \mathbf{r})$ denotes a scalar field. When assuming that the source term $f(\mathbf{r}, t)$ has the properties of a white noise, it can be demonstrated that the correlation C between the fields at A and B verifies

$$\frac{\mathrm{d}}{\mathrm{d}\tau}\,C(\tau, \mathbf{r}_A, \mathbf{r}_B) = \frac{-\sigma^2}{4\,a}(G_a(\tau, \mathbf{r}_A, \mathbf{r}_B) - G_a(-\tau, \mathbf{r}_A, \mathbf{r}_B)). \tag{12.6}$$

In this expression τ is the correlation time, σ is the variance of the source distribution, and G_a is the *exact* Green's function of the medium. Equation (12.6) can be extended to the case of elastic waves in an arbitrary inhomogeneous medium and generalizes the results of Lobkis and Weaver (2001) for a finite body and of Roux et al. (2005) for an infinite medium. Equation (12.6) is reminiscent of the fluctuation–dissipation theorem and has been verified for various kinds of waves in very different frequency bands. Weaver and Lobkis (2001)showed experimentally that the cross-correlation of the thermal noise recorded at two piezoelectric sensors at the surface of an aluminum sample leads to the complete Green's function between these two points. Local helioseismology (e.g., Duvall et al. 1993, Gizon & Birch 2004, 2005) uses cross-correlation of dopplergrams to reconstruct acoustic body waves propagating in the Sun.

The direct application of expression (12.6) to seismology seems rather limited because evenly distributed random sources do not exist in the Earth. For instance, earthquakes occur on localized fault systems. Thanks to modern seismological instrumentation, a huge quantity of continuous records are now available. In absence of earthquakes, a permanent motion of the ground is recorded. We refer to it as "ambient noise." For periods larger than 1 s, this background "noise" results from the mechanical coupling between the solid Earth and its fluid envelopes. The oceanic wave activity is a prominent cause for periods between 1 and 40 s. Numerous observations suggest that the sources of seismic noise in the period band of 5–40 s change position with time, exhibiting a clear seasonality (e.g., Stehly et al. 2006). This time-varying distribution of sources makes it possible to reconstruct efficiently the surface wave part of the Green's function by correlating long time series of noise (Shapiro & Campillo 2004). Interestingly, the long-range correlations of coda waves were also proved to contain the surface wave part of the Green's function (Campillo & Paul 2003, Paul et al. 2005). In this last case, it is the multiple scattering process that generates the random distribution of sources. The reconstructed Rayleigh waves can in turn be used for tomographic studies as shown by for example Shapiro et al. (2005). We refer to Campillo (2006), Larose, Margerin, Derode, van Tiggelen, Campillo, M. Shapiro, Paul, Stehly & Tanter (2006) and Gouédard et al. (2008) for further details on the correlation properties of seismic noise and coda waves.

Since 2005, the use of noise correlation for seismic tomography has become a rapidly growing field. There are nevertheless several issues that must be clarified. The reconstruction of the Green's function is not perfect in practice, even when the correlation is computed for very long time series of noise (up to several years). The fluctuations of the correlation around the Green's function remain significant. It is therefore difficult to identify arrivals other than the predominant surface wave. The convergence towards the Green's function, that is, the ratio between the Green's function amplitude and the residual fluctuations, is expected to evolve as the square root of the amount of data used in the cross-correlation (Larose, Margerin, van Tiggelen & Campillo 2004, Snieder 2004, Sabra, Roux & Kuperman 2005, Weaver & Lobkis 2005). With actual data, the perfect convergence cannot be achieved. This is caused by some deterministic and permanent structures of the seismic noise that remain after averaging, such as spatially localized sources and temporal correlations of the excitation. How can we demonstrate that waves other than direct surface waves are also present in the correlations, as expected from Equation (12.6)? We performed a specific experiment with real data to demonstrate the presence of multiply scattered waves in the Green's function obtained from noise correlation (Stehly et al. 2008). The basic idea is that if diffuse coda waves are present in the correlations, their correlations should in turn contain the Green's function. In other words the correlation of coda of correlations between two stations A and B – referred to as the C^3 function – contains the direct waves between stations A and B. To illustrate this idea, we used data from European stations located in and around the Alps. Let us consider two stations EMV and GIMEL whose location is shown in Figure 12.8. We compute noise correlations between the station EMV (resp. GIMEL) and all other stations of the network located at regional distances (see Figure 12.8). Formula 12.6 indicate that these correlations must contain direct and coda waves albeit with various signal-to-noise ratios. We consider these correlation functions as equivalents of seismograms produced by sources acting at regional distances and recorded at EMV and GIMEL. We select time windows corresponding to coda waves and compute the average correlation between the virtual seismograms at EMV and GIMEL. Just like for earthquake data (Campillo & Paul 2003, Paul et al. 2005), after averaging over the 100 stations of the network, the C^3 function clearly exhibits the direct arrivals of the Green's function (see Figure 12.8). It is therefore possible to extract clear direct arrivals from late time windows of the correlation function, which are apparently dominated by random fluctuations. This experimental result shows that noise correlations actually contain multiply scattered waves.

12.5 Tomography and Temporal Changes from Seismic Noise

Even though only surface waves have been reconstructed from noise so far, it is a valuable result because dispersive surface waves are widely used for imaging crustal or lithospheric shear velocity structures. Shapiro et al. (2005) were the first to use seismic noise to map the distribution of Rayleigh wave group velocities across California. After this first successful application, the noise-based surface wave tomography has been rapidly emerging as a powerful method for imaging the Earth's crust and uppermost mantle at local and regional scales. Numerous applications can be cited: Sabra, Gerstoft, Roux, Kuperman & Fehler (2005) and Moschetti et al. (2007) in the Western United States; Yang et al. (2007) and Villasenor et al. (2007) in Europe; Lin et al. (2007) in New Zealand; Kang & Shin (2006) in Korea; Yao et al. (2006) in China; Nishida et al. (2008) in Japan. In recent work Stehly et al. (2009)

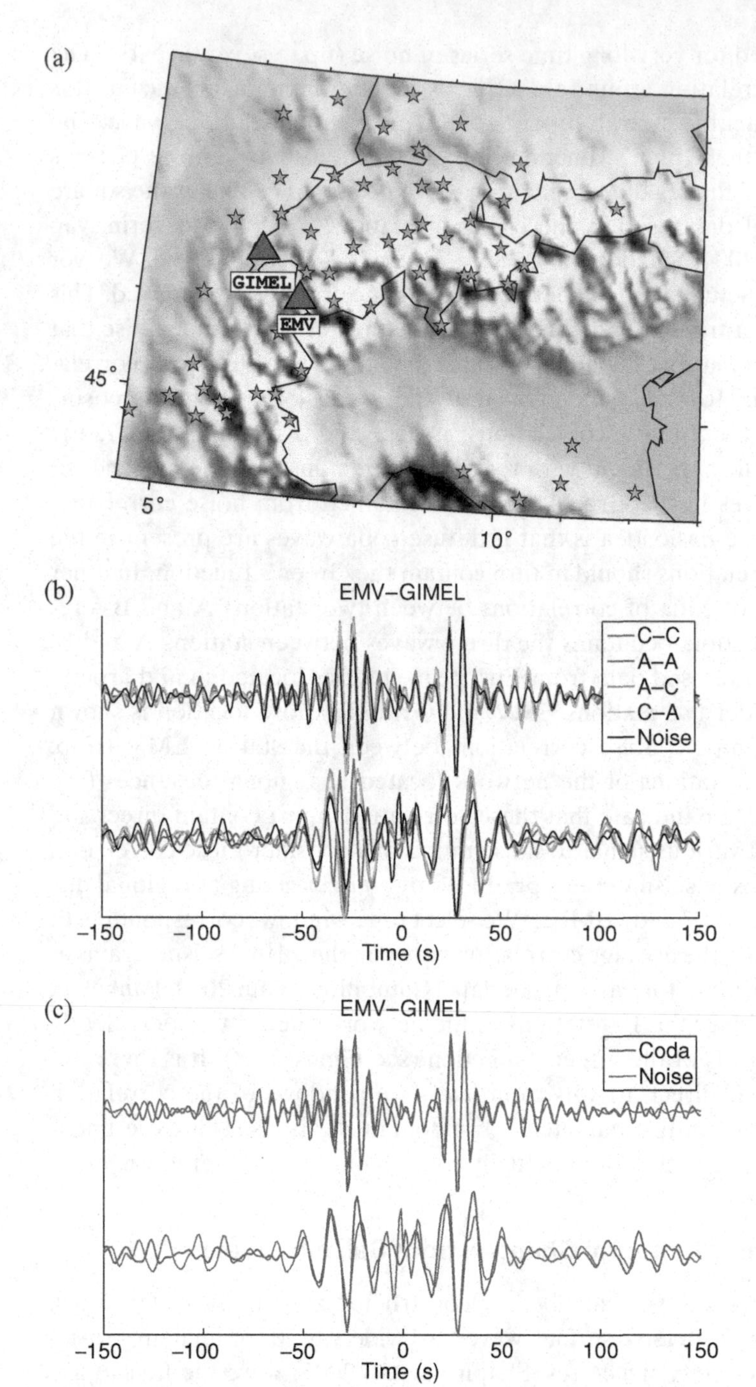

Figure 12.8. (a) location map of the permanent seismic network in the Alps. (b) Green's function between stations EMV and GIMEL (shown in (a)) reconstructed by correlating one year of noise records (black) and by correlating coda waves reconstructed by noise correlations (C3, in blue). We show with different colors the C3 functions, where the coda was selected on the positive noise correlation time, negative time, or a mix of the two. (c) noise correlation function between EMV and GIMEL (red) and the stack of the 4 (C3) functions. We show the results for two period bands: (b) 5–10 s and 10–20 s. From Stehly et al. (c) (n.d.). See http://www.cambridge.org/wrightandweaver for a colour version of this figure.

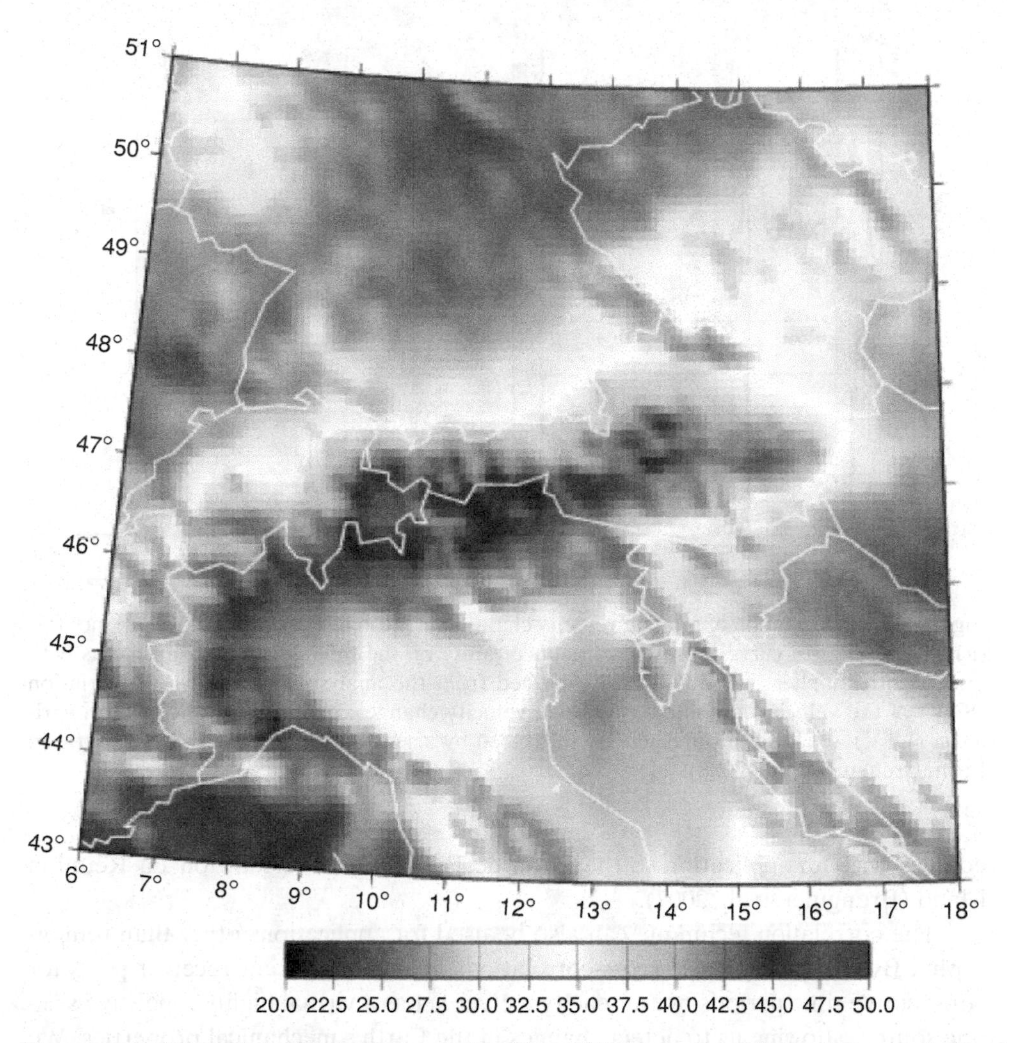

Figure 12.9. View of the Moho beneath the Alps. This image was obtained from seismic noise only. Records from broadband European network were cross-correlated to construct the Green's function between each pair of stations, and a three-dimensional tomography was performed based on local Rayleigh wave dispersion properties. Modified from Stehly et al. (2009). Copyright 2008 Wiley-Blackwell. See http://www.cambridge.org/wrightandweaver for a colour version of this figure.

applied the noise-based Rayleigh wave tomography to image the structure beneath the Alpine region with the European seismic network. They used the correlation of the seismic ambient noise to study the lithosphere. Cross-correlation of one year of noise recorded at 150 broadband stations yields more than 3,000 Rayleigh group velocity measurements. These measurements are used to construct Rayleigh group velocity maps of the Alpine region and the surrounding areas in the period band between 5 and 80 s. The local dispersion curves are then inverted to obtain depth-dependent shear wave velocity. The results of this processing are illustrated in Figure 12.9 where the depth of the crust-mantle boundary is presented. The thickening of the crust in the axis of the Alps is well imaged with this passive imaging approach.We also demonstrated that the noise-based imaging can be applied to relatively short-period (1–5 s) Rayleigh waves to study internal structure of volcanic

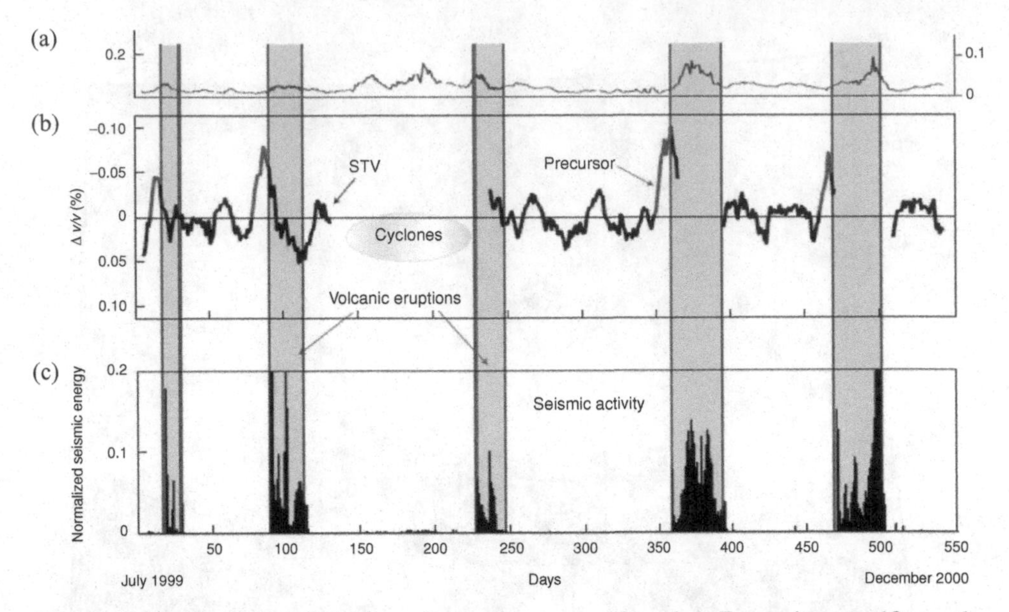

Figure 12.10. Evolution of relative velocity changes on Piton de la Fournaise over 18 months. (a) The blue error curve represents the uncertainty on the time delay. Measurements with uncertainties higher than 0.04% are excluded from the analysis. (b) Short-term variation (STV) of the velocity computed as relative velocity changes corrected for a long-term variation (LTV). (c) Daily seismic activity measured by a sensor located near Dolomieu crater. From Brenguier et al. (2008).

edifices with an application to the Piton de la Fournaise volcano on La Réunion island (Brenguier et al., 2007).

The correlation technique can also be used for applications other than tomography. By computing noise cross-correlations between different receiver pairs for consecutive time periods, we let each receiver act as a virtual highly repetitive seismic source, allowing us to detect changes in the Earth's mechanical properties. We have already made a series of applications (Stehly et al. 2007, Brenguier et al. 2008). In Stehly et al. (2007) we selected a series of records obtained during long periods (more than 10 years). This analysis allowed us to determine the duration of averaging required to obtain stable measurements. This point is strongly linked to the question of the origin of the seismic noise. Stehly et al. (2007) proposed a method to correct for instrumental errors, which is based on the property of time symmetry of the wave equation. We then applied the repetitive noise-based measurements to study the Piton de la Fournaise volcano on La Réunion island and demonstrated that the associated reconstructed seismic waves (Green's functions) can be used to detect temporal perturbations associated with small velocity changes. The accuracy of the measurements of relative velocity change is better than 0.1% (Brenguier et al. 2008). In particular we observed a new type of precursory phenomenon. The preeruptive dilatations of the Piton de la Fournaise volcano provokes a slight decreases of seismic velocity a few days prior to eruptions. Noise-based monitoring is precise enough to detect such changes of velocity (Figure 12.10). The measurements reach a high accuracy level because we associate the noise-based reconstruction with a doublet analysis that takes advantage of the presence of scattered waves

(Poupinet et al. 1984, Grêt et al. 2005, Sens-Schönfelder & Wegler 2006, Wegler & Sens-Schönfelder 2007). The technique is similar to the diffuse wave spectroscopy used in optics and acoustics (see Snieder & Page 2007 for a review of recent results).

12.6 Conclusion

We have tried to demonstrate the usefulness of multiple scattering and mesoscopic concepts in seismology. We have put emphasis on the experimental illustration of the closely related phenomena of equipartition and long-range field correlations. A large number of applications such as seismic tomography with ambient noise are currently being developed. Recently, the possibility to monitor local, small temporal variations in active regions with seismic noise has been demonstrated (Wegler & Sens-Schönfelder 2007, Brenguier et al. 2008), which opens new ways for the investigation of the Earth dynamics. This illustrates the possibility to do mesoscopic physics with signals that are themselves the output of mesoscopic seismic experiments. Long-range correlations (Stehly et al. 2008) and weak localization (Larose, Lobkis & Weaver 2006) have also been observed in the coda of signals resulting from the cross-correlation of ambient or thermal noise. The far-reaching implications of mesoscopic concepts for seismology are still to be further explored.

13 Random Matrices in Structural Acoustics

Christian Soize

Laboratoire de Modélisation et Simulation Multi-Echelle, Université Paris-Est, Paris, France

This chapter is devoted to the predictions of complex structural–acoustic systems in the low and medium frequencies for which computational structural–acoustic models are required. The presentation is limited to a bounded structure coupled with bounded internal acoustic cavities. In order to simplify the presentation, the acoustic coupling of the structure with an unbounded external acoustic fluid is not considered here but can be taken into account without any difficulties. For complex systems, the main problem induced by such predictions is due to the incapacity of computational models (and of any another approaches) to represent a real complex system even if the model used is very sophisticated (multiscale modeling, very large number of degrees of freedom used in the finite element model, etc.). This problem is induced by the presence of both the *system parameter uncertainties* and the *model uncertainties* in the computational model and by the *variabilities* of the real complex system with respect to the design system. The objectives of this chapter is to model the uncertainties in the structural–acoustic computational model by using the random matrix theory and also to present a methodology to perform an experimental identification of the stochastic model and to present an experimental validation.

The designed structural–acoustic system is the system conceived by the designers and analysts. A designed structural–acoustic system, made up of a structure coupled with an internal acoustic cavity, is defined by geometrical parameters, by the choice of materials, and by many other parameters. The real structural–acoustic system is a manufactured version of the system realized from the designed structural–acoustic system. Consequently, the real structural–acoustic system is a man-made physical system that is never exactly known because of the variability induced by the manufacturing process and by small differences in the configurations. The objective is to predict the behaviour of the real structural–acoustic system including its variabilities, using a unique computational model. We then must predict the output $(\mathbf{u}_{\text{exp}}, p_{\text{exp}})$ of the real structural–acoustic system because of a given input $\mathbf{f}_{\text{exp}}$, in which $\mathbf{u}_{\text{exp}}$ is the response in displacement of the structure and p_{exp} is the acoustic pressure inside the acoustic cavity. Such a computational model is constructed by developing a mathematical–physical model of the designed structural–acoustic system for a given input (see Figure 13.1). Consequently, the mean computational model is defined by an input $\mathfrak{f}$ that is a model of $\mathbf{f}_{\text{exp}}$ and by an output $(\mathfrak{u}, \mathfrak{p})$ that is a model of $(\mathbf{u}_{\text{exp}}, p_{\text{exp}})$ and that exhibits a vector-valued parameter $\mathbf{w}$ for which

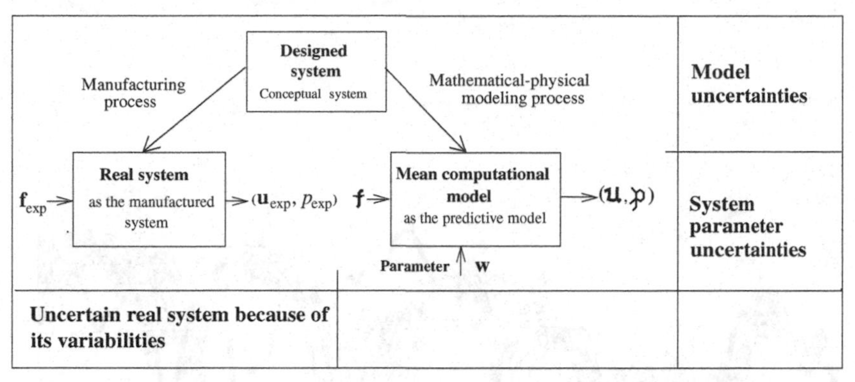

Figure 13.1. Designed system, real system, and mean computational model as the predictive model of the real system.

data have to be given. The errors are related to the construction of the approximation $(\mathfrak{u}, \mathfrak{p})$ of the output $(\mathbf{u}, p)$ of the mean boundary value problem for a given input $\mathbf{f}$ and for a given parameter $\mathbf{w}$. These errors are induced by the approximation of the mean boundary value problem by the mean computational model using, for instance, the finite element method or another numerical method. Clearly, such errors have to be reduced and controlled using adapted methods developed in applied mathematics and in numerical analysis and must not be considered as uncertainties. The mathematical physical modeling process of the designed structural–acoustic system introduces two fundamental types of uncertainties: the system parameter uncertainties and the model uncertainties. The manufacturing process induces variabilities in the real structural–acoustic system. Figure 13.1 displays a scheme showing the relationships between the different notions introduced above. Figure 13.2 gives an experimental illustration of the role played by the variabilities, the system parameter uncertainties, and the model uncertainties for 20 cars of the same type with optional extras (small differences in the configurations). This figure shows the variabilities in the real system induced by the manufacturing process and optional extras. In addition, Figure 13.2 also shows the comparison between the measurements and the prediction constructed with a sophisticated mean computational model. The significant differences between measurements and numerical simulations are due to the uncertainties in the computational model. This figure clearly shows that the system parameter uncertainties and the model uncertainties have to be taken into account in the mean computational structural–acoustic model in order to improve the predictability and the robustness of the predictions. Consequently, one needs to model the uncertainties in the structure, in the acoustic cavity, and for the structure-acoustic coupling interface. It is known that several approaches can be used to take into account the uncertainties in the computational models of complex structural–acoustic systems for the low and medium frequencies (method of intervals, fuzzy sets, probabilistic approach, etc). The most efficient and the most powerful mathematical tool adapted to model the uncertainties is the probabilistic approach as soon as the probability theory can be used. The *parametric probabilistic approach*, which includes the stochastic finite element method, is the most efficient method to address the system parameter uncertainties in the predictive models. Such an approach consists of modeling the vector-valued parameter $\mathbf{w}$ by a vector-valued

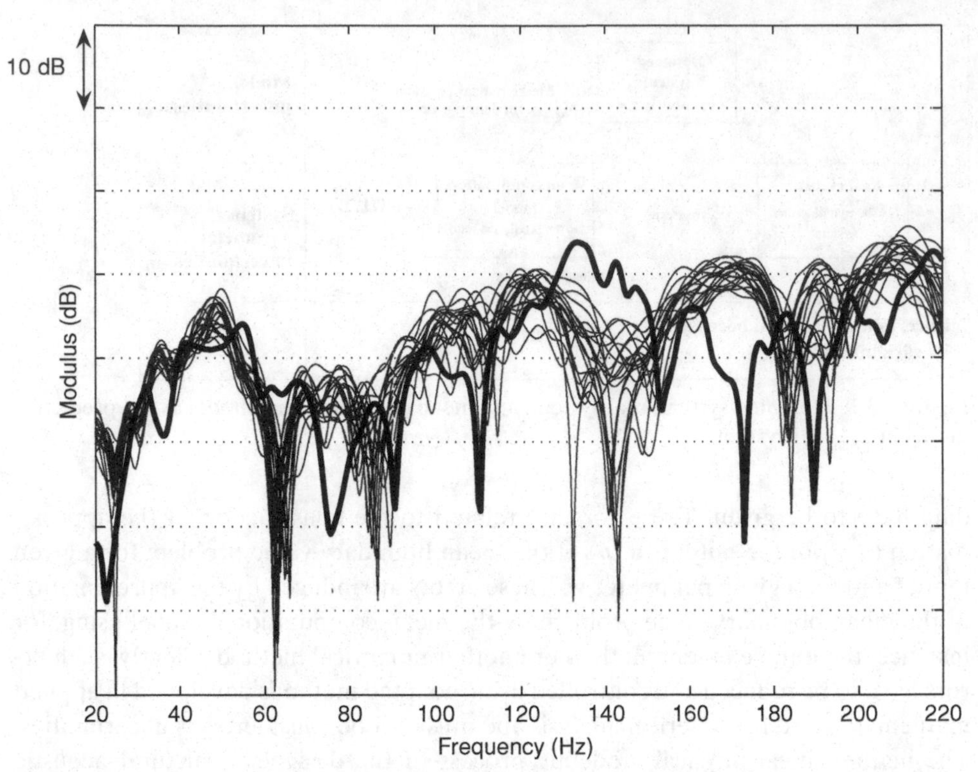

Figure 13.2. Modulus of the experimental frequency response function for the structural displacement response at a given observation point, because of the engine excitation and for 20 cars of the same type with optional extras (20 gray thin solid lines corresponding to measurements). Mean computational model predictions (thick solid line) (from Durand (2007) and reprinted with permission from Durand et al. (2008)). Copyright 2008, Acoustical Society of America.

random variable **W** for which the probability distribution must be constructed and identified with experiments. However, the parametric probabilistic approach cannot address the model uncertainties.This difficulty is illustrated in Figure 13.3, which is devoted to the frequency response function for the transversal displacement in a point on the neutral fiber of a slender elastic body with dimensions $10 \times 1 \times 1.5$ m, simply supported and driven by a point force. The mean computational model is constructed using one-dimensional Euler beam finite elements. This figure shows significant differences between the reference solution, which is constructed from a large number of three-dimensional finite elements, and the prediction given by the mean computational model. The parametric probabilistic approach for the system parameter uncertainties has been used. The random variables are the mass density, the geometric parameters, the Young's modulus, and the damping rate. This figure also displays the confidence region calculated with the parametric probabilistic approach for a probability value equal to 0.98. It can be seen that the prediction of the mean model is good for frequencies lower than 120 Hz and is locally different for frequencies greater than 120 Hz. In the low-frequency range 0–100 Hz, the computational model is relatively robust with respect to the model uncertainties. The results are different in the frequency band 100–1,000 Hz. The parametric probabilistic approach allows the system parameter uncertainties to be taken into account

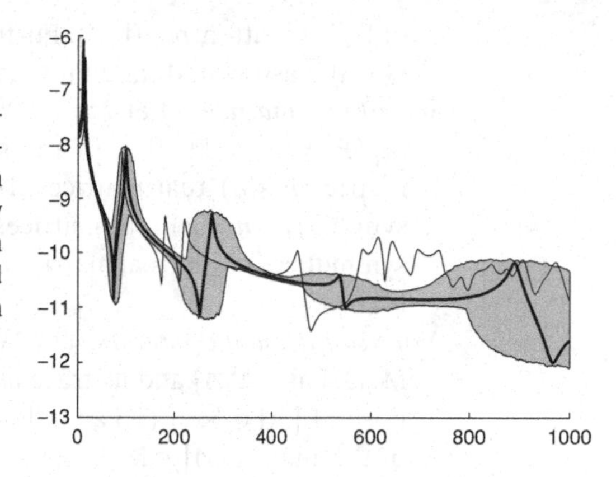

Figure 13.3. Graph of the modulus of the frequency response function in log scale for transversal displacement of a slender elastic body at a given observation point as a function of the frequency in hertz (horizontal axis): reference solution (thin solid line); mean computational model (thick solid line), and confidence region (gray region) (from Soize 2005b).

but cannot address the model uncertainties, which become more significant when frequency increases. In addition, the use of the parametric probabilistic approach of the system parameter uncertainties for structural–acoustic analysis of complex system generally requires the introduction of a very large number of random variables. This is due to the fact that structural–acoustic responses can be very sensitive to many parameters relative to the boundary conditions or to the geometry (such as the thicknesses of plates or the curvatures of panels), among others. Typically, several tens of thousands of random variables can be necessary for a complex system. It should be noted that the construction of the probabilistic model for a large number of parameters is not so easy to carry out. In addition, the experimental identification of a large number of probability distributions using measurements of the structural–acoustic system responses (solving an inverse stochastic problem and, consequently, solving an optimization problem) can become unrealistic.

We therefore need new concepts for taking into account the two types of uncertainties. The *nonparametric probabilistic approach* recently proposed and based on the use of the random matrix theory is a way to address both the model uncertainties and the system parameter uncertainties. We will see that the nonparametric approach proposed introduces a very small number of parameters (typically seven parameters for a structural–acoustic system), which allows the level of uncertainties to be controlled. In these conditions, the experimental identification of these parameters is realistic and can be performed by solving the stochastic inverse problem using adapted mathematical statistical tools.

In this chapter, the following algebraic notations are used:

Euclidean and Hermitian spaces. Let $\mathbf{x} = (x_1, \ldots, x_n)$ be a vector in $\mathbb{K}^n$ with $\mathbb{K} = \mathbb{R}$ or $\mathbb{K} = \mathbb{C}$. The Euclidean space $\mathbb{R}^n$ (or the Hermitian space $\mathbb{C}^n$) is equipped with the usual inner product $\langle \mathbf{x}, \mathbf{y} \rangle = \sum_{j=1}^{n} x_j y_j^*$ and the associated norm $\|\mathbf{x}\| = \langle \mathbf{x}, \mathbf{x} \rangle^{1/2}$ in which y_j^* is the complex conjugate of the complex number y_j and where $y_j^* = y_j$ when y_j is a real number.

Matrices and vectors. In this chapter, $[\mathscr{A}]$ denotes a deterministic matrix related to the computational model in physical coordinates; $[\underline{A}]$ denotes a deterministic matrix related to the mean reduced computational model in generalized coordinates; $[\mathbf{A}]$ denotes a random matrix whose mean value is $[\underline{A}]$; and $[A]$ is a deterministic matrix associated with $[\mathbf{a}]$. The probability density function of $[\mathbf{A}]$ is written $p_{[\mathbf{A}]}([A])$. Furthermore $\mathbf{q}$ denotes a deterministic vector and $\mathbf{Q}$ is the associated random vector.

Sets of real matrices. Let $\mathbb{M}_{n,m}(\mathbb{R})$ be the set of all the $(n \times m)$ real matrices, $\mathbb{M}_n(\mathbb{R}) = \mathbb{M}_{n,n}(\mathbb{R})$ that of the square matrices, $\mathbb{M}_n^S(\mathbb{R})$ that of all the symmetric $(n \times n)$ real matrices, $\mathbb{M}_n^{+0}(\mathbb{R})$ that of all the semi-positive definite symmetric $(n \times n)$ real matrices, and $\mathbb{M}_n^+(\mathbb{R})$ that of all the positive definite symmetric $(n \times n)$ real matrices. We then have $\mathbb{M}_n^+(\mathbb{R}) \subset \mathbb{M}_n^{+0}(\mathbb{R}) \subset \mathbb{M}_n^S(\mathbb{R}) \subset \mathbb{M}_n(\mathbb{R})$.

Norms and usual operators. (1) We denote the determinant of matrix $[A] \in \mathbb{M}_n(\mathbb{R})$ as $\det[A]$ and its trace as $\mathrm{tr}[A] = \sum_{j=1}^n [A]_{jj}$; we denote (2) the transpose of $[A] \in \mathbb{M}_{n,m}(\mathbb{R})$ as $[A]^T \in \mathbb{M}_{m,n}(\mathbb{R})$; we denote (3) the operator norm of the matrix $[A] \in \mathbb{M}_{n,m}(\mathbb{R})$ as $\|A\| = \sup_{\|\mathbf{x}\| \leq 1} \|[A]\mathbf{x}\|$ for all $\mathbf{x}$ in $\mathbb{R}^m$, which is such that $\|[A]\mathbf{x}\| \leq \|A\| \, \|\mathbf{x}\|$ for all $\mathbf{x}$ in $\mathbb{R}^m$; (4) for $[A]$ and $[B] \in \mathbb{M}_{n,m}(\mathbb{R})$, we denote $\langle\!\langle [A], [B] \rangle\!\rangle = \mathrm{tr}\{[A]^T [B]\}$ and the Frobenius norm (or Hilbert–Schmidt norm) $\|A\|_F$ of $[A]$ is such that $\|A\|_F^2 = \langle\!\langle [A], [A] \rangle\!\rangle = \mathrm{tr}\{[A]^T [A]\} = \sum_{j=1}^n \sum_{k=1}^m [A]_{jk}^2$, which is such that $\|A\| \leq \|A\|_F \leq \sqrt{n}\|A\|$.

13.1 Reduced Mean Computational structural–acoustic Model

In this section, we present the construction of the reduced mean computational structural–acoustic model formulated in the frequency domain. Such a reduced model is necessary to implement the nonparametric probabilistic approach of uncertainties. The first step of this construction is the physical modeling of the structural–acoustic system for the low and medium frequencies, which allows the mean boundary value problem to be built. This model is developed in the context of the three-dimensional linear elastoacoustics for a structural–acoustic system made up of a damped elastic structure coupled with a closed internal acoustic cavity filled with a dissipative acoustic fluid. The linear responses of the structural–acoustic system are studied around a static equilibrium state, which is taken as natural state at rest. The external acoustic fluid is a gas, and its effects on the structural–acoustic system are assumed to be negligible in the frequency band of analysis. The mean computational structural–acoustic system is derived from the mean boundary value problem using the finite element method. Finally, the last step consists of constructing the reduced mean computational structural–acoustic model using modal analysis.

13.1.1 Mean Boundary Value Problem

Let Ω_s be a fixed damped structure subjected to external loads. The structure is coupled with its internal cavity Ω_a filled with a dissipative acoustic fluid. The frequency band of analysis is denoted by $\mathcal{F} = [\omega_{\min}, \omega_{\max}]$ rad/s. The three-dimensional space is referred to a Cartesian system, and its generic point is denoted by $\mathbf{x} = (x_1, x_2, x_3)$. The system is defined in Figure 13.4. Let $\partial\Omega_s = \Gamma_s \cup \Gamma_0 \cup \Gamma_a$ be the boundary of Ω_s. The outward unit normal to $\partial\Omega_s$ is denoted by $\mathbf{n}_s = (n_{s,1}, n_{s,2}, n_{s,3})$. The displacement field in Ω_s is denoted by $\mathbf{u}(\mathbf{x}, \omega) = (u_1(\mathbf{x}, \omega), u_2(\mathbf{x}, \omega), u_3(\mathbf{x}, \omega))$. The structure is assumed to be fixed on the part Γ_0 of the boundary $\partial\Omega_s$. The internal acoustic cavity Ω_a is a bounded domain filled with a dissipative acoustic fluid. The boundary $\partial\Omega_a$ of Ω_a is Γ_a. The outward unit normal to $\partial\Omega_a$ is $\mathbf{n}_a = (n_{a,1}, n_{a,2}, n_{a,3})$ such that

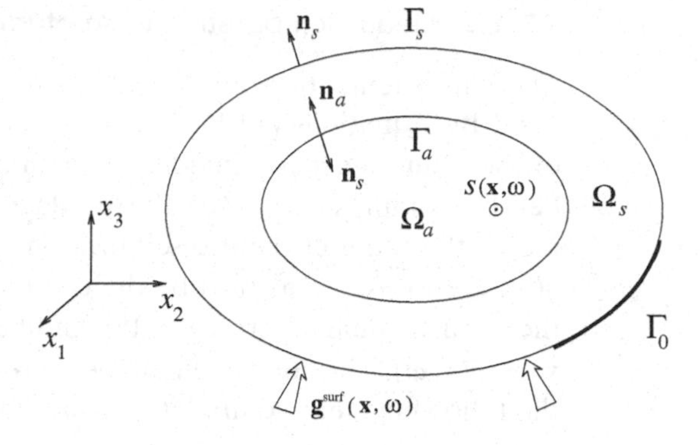

Figure 13.4. Scheme of the structural–acoustic system.

$\mathbf{n}_a = -\mathbf{n}_s$ on $\partial\Omega_a$. The pressure field in Ω_a is denoted by $p(\mathbf{x},\omega)$. For all ω in $\mathcal{F}$, the equation for the structure is written as

$$-\omega^2 \rho_s u_i - \frac{\partial \sigma_{ij}}{\partial x_j} = g_i^{\text{vol}} \quad \text{in} \quad \Omega_s, \tag{13.1}$$

with the convention for summations over repeated Latin indices, in which σ_{ij} is the stress tensor, $\mathbf{g}^{\text{vol}}(\mathbf{x},\omega) = (g_1^{\text{vol}}(\mathbf{x},\omega), g_2^{\text{vol}}(\mathbf{x},\omega), g_3^{\text{vol}}(\mathbf{x},\omega))$ is the body force field applied to the structure, and ρ_s is the mass density. For $i = 1, 2$, or 3, the boundary conditions are

$$\sigma_{ij}(\mathbf{u}) n_{s,j} = g_i^{\text{sur}} \quad \text{on} \quad \Gamma_s, \quad \sigma_{ij}(\mathbf{u}) n_{s,j} = -p n_{s,i} \quad \text{on} \quad \Gamma_a, \tag{13.2}$$

$$u_i = 0 \quad \text{on} \quad \Gamma_0, \tag{13.3}$$

in which $\mathbf{g}^{\text{sur}}(\mathbf{x},\omega) = (g_1^{\text{sur}}(\mathbf{x},\omega), g_2^{\text{sur}}(\mathbf{x},\omega), g_3^{\text{sur}}(\mathbf{x},\omega))$ is the surface force field applied to the boundary Γ_s of the structure. The damped structure is made up of a linear nonhomogeneous anisotropic viscoelastic material. The constitutive equation is then written as $\sigma_{ij} = a_{ijkh}(\mathbf{x})\epsilon_{kh}(\mathbf{u}) + i\omega b_{ijkh}(\mathbf{x},\omega)\epsilon_{kh}(\mathbf{u})$. The tensor $a_{ijkh}(\mathbf{x})$ of the elastic coefficients depends on $\mathbf{x}$ but is approximated by a function independent of ω for all ω in the frequency band of analysis $\mathcal{F}$. The tensor $b_{ijkh}(\mathbf{x},\omega)$ of the damping coefficients of the material depends on $\mathbf{x}$ and ω. These two tensors have the usual properties of symmetry and positive definiteness. The strain tensor $\epsilon_{kh}(\mathbf{u})$ is related to the displacement field $\mathbf{u}$ by $\epsilon_{kh}(\mathbf{u}) = (\partial u_k/\partial x_h + \partial u_h/\partial x_k)/2$. Concerning the internal dissipative acoustic fluid, the formulation in pressure is used. The equation governing the vibration of the dissipative acoustic fluid occupying domain Ω_a is written as

$$-\frac{\omega^2}{\rho_a c_a^2} p - i\omega \frac{\tau}{\rho_a} \nabla^2 p - \frac{1}{\rho_a} \nabla^2 p = -\frac{\tau}{\rho_a} c_a^2 \nabla^2 s + i\frac{\omega}{\rho_a} s \quad \text{in} \quad \Omega_a, \tag{13.4}$$

for which the boundary conditions are

$$\frac{1}{\rho_a}(1 + i\omega\tau)\frac{\partial p}{\partial \mathbf{n}_a} = \omega^2 \mathbf{u} \cdot \mathbf{n}_a + \tau \frac{c_a^2}{\rho_a} \frac{\partial s}{\partial \mathbf{n}_a} \quad \text{on} \quad \Gamma_a, \tag{13.5}$$

where ρ_a is the mass density of the acoustic fluid at equilibrium, c_a is the speed of sound, τ is the coefficient owing to the viscosity of the fluid that can depend on ω (the coefficient owing to thermal conduction is neglected) and where $s(\mathbf{x},\omega)$ is the acoustic source density.

13.1.2 Mean Computational structural–acoustic Model

The finite element method is used to approximate the boundary value problem defined by Equations (13.1)–(13.5). A finite element mesh of the structure Ω_s and of the internal acoustic fluid Ω_a is then introduced. Let $\mathbf{u}(\omega) = (\mathrm{u}_1(\omega), \ldots, \mathrm{u}_{n_s}(\omega))$ be the complex vector of the n_s degrees of freedom of the structure according to the finite element discretization of the displacement field $\mathbf{x} \mapsto \mathbf{u}(\mathbf{x}, \omega)$. Let $\mathfrak{p}(\omega) = (\mathfrak{p}_1(\omega), \ldots, \mathfrak{p}_{n_a}(\omega))$ be the complex vector of the n_a degrees of freedom of the acoustic fluid according to the finite element discretization of the pressure field $\mathbf{x} \mapsto p(\mathbf{x}, \omega)$. The finite element discretization of the boundary value problem yields the following mean computational structural–acoustic model:

$$\begin{pmatrix} [\mathscr{A}^s(\omega)] & [\mathscr{H}] \\ [\mathscr{H}]^T & [\mathscr{A}^a(\omega)] \end{pmatrix} \begin{pmatrix} \mathbf{u}(\omega) \\ \mathfrak{p}(\omega) \end{pmatrix} = \begin{pmatrix} \mathfrak{f}^s(\omega) \\ \mathfrak{f}^a(\omega) \end{pmatrix}, \tag{13.6}$$

where $[\mathscr{A}^s(\omega)]$ is the dynamical stiffness matrix of the damped structure in vacuo which is a symmetric $(n_s \times n_s)$ complex matrix such that

$$[\mathscr{A}^s(\omega)] = -\omega^2[\mathscr{M}^s] + i\omega[\mathscr{D}^s(\omega)] + [\mathscr{K}^s], \tag{13.7}$$

in which $[\mathscr{M}^s]$, $[\mathscr{D}^s(\omega)]$, and $[\mathscr{K}^s]$ are the mass, damping, and stiffness matrices of the structure that belong to $\mathbb{M}^+_{n_s}(\mathbb{R})$. In Equation (13.6), $[\mathscr{A}^a(\omega)]$ is the dynamical stiffness matrix of the dissipative acoustic fluid that is a symmetric $(n_a \times n_a)$ complex matrix such that

$$[\mathscr{A}^a(\omega)] = -\omega^2[\mathscr{M}^a] + i\omega[\mathscr{D}^a(\omega)] + [\mathscr{K}^a], \tag{13.8}$$

in which $[\mathscr{M}^a]$, $[\mathscr{D}^a(\omega)]$, and $[\mathscr{K}^a]$ are the mass, damping, and stiffness matrices of the acoustic cavity with fixed coupling interface. The matrix $[\mathscr{M}^a]$ belongs to $\mathbb{M}^+_{n_a}(\mathbb{R})$, and $[\mathscr{D}^a(\omega)]$ and $[\mathscr{K}^a]$ belong to $\mathbb{M}^{+0}_{n_a}(\mathbb{R})$ for which the rank is $n_a - 1$. The matrix $[\mathscr{H}]$ is the structural–acoustic coupling matrix that belongs to $\mathbb{M}_{n_s,n_a}(\mathbb{R})$.

13.1.3 Reduced Mean Computational structural–acoustic Model

The projection of the mean computational structural–acoustic Equation (13.6) on the structural modes in vacuo and on the acoustic modes of the acoustic cavity with fixed coupling interface yields the reduced mean computational structural–acoustic model. For a fixed value of the parameter $\mathbf{w}$ (a given nominal value), the structural modes in vacuo and the acoustic modes of the cavity with fixed coupling interface are calculated by solving the two generalized eigenvalue problems

$$[\mathscr{K}^s]\boldsymbol{\phi}^s = \lambda^s[\mathscr{M}^s]\boldsymbol{\phi}^s, \quad [\mathscr{K}^a]\boldsymbol{\phi}^a = \lambda^a[\mathscr{M}^a]\boldsymbol{\phi}^a, \tag{13.9}$$

in which $[\Phi^s]$ is the $(n_s \times n)$ real matrix whose columns are the n structural modes associated with the n first positive eigenvalues $0 < \lambda^s_1 \leq \lambda^s_2 \leq \cdots \leq \lambda^s_n$ (the n first structural eigenfrequencies are $\omega^s_\alpha = \sqrt{\lambda^s_\alpha}$). In Equation (13.9), $[\Phi^a]$ is the $(n_a \times m)$ real matrix whose columns are constituted (1) of the constant pressure mode associated with the zero eigenvalue $\lambda_1 = 0$ and (2) of the acoustic modes associated with the positive eigenvalues $0 < \lambda^a_2 \leq \lambda^a_3 \leq \cdots \leq \lambda^a_m$ (the $m - 1$ first acoustical eigenfrequencies are $\omega^a_\beta = \sqrt{\lambda^a_\beta}$). It should be noted that the constant pressure mode is kept in order to model the quasi-static variation of the internal fluid pressure induced by the deformation of the coupling interface. The eigenvectors satisfied the orthogonality conditions

$$[\Phi^s]^T[\mathcal{M}^s][\Phi^s] = [I_n], \quad [\Phi^s]^T[\mathcal{K}^s][\Phi^s] = [\lambda^s], \tag{13.10}$$

$$[\Phi^a]^T[\mathcal{M}^a][\Phi^a] = [I_m], \quad [\Phi^a]^T[\mathcal{K}^a][\Phi^a] = [\lambda^a], \tag{13.11}$$

in which $[\lambda^s]_{\alpha\alpha'} = \lambda^s_\alpha \delta_{\alpha\alpha'}$ and $[\lambda^a]_{\beta\beta'} = \lambda^a_\beta \delta_{\beta\beta'}$ are diagonal matrices, where $[I_n]$ and $[I_m]$ are the identity matrices of dimension n and m respectively and where $\delta_{\alpha\alpha} = 1$ and $\delta_{\alpha\alpha'} = 0$ for $\alpha \neq \alpha'$. Using Equations (13.10) and (13.11), the projection of Equation (13.6) yields the reduced mean matrix model of the structural–acoustic system (also called the reduced mean computational structural–acoustic model). For all ω in the frequency band $\mathcal{F}$, the structural displacement $\mathbf{u}^n(\omega)$ and the internal acoustic pressure $\mathbf{p}^m(\omega)$ are written as

$$\mathbf{u}^n(\omega) = [\Phi^s]\mathbf{q}^s(\omega), \quad \mathbf{p}^m(\omega) = [\Phi^a]\mathbf{q}^a(\omega), \tag{13.12}$$

in which the complex vectors $\mathbf{q}^s(\omega)$ and $\mathbf{q}^a(\omega)$ are such that

$$\begin{pmatrix} [\underline{A}^s(\omega)] & [\underline{H}] \\ [\underline{H}]^T & [\underline{A}^a(\omega)] \end{pmatrix} \begin{pmatrix} \mathbf{q}^s(\omega) \\ \mathbf{q}^a\omega) \end{pmatrix} = \begin{pmatrix} \mathbf{f}^s(\omega) \\ \mathbf{f}^a(\omega) \end{pmatrix}, \tag{13.13}$$

where $[\underline{H}] = [\Phi^s]^T[\mathcal{H}][\Phi^a]$ is in $\mathbb{M}_{n,m}(\mathbb{R})$ and $\mathbf{f}^s(\omega) = [\Phi^s]^T\mathfrak{f}^s(\omega)$ and $\mathbf{f}^a(\omega) = [\Phi^a]^T\mathfrak{f}^a(\omega)$ are complex vectors. The generalized dynamical stiffness matrix $[\underline{A}^s(\omega)]$ of the damped structure is the $(n_s \times n_s)$ complex matrix written as

$$[\underline{A}^s(\omega)] = -\omega^2[\underline{M}^s] + i\omega[\underline{D}^s(\omega)] + [\underline{K}^s], \tag{13.14}$$

in which $[\underline{M}^s] = [\Phi^s]^T[\mathcal{M}^s][\Phi^s] = [I_n]$ and $[\underline{K}^s] = [\Phi^s]^T[\mathcal{K}^s][\Phi^s] = [\lambda^s]$ are diagonal matrices belonging to $\mathbb{M}_n^+(\mathbb{R})$ and $[\underline{D}^s(\omega)] = [\Phi^s]^T[\mathcal{D}^s(\omega)][\Phi^s]$ is a full matrix belonging to $\mathbb{M}_n^+(\mathbb{R})$. The generalized dynamical stiffness matrix $[\underline{A}^a(\omega)]$ of the dissipative acoustic fluid is the $(n_a \times n_a)$ complex matrix written as

$$[\underline{A}^a(\omega)] = -\omega^2[\underline{M}^a] + i\omega[\underline{D}^a(\omega)] + [\underline{K}^a], \tag{13.15}$$

in which $[\underline{M}^a] = [\Phi^a]^T[\mathcal{M}^a][\Phi^a] = [I_m]$ is a diagonal matrix belonging to $\mathbb{M}_m^+(\mathbb{R})$ and $[\underline{D}^a(\omega)] = [\Phi^a]^T[\mathcal{D}^a(\omega)][\Phi^a] = \tau(\omega)[\lambda^a]$ and $[\underline{K}^a] = [\Phi^a]^T[\mathcal{K}^a][\Phi^a] = [\lambda^a]$ are diagonal matrices belonging to $\mathbb{M}_m^{+0}(\mathbb{R})$ of rank $m-1$.

13.2 Parametric Probabilistic Approach of the System Parameter Uncertainties

In order to explain why the parametric probabilistic approach cannot address model uncertainties, this approach is briefly presented next. The parameter $\mathbf{w}$ of the system represents data such as geometrical parameters, mass density, coefficients of the constitutive equation, and parameters describing the boundary conditions. As explained in the beginning of this chapter, the methodology of the parametric probabilistic approach of system parameter uncertainties consists of modeling the parameter $\mathbf{w}$ with values in an admissible set $\mathcal{W}$ by a random variable $\mathbf{W}$ for which the probability distribution $P_{\mathbf{W}}$ has a support that is $\mathcal{W}$. It should be noted that $P_{\mathbf{W}}$ must be explicitly constructed either by using the nonparametric statistics if a sufficient amount of data are available (which is seldom the case) or by using the parametric statistics coupled with the use of the maximum entropy principle when a small amount of data are available. Concerning the construction of the reduced mean computational model for the implementation of the parametric probabilistic approach, two methods can be used. The first one consists of solving the generalized eigenvalue problems defined by Equation (13.9) for each value of $\mathbf{w}$ in $\mathscr{W}$, and then $[\Phi^s]$ and $[\Phi^a]$ become functions of $\mathbf{w}$. The second one consists of keeping the matrices $[\Phi^s]$ and $[\Phi^a]$ calculated for the nominal value of the parameter $\mathbf{w}$ (see Equation (13.9)). For each value of $\mathbf{w}$ in $\mathscr{W}$, the projection of Equation (13.6) is then carried out using these bases $[\Phi^s]$ and $[\Phi^a]$ that are independent of $\mathbf{w}$. Both methods are perfectly correct and converge as n and m increase. Next, we choose the second one. Consequently, each matrix appearing in the reduced mean computational structural–acoustic model defined by Equations (13.13)–(13.15) depends on parameter $\mathbf{w}$ and can then be rewritten as $[B(\mathbf{w})]$ in which $\mathbf{w} \mapsto [B(\mathbf{w})]$ is a matrix-valued deterministic mapping on $\mathscr{W}$, which is perfectly defined and known. The parametric probabilistic approach then consists of substituting each deterministic matrix $[B(\mathbf{w})]$ by the random matrix $[\mathbf{B}_{\text{par}}] = [B(\mathbf{W})]$ for which the probability distribution $P_{[\mathbf{B}_{\text{par}}]}$ is deduced from the deterministic mapping $\mathbf{w} \mapsto [B(\mathbf{w})]$ and the probability distribution $P_{\mathbf{W}}$ of the random parameter $\mathbf{W}$. For all ω in the frequency band $\mathscr{F}$, the random structural displacement $\mathbf{U}^n_{\text{par}}(\omega)$ and the random internal acoustic pressure $\mathbf{P}^m_{\text{par}}(\omega)$ are then written as

$$\mathbf{U}^n_{\text{par}}(\omega) = [\Phi^s]\mathbf{Q}^s_{\text{par}}(\omega), \quad \mathbf{P}^m_{\text{par}}(\omega) = [\Phi^a]\mathbf{Q}^a_{\text{par}}(\omega), \tag{13.16}$$

in which the complex random vectors $\mathbf{Q}^s_{\text{par}}(\omega)$ and $\mathbf{Q}^a_{\text{par}}(\omega)$ are solution of the random matrix equation,

$$\begin{pmatrix} [\mathbf{A}^s_{\text{par}}(\omega)] & [\mathbf{H}_{\text{par}}] \\ [\mathbf{H}_{\text{par}}]^T & [\mathbf{A}^a_{\text{par}}(\omega)] \end{pmatrix} \begin{pmatrix} \mathbf{Q}^s_{\text{par}}(\omega) \\ \mathbf{Q}^a_{\text{par}}(\omega) \end{pmatrix} = \begin{pmatrix} \mathbf{f}^s(\omega) \\ \mathbf{f}^a(\omega) \end{pmatrix}, \tag{13.17}$$

where $[\mathbf{H}_{\text{par}}] = [\Phi^s]^T[\mathscr{H}(\mathbf{W})][\Phi^a]$ is a random matrix and where the random generalized dynamical stiffness matrix of the damped structure is then written as

$$[\mathbf{A}^s_{\text{par}}(\omega)] = -\omega^2[\mathbf{M}^s_{\text{par}}] + i\omega[\mathbf{D}^s_{\text{par}}(\omega)] + [\mathbf{K}^s_{\text{par}}], \tag{13.18}$$

in which $[\mathbf{M}^s_{\text{par}}] = [\Phi^s]^T[\mathscr{M}^s(\mathbf{W})][\Phi^s]$, $[\mathbf{D}^s_{\text{par}}(\omega)] = [\Phi^s]^T[\mathscr{D}^s(\omega, \mathbf{W})][\Phi^s]$, and $[\mathbf{K}^s_{\text{par}}] = [\Phi^s]^T[\mathscr{K}^s(\mathbf{W})][\Phi^s]$ are full random matrices with values in the set $\mathbb{M}^+_n(\mathbb{R})$. The random generalized dynamical stiffness matrix of the dissipative acoustic fluid is also written as

$$[\mathbf{A}^a_{\text{par}}(\omega)] = -\omega^2[\mathbf{M}^a_{\text{par}}] + i\omega[\mathbf{D}^a_{\text{par}}(\omega)] + [\mathbf{K}^a_{\text{par}}], \tag{13.19}$$

in which $[\mathbf{M}^a_{\text{par}}] = [\Phi^a]^T[\mathscr{M}^a(\mathbf{W})][\Phi^a]$ is a full random matrix with values in $\mathbb{M}^+_m(\mathbb{R})$ and $[\mathbf{D}^a_{\text{par}}(\omega)] = [\Phi^a]^T[\mathscr{D}^a(\omega, \mathbf{W})][\Phi^a]$ and $[\mathbf{K}^a_{\text{par}}] = [\Phi^a]^T[\mathscr{K}^a(\mathbf{W})][\Phi^a]$ are random matrices with values in the set $\mathbb{M}^{+0}_m(\mathbb{R})$ (note that the rank of these two random matrices is fixed and is equal to $m - 1$).

In order to finish the explanation concerning the limitation of the parametric probabilistic approach, let us consider the random matrix $[\mathbf{B}_{\text{par}}] = [B(\mathbf{W})]$ corresponding, for instance, to the generalized stiffness matrix of the structure, that is, $[B(\mathbf{w})] = [\Phi^s]^T[\mathscr{K}^s(\mathbf{w})][\Phi^s]$. In this case, $\mathbf{w} \mapsto [B(\mathbf{w})]$ is a mapping from $\mathscr{W}$ into $\mathbb{M}^+_n(\mathbb{R})$. Therefore, when $\mathbf{w}$ runs through the set $\mathscr{W}$, the family $\{[B(\mathbf{w})], \mathbf{w} \in \mathscr{W}\}$ does

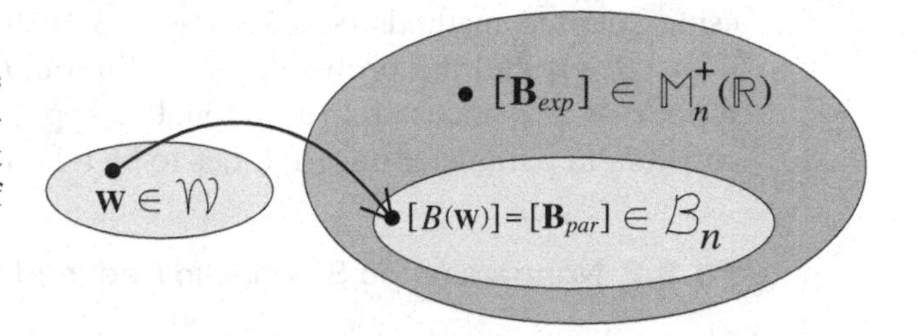

Figure 13.5. Admissible set $\mathscr{W}$ of the parameter $\mathbf{w}$, subset $\mathscr{B}_n$ of all the possible values of $[\mathbf{B}_{\text{par}}] = [B(\mathbf{W})]$, and set $\mathbb{M}_n^+(\mathbb{R})$ of all the possible values of $[\mathbf{B}_{\text{exp}}]$.

not span all the set $\mathbb{M}_n^+(\mathbb{R})$ but spans a subset $\mathscr{B}_n$ of $\mathbb{M}_n^+(\mathbb{R})$ (see Figure 13.5). Because the support of the probability distribution $P_{\mathbf{W}}$ is $\mathscr{W}$, the support of the probability distribution $P_{[\mathbf{B}_{\text{par}}]}$ of the random matrix $[\mathbf{B}_{\text{par}}] = [B(\mathbf{W})]$ is $\mathscr{B}_n \subset \mathbb{M}_n^+(\mathbb{R})$. Clearly, if the corresponding experimental matrix $[\mathbf{B}_{\text{exp}}]$ belongs to the subset $\mathscr{B}_n$, this means that there are only system parameter uncertainties, and the probabilistic approach is the most efficient approach. In opposite, if $[\mathbf{B}_{\text{exp}}]$ does not belong to the subset $\mathscr{B}_n$ but will necessarily belong to $\mathbb{M}_n^+(\mathbb{R})$, this means that there are model uncertainties. In this case, the distance between $[\mathbf{B}_{\text{exp}}]$ and the subset $\mathscr{B}_n$ cannot be reduced, and the parametric probabilistic approach cannot clearly address model uncertainties. The nonparametric probabilistic approach presented in the next section will be a way to address both model uncertainties and system parameter uncertainties.

13.3 Nonparametric Probabilistic Approach of Uncertainties

In this section, we introduce the concept and the methodology of the nonparametric probabilistic approach of both model uncertainties and system parameter uncertainties.

13.3.1 Concept of the Nonparametric Probabilistic Approach

The concept of the nonparametric probabilistic approach of uncertainties consists, for each matrix $[\underline{B}]$ in the reduced mean computational structural–acoustic model defined by Equations (13.13)–(13.15), of directly constructing the random matrix $[\mathbf{B}]$ with a probability distribution $P_{[\mathbf{B}]}$ such that its support is the largest set of all the possible values of $[B]$. For instance, if $[\underline{B}] = [\underline{K}^s]$ is the generalized stiffness matrix of the structure that must belong to $\mathbb{M}_n^+(\mathbb{R})$, then the random matrix $[\mathbf{B}]$ is constructed to be with values in $\mathbb{M}_n^+(\mathbb{R})$. This means that the support of the probability distribution $P_{[\mathbf{B}]}$ will be the set $\mathbb{M}_n^+(\mathbb{R})$, which contains the subset $\mathscr{B}_n$ (see Figure 13.5). It should be noted that $P_{[\mathbf{B}]} = P_{[\mathbf{B}_{\text{par}}]}$ could be chosen because the support $\mathscr{B}_n$ of $P_{[\mathbf{B}_{\text{par}}]}$ is included in $\mathbb{M}_n^+(\mathbb{R})$. In this case, one would have $[\mathbf{B}] = [\mathbf{B}_{\text{par}}]$ which proves that the nonparametric probabilistic approach has the capability to take into account system parameter uncertainties. In addition, for the nonparametric approach, because the support of $P_{[\mathbf{B}]}$ is $\mathbb{M}_n^+(\mathbb{R})$ with $\mathscr{B}_n \subset \mathbb{M}_n^+(\mathbb{R})$, the nonparametric probabilistic approach allows a larger class of random matrices to be constructed and consequently has, a priori, the capability to take into account the model uncertainties. Nevertheless, the probability distribution $P_{[\mathbf{B}]}$ of random matrix $[\mathbf{B}]$ whose support is $\mathbb{M}_n^+(\mathbb{R})$ cannot arbitrarily be chosen but must be constructed using adapted methods (see Section 13.3.2). In such a construction, the mapping $\mathbf{w} \mapsto [B(\mathbf{w})]$ and the probability distribution $P_{\mathbf{W}}$ of the random variable $\mathbf{W}$ with support $\mathscr{W}$ are not used. In addition, this approach allows both model uncertainties and system parameter uncertainties to be taken into account.

13.3.2 Nonparametric Stochastic Reduced Model

The application of the concept introduced in Section 13.3.1 leads us to the following stochastic reduced computational structural–acoustic model, allowing both model uncertainties and system parameter uncertainties to be taken into account. For all ω in $\mathscr{F}$, the random vector $\mathbf{U}^n(\omega)$ of the structural displacements and the random vector $\mathbf{P}^m(\omega)$ of the acoustic pressures are given by

$$\mathbf{U}^n(\omega) = [\Phi^s]\mathbf{Q}^s(\omega), \quad \mathbf{P}^m(\omega) = [\Phi^a]\mathbf{Q}^a(\omega), \tag{13.20}$$

in which the random vectors $\mathbf{Q}^s(\omega)$ and $\mathbf{Q}^a(\omega)$ are such that

$$\begin{pmatrix} [\mathbf{A}^s(\omega)] & [\mathbf{H}] \\ [\mathbf{H}]^T & [\mathbf{A}^a(\omega)] \end{pmatrix} \begin{pmatrix} \mathbf{Q}^s(\omega) \\ \mathbf{Q}^a(\omega) \end{pmatrix} = \begin{pmatrix} \mathbf{f}^s(\omega) \\ \mathbf{f}^a(\omega) \end{pmatrix}, \tag{13.21}$$

where $[\mathbf{H}]$ is a random matrix with values in the set $\mathbb{M}_{n,m}(\mathbb{R})$. The random generalized dynamical stiffness matrix of the damped structure is written as

$$[\mathbf{A}^s(\omega)] = -\omega^2[\mathbf{M}^s] + i\omega[\mathbf{D}^s(\omega)] + [\mathbf{K}^s], \tag{13.22}$$

in which $[\mathbf{M}^s]$, $[\mathbf{D}^s(\omega)]$, and $[\mathbf{K}^s]$ are random matrices with values in $\mathbb{M}_n^+(\mathbb{R})$. The random generalized dynamical stiffness matrix of the dissipative acoustic fluid is written as

$$[\mathbf{A}^a(\omega)] = -\omega^2[\mathbf{M}^a] + i\omega[\mathbf{D}^a(\omega)] + [\mathbf{K}^a], \tag{13.23}$$

in which $[\mathbf{M}^a]$ is a random matrix with values in $\mathbb{M}_m^+(\mathbb{R})$ and $[\mathbf{D}^a(\omega)]$ and $[\mathbf{K}^a]$ are random matrices with values in $\mathbb{M}_m^{+0}(\mathbb{R})$. The rank of the two random matrices $[\mathbf{D}^a(\omega)]$ and $[\mathbf{K}^a]$ is fixed and is equal to $m - 1$.

The development of the nonparametric probabilistic approach then requires precise definition of the available information for the above-given random matrices in order to construct their probability distribution using the maximum entropy principle, which allows to maximize uncertainties. For all ω fixed in $\mathscr{F}$, this available information is defined as follows:

(i) The random matrices $[\mathbf{H}]$, $[\mathbf{M}^s]$, $[\mathbf{D}^s(\omega)]$, $[\mathbf{K}^s]$, $[\mathbf{M}^a]$, $[\mathbf{D}^a(\omega)]$, and $[\mathbf{K}^a]$ are defined on a probability space $(\Theta, \mathscr{T}, \mathscr{P})$, and for $\mathscr{P}$-almost all θ in Θ (almost surely property), one has

$$[\mathbf{H}(\theta)] \in \mathbb{M}_{n,m}(\mathbb{R}), \tag{13.24}$$

$$[\mathbf{M}^s(\theta)],\ [\mathbf{D}^s(\omega,\theta)],\ [\mathbf{K}^s(\theta)] \in \mathbb{M}_n^+(\mathbb{R}), \tag{13.25}$$

$$[\mathbf{M}^a(\theta)] \in \mathbb{M}_m^+(\mathbb{R}); \quad [\mathbf{D}^a(\omega,\theta)],\ [\mathbf{K}^a(\theta)] \in \mathbb{M}_m^{+0}(\mathbb{R}). \tag{13.26}$$

(ii) The mean value of these random matrices must be equal to the matrices of the reduced mean computational structural–acoustic model defined in Section 13.1.3,

$$E\{[\mathbf{H}]\} = [\underline{H}], \tag{13.27}$$

$$E\{[\mathbf{M}^s]\} = [\underline{M}^s], \quad E\{[\mathbf{D}^s(\omega)]\} = [\underline{D}^s(\omega)], \quad E\{[\mathbf{K}^s]\} = [\underline{K}^s], \tag{13.28}$$

$$E\{[\mathbf{M}^a]\} = [\underline{M}^a], \quad E\{[\mathbf{D}^a(\omega)]\} = [\underline{D}^a(\omega)], \quad E\{[\mathbf{K}^a]\} = [\underline{K}^a], \tag{13.29}$$

in which $E\{[\mathbf{B}]\}$ denotes the mathematical expectation of random matrix $[\mathbf{B}]$, which is such that $E\{[\mathbf{B}]\} = \int_\Theta [\mathbf{B}(\theta)]\, d\mathscr{P}(\theta)$.

(iii) For all ω fixed in $\mathscr{F}$, the probability distributions of the random matrices $[\mathbf{H}]$, $[\mathbf{M}^s]$, $[\mathbf{D}^s(\omega)]$, $[\mathbf{K}^s]$, $[\mathbf{M}^a]$, $[\mathbf{D}^a(\omega)]$, and $[\mathbf{K}^a]$ must be such that the random solution $(\mathbf{U}^n(\omega), \mathbf{P}^m(\omega))$ of Equations (13.20)–(13.23) is a second-order random vector,

$$E\{\|\mathbf{U}^n(\omega)\|^2\} < +\infty, \quad E\{\|\mathbf{P}^m(\omega)\|^2\} < +\infty. \tag{13.30}$$

13.3.3 Normalization of Random Matrices

In the construction of the nonparametric probabilistic approach, the random matrices introduced in Section 13.3.2 are normalized as follows.

13.3.3.1 Rectangular Random Matrix

Let us assume that $n \geq m$. The elastic modes of the structure and the acoustic modes of the internal acoustic cavity are selected in order for the matrix $[\underline{H}]$ that belongs to $\mathbb{M}_{n,m}(\mathbb{R})$ to be such that $[\underline{H}]\mathbf{q}^a = 0$ implies $\mathbf{q}^a = 0$. It should be noted that if $m \geq n$, the following construction must be applied to $[\underline{H}]^T$ instead of $[\underline{H}]$. Then, using the singular value decomposition of the rectangular matrix $[\underline{H}]$, one can write $[\underline{H}] = [\underline{C}][\underline{T}]$ in which the rectangular matrix $[\underline{C}]$ in $\mathbb{M}_{n,m}(\mathbb{R})$ is such that $[\underline{C}]^T[\underline{C}] = [I_m]$ and the symmetric square matrix $[\underline{T}]$ belongs to $\mathbb{M}_m^+(\mathbb{R})$ and can then be written (using the Cholesky decomposition) as $[\underline{T}] = [\underline{L}_T]^T[\underline{L}_T]$, where $[\underline{L}_T]$ is an upper triangular matrix in $\mathbb{M}_m(\mathbb{R})$. Because $[\underline{T}]$ belongs to $\mathbb{M}_m^+(\mathbb{R})$, we introduce the random matrix $[\mathbf{T}]$ with values in $\mathbb{M}_m^+(\mathbb{R})$ such that $E\{[\mathbf{T}]\} = [\underline{T}]$. Using Equation (13.27), the random matrix $[\mathbf{H}]$ with values in $\mathbb{M}_{n,m}(\mathbb{R})$ can then be written as

$$[\mathbf{H}] = [\underline{C}][\mathbf{T}], \quad [\mathbf{T}] = [\underline{L}_T]^T[\mathbf{G}_H][\underline{L}_T], \tag{13.31}$$

with $[\mathbf{G}_H]$ a random matrix with values in $\mathbb{M}_m^+(\mathbb{R})$ such that $E\{[\mathbf{G}_H]\} = [I_m]$.

13.3.3.2 Positive and Semi-Positive Definite Random Matrices

For all ω in $\mathscr{F}$, it can be seen that the positive definite matrices $[\underline{M}^s]$, $[\underline{K}^s]$, $[\underline{D}^s(\omega)]$, $[\underline{M}^a]$ and the semi-positive definite matrices $[\underline{K}^a]$, $[\underline{D}^a(\omega)]$ can be written as

$$[\underline{M}^s] = [L_{M^s}]^T[L_{M^s}], \quad [\underline{K}^s] = [L_{K^s}]^T[L_{K^s}], \tag{13.32}$$

$$[\underline{D}^s(\omega)] = [L_{D^s}(\omega)]^T[L_{D^s}(\omega)], \quad [\underline{M}^a] = [L_{M^a}]^T[L_{M^a}], \tag{13.33}$$

$$[\underline{K}^a] = [L_{K^a}]^T[L_{K^a}], \quad [\underline{D}^a(\omega)] = [L_{D^a}(\omega)]^T[L_{D^a}(\omega)], \tag{13.34}$$

in which $[L_{M^s}]$ is the unity matrix $[I_n]$; $[L_{K^s}]$ is the diagonal matrix $[\lambda^s]^{1/2}$ in $\mathbb{M}_n(\mathbb{R})$; $[L_{D^s}(\omega)]$ is an upper triangular matrix in $\mathbb{M}_n(\mathbb{R})$ (Cholesky's decomposition); $[L_{M^a}]$ is the unity matrix $[I_m]$; $[L_{K^a}]$ is a rectangular matrix in $\mathbb{M}_{m-1,m}(\mathbb{R})$ such that for all $\beta = 1, \ldots, m-1$, one has, for $\beta' = 1$, $[L_{K^a}]_{\beta 1} = 0$, and for $\beta' = 2, \ldots, m$, $[L_{K^a}]_{\beta\beta'} = \sqrt{\lambda^a_{\beta'}}\, \delta_{(\beta+1)\beta'}$; and finally, $[L_{D^a}(\omega)] = \sqrt{\tau(\omega)}\,[L_{K^a}]$ is a rectangular matrix in $\mathbb{M}_{m-1,m}(\mathbb{R})$. Using Equations (13.25) and (13.28), the random matrices $[\mathbf{M}^s]$, $[\mathbf{K}^s]$, and $[\mathbf{D}^s(\omega)]$ can then be written as

$$[\mathbf{M}^s] = [L_{M^s}]^T[\mathbf{G}_{M^s}][L_{M^s}], \quad [\mathbf{K}^s] = [L_{K^s}]^T[\mathbf{G}_{K^s}][L_{K^s}], \tag{13.35}$$

$$[\mathbf{D}^s(\omega)] = [L_{D^s}(\omega)]^T[\mathbf{G}_{D^s}][L_{D^s}(\omega)], \tag{13.36}$$

in which $[\mathbf{G}_{M^s}]$, $[\mathbf{G}_{K^s}]$, and $[\mathbf{G}_{D^s}]$ are random matrices with values in $\mathbb{M}^+_n(\mathbb{R})$, independent of ω and such that $E\{[\mathbf{G}_{M^s}]\} = E\{[\mathbf{G}_{K^s}]\} = E\{[\mathbf{G}_{D^s}]\} = [I_n]$. Similarly, using Equations (13.26) and (13.29), the random matrices $[\mathbf{M}^a]$, $[\mathbf{K}^a]$, and $[\mathbf{D}^a(\omega)]$ can then be written as

$$[\mathbf{M}^a] = [L_{M^a}]^T[\mathbf{G}_{M^a}][L_{M^a}], \quad [\mathbf{K}^a] = [L_{K^a}]^T[\mathbf{G}_{K^a}][L_{K^a}], \tag{13.37}$$

$$[\mathbf{D}^a(\omega)] = [L_{D^a}(\omega)]^T[\mathbf{G}_{D^a}][L_{D^a}(\omega)], \tag{13.38}$$

in which $[\mathbf{G}_{M^a}]$ is a random matrix with values in $\mathbb{M}^+_m(\mathbb{R})$, independent of ω and such that $E\{[\mathbf{G}_{M^a}]\} = [I_m]$, and $[\mathbf{G}_{K^a}]$ and $[\mathbf{G}_{D^a}]$ are random matrices with values in $\mathbb{M}^+_{m-1}(\mathbb{R})$, independent of ω and such that $E\{[\mathbf{G}_{K^a}]\} = E\{[\mathbf{G}_{D^a}]\} = [I_{m-1}]$.

Taking into account Equations (13.31)–(13.38), the stochastic reduced computational model is completely defined as soon as the joint probability distribution of the random matrices $[\mathbf{G}_H]$, $[\mathbf{G}_{M^s}]$, $[\mathbf{G}_{D^s}]$, $[\mathbf{G}_{K^s}]$, $[\mathbf{G}_{M^a}]$, $[\mathbf{G}_{D^a}]$, and $[\mathbf{G}_{K^a}]$ are constructed. Let $[\mathbf{G}]$ be any one of these random matrices with values, in $\mathbb{M}^+_\mu(\mathbb{R})$ for which μ is equal to n, m, or $m-1$. We then have $E\{[\mathbf{G}]\} = [I_\mu]$, and it has been proven that $E\{\|[\mathbf{G}]^{-1}\|_F^2\}$ must be finite in order that Equation (13.30) be verified. This last condition is necessary but is not sufficient to get a second-order solution (see Equation (13.30)) when dimension n_s and n_a go to infinity (that is to say, for the continuous structural–acoustic system in infinite dimension) and, then, when n and m go to infinity. To obtain such a fundamental property, it is necessary that for all $\mu \geq 1$, one has $E\{\|[\mathbf{G}]^{-1}\|^2\} \leq C < +\infty$ in which C is a positive finite constant independent of μ. Such a construction is performed in the next section using the maximum entropy principle with the constraints defined by the above-mentioned available information.

Finally, because the construction proposed depends on the reduced-order model n and m, a mean square convergence analysis of the stochastic computational model must systematically be performed in studying the function $(n, m) \mapsto E\{\int_{\mathcal{F}}(\|\mathbf{U}^n(\omega)\|^2 + \|\mathbf{P}^m(\omega)\|^2)\,\mathrm{d}\omega\}$. With the probabilistic model constructed, it is mathematically proven that convergence can always be obtained for finite and infinite dimensions.

13.4 Random Matrix Theory

The objective of this section is to construct an ensemble SG^+ of random matrices $[\mathbf{G}]$ for which the properties introduced in Section 13.3.3 are satisfied. This means that such a random matrix $[\mathbf{G}]$ is defined on the probability space $(\Theta, \mathcal{T}, \mathcal{P})$, is with values in $\mathbb{M}_\mu^+(\mathbb{R})$, and must be such that $E\{[\mathbf{G}]\} = [I_\mu]$ and $E\{\|[\mathbf{G}]^{-1}\|^2\} \leq C < +\infty$ with C independent of μ.

13.4.1 Volume Element and Probability Density Function on $\mathbb{M}_\mu^+(\mathbb{R})$

Let $[\mathbf{G}]$ be a random matrix with values in $\mathbb{M}_\mu^+(\mathbb{R}) \subset \mathbb{M}_\mu^S(\mathbb{R})$ whose probability distribution $P_{[\mathbf{G}]} = p_{[\mathbf{G}]}([G])\,\tilde{d}G$ is defined by a probability density function $[G] \mapsto p_{[\mathbf{G}]}([G])$ from $\mathbb{M}_\mu^+(\mathbb{R})$ into $\mathbb{R}^+ = [0, +\infty[$ with respect to the volume element $\tilde{d}G$ on $\mathbb{M}_\mu^S(\mathbb{R})$ defined below. This probability density function can then verify the normalization condition,

$$\int_{\mathbb{M}_\mu^+(\mathbb{R})} p_{[\mathbf{G}]}([G])\,\tilde{d}G = 1, \tag{13.39}$$

In this section, the volume element $\tilde{d}G$ on $\mathbb{M}_\mu^S(\mathbb{R})$ is clearly defined.

13.4.1.1 Volume Element on Euclidean Space $\mathbb{R}^\mu$

Let $\{\mathbf{e}_1, \ldots, \mathbf{e}_\mu\}$ be the orthonormal basis of $\mathbb{R}^\mu$ such that $\mathbf{e}_j = (0, \ldots, 1, \ldots, 0)$ is the null vector with 1 in position j. Consequently, one has $\langle \mathbf{e}_j, \mathbf{e}_k \rangle = \delta_{jk}$ in which $\delta_{jk} = 0$ if $j \neq k$ and $\delta_{jj} = 1$. Any vector $\mathbf{x} = (x_1, \ldots, x_\mu)$ in $\mathbb{R}^\mu$ can then be written as $\mathbf{x} = \sum_{j=1}^\mu x_j \mathbf{e}_j$. This Euclidean structure on $\mathbb{R}^\mu$ defines the volume element $d\mathbf{x}$ on $\mathbb{R}^\mu$ such that $d\mathbf{x} = \prod_{j=1}^\mu dx_j$.

13.4.1.2 Volume Element on Euclidean Space $\mathbb{M}_\mu(\mathbb{R})$

Similarly, let $\{[b_{jk}]\}_{jk}$ be the orthonormal basis of $\mathbb{M}_\mu(\mathbb{R})$ such that $[b_{jk}] = \mathbf{e}_j \mathbf{e}_k^T$. Consequently, one has $\langle\!\langle [b_{jk}], [b_{j'k'}] \rangle\!\rangle = \delta_{jj'}\delta_{kk'}$. Any matrix $[G]$ in $\mathbb{M}_\mu(\mathbb{R})$ can be written as $[G] = \sum_{j,k=1}^\mu G_{jk}[b_{jk}]$ in which $G_{jk} = [G]_{jk}$. This Euclidean structure on $\mathbb{M}_\mu(\mathbb{R})$ defines the volume element dG on $\mathbb{M}_\mu(\mathbb{R})$ such that $dG = \prod_{j,k=1}^\mu dG_{jk}$.

13.4.1.3 Volume Element on Euclidean Space $\mathbb{M}_\mu^S(\mathbb{R})$

Let $\{[\tilde{b}_{jk}], 1 \leq j \leq k \leq \mu\}$ be the orthonormal basis of $\mathbb{M}_\mu^S(\mathbb{R})$ such that $[\tilde{b}_{jj}] = \mathbf{e}_j \mathbf{e}_j^T$ and $[\tilde{b}_{jk}] = 2^{-1/2}(\mathbf{e}_j \mathbf{e}_k^T + \mathbf{e}_k \mathbf{e}_j^T)$ if $j < k$. Consequently, one has $\langle\!\langle [\tilde{b}_{jk}], [\tilde{b}_{j'k'}] \rangle\!\rangle = \delta_{jj'}\delta_{kk'}$ for $j \leq k$ and $j' \leq k'$. Any symmetric matrix $[G]$ in $\mathbb{M}_n^S(\mathbb{R})$ can be written as $[G] = \sum_{1 \leq j \leq k \leq \mu} \widetilde{G}_{jk}[\tilde{b}_{jk}]$ in which $\widetilde{G}_{jj} = G_{jj}$ and $\widetilde{G}_{jk} = \sqrt{2}\,G_{jk}$ if $j < k$. This Euclidean structure on $\mathbb{M}_n^S(\mathbb{R})$ defines the volume element $\tilde{d}G$ on $\mathbb{M}_\mu^S(\mathbb{R})$ such that $\tilde{d}G = \prod_{1 \leq j \leq k \leq \mu} d\widetilde{G}_{jk}$. We then deduce that

$$\tilde{d}G = 2^{\mu(\mu-1)/4} \prod_{1 \leq j \leq k \leq \mu} dG_{jk}. \tag{13.40}$$

13.4.2 Why the Gaussian Orthogonal Ensemble Cannot Be Used in the Low- and Medium-Frequency Ranges

As it as been explained in the previous chapters, the use of the Gaussian orthogonal ensemble (GOE) is perfectly adapted to the description of the statistical fluctuations in the high-frequency range. In the low-frequency range (presence of global structural and acoustical modes) and for a large part of the medium-frequency range (mixing of global structural modes with local structural modes) the use of GOE is not well adapted as explained next and in the bibliographical comments at the end of this chapter. If the GOE is chosen to construct the random matrix $[\mathbf{G}]$ (denoted by $[\mathbf{G}^{\mathrm{GOE}}]$), then the random matrix $[\mathbf{G}^{\mathrm{GOE}}]$ should be written as $[\mathbf{G}^{\mathrm{GOE}}] = [I_\mu] + [\mathbf{X}^{\mathrm{GOE}}]$ in which $[\mathbf{X}^{\mathrm{GOE}}]$ would belong to the GOE and would consequently be a second-order centered random matrix with values in $\mathbb{M}_\mu^S(\mathbb{R})$ such that $E\{[\mathbf{X}^{\mathrm{GOE}}]\} = [0]$ and $E\{\|[\mathbf{X}^{\mathrm{GOE}}]\|_F^2\} < +\infty$ and for which the probability density function with respect to the volume element $\tilde{d}[X] = 2^{\mu(\mu-1)/4} \prod_{1\leq i\leq j\leq\mu} d[X]_{ij}$ is written as

$$p_{[\mathbf{X}^{\mathrm{GOE}}]}([X]) = C_\mu \times \exp\left\{-\frac{(\mu+1)}{4\delta^2}\,\mathrm{tr}\{[X]^2\}\right\}, \tag{13.41}$$

where C_μ is the constant of normalization, which can easily be calculated, and δ is the coefficient of variation of the random matrix $[\mathbf{G}^{\mathrm{GOE}}]$, which is such that $\delta^2 = \mu^{-1} E\{\|[\mathbf{G}^{\mathrm{GOE}}] - [I_\mu]\|_F^2\}$ because $\|[I_\mu]\|_F^2 = \mu$. Equation (13.39) shows that the real-valued random variables $\{[\mathbf{X}^{\mathrm{GOE}}]_{jk},\, j \leq k\}$ are statistically independent, second order, centered, and Gaussian. It can be seen that $[\mathbf{G}^{\mathrm{GOE}}]$ is with values in $\mathbb{M}_\mu^S(\mathbb{R})$ but not in $\mathbb{M}_\mu^+(\mathbb{R})$, and in addition, $E\{\|[\mathbf{G}^{\mathrm{GOE}}]^{-1}\|^2\} = +\infty$. This last result can easily be understood in dimension $\mu = 1$ because $\mathbf{G}^{\mathrm{GOE}} = 1 + \mathbf{X}^{\mathrm{GOE}}$ and then

$$E\left\{\frac{1}{(\mathbf{G}^{\mathrm{GOE}})^2}\right\} = \frac{1}{\sqrt{2\pi}\,\delta}\int_{\mathbb{R}} \frac{1}{(1+x)^2}\exp\left(-\frac{x^2}{2\delta^2}\right) dx = +\infty. \tag{13.42}$$

Consequently, $[\mathbf{G}^{\mathrm{GOE}}]$ does not belong to SG^+, and such a construction would not be acceptable for the low- and medium-frequency ranges that are driven by a boundary value problem for which the boundary conditions play an important role. Consider for instance the static problem for the structure associated with Equation (13.21). We then have to find the random vector $\mathbf{Q}^s$ such that $[\mathbf{K}^s]\mathbf{Q}^s = \mathbf{f}^s$ in which $[\mathbf{K}^s] = [L_{K^s}]^T[\mathbf{G}^{\mathrm{GOE}}][L_{K^s}]$ (see Equation (13.35)). We then have $E\{\|\mathbf{Q}^s\|^2\} = +\infty$. As explained at the end of Section 13.3.3, the random matrices $[\mathbf{G}_H]$, $[\mathbf{G}_{M^s}]$, $[\mathbf{G}_{D^s}]$, $[\mathbf{G}_{K^s}]$, $[\mathbf{G}_{M^a}]$, $[\mathbf{G}_{D^a}]$, and $[\mathbf{G}_{K^a}]$ must belong to ensemble SG^+ in order that Equation (13.30) be verified for all ω in $\mathcal{F}$. This means that the GOE cannot be used and that a new ensemble SG^+ must be constructed (see the beginning of Section 13.4).

13.4.3 Construction of the Ensemble SG^+

13.4.3.1 Definition of the Ensemble SG^+

The ensemble SG^+ is defined as the set of all the random matrices $[\mathbf{G}]$, defined on a probability space $(\Theta, \mathcal{T}, \mathcal{P})$, with values in $\mathbb{M}_\mu^+(\mathbb{R})$, whose probability distribution $P_{[\mathbf{G}]} = p_{[\mathbf{G}]}([G])\tilde{d}G$ is such that $E\{[\mathbf{G}]\} = [I_\mu]$ and $E\{\log(\det[\mathbf{G}])\} = v$ with $|v| < +\infty$. The available information can then be rewritten as

$$h_0(p_{[\mathbf{G}]}) = 0, \quad [h_1(p_{[\mathbf{G}]})] = [0], \quad h_2(p_{[\mathbf{G}]}) = 0, \tag{13.43}$$

in which the functions h_0 (normalization condition), $[h_1]$ (mean value), and h_2 (decreasing condition in the neighborhood of the origin) are defined by

$$h_0(p_{[\mathbf{G}]}) = \int_{\mathbb{M}_\mu^+(\mathbb{R})} p_{[\mathbf{G}]}([G])\, \tilde{d}G - 1, \tag{13.44}$$

$$[h_1(p_{[\mathbf{G}]})] = \int_{\mathbb{M}_\mu^+(\mathbb{R})} [G] p_{[\mathbf{G}]}([G])\, \tilde{d}G - [I_\mu], \tag{13.45}$$

$$h_2(p_{[\mathbf{G}]}) = \int_{\mathbb{M}_\mu^+(\mathbb{R})} \log(\det[G]) p_{[\mathbf{G}]}([G])\, \tilde{d}G - v, \quad |v| < +\infty. \tag{13.46}$$

13.4.3.2 Effective Construction by Using the Maximum Entropy Principle

The function $[G] \mapsto p_{[\mathbf{G}]}([G])$ from $\mathbb{M}_n^+(\mathbb{R})$ into $\mathbb{R}^+$, which satisfies Equations (13.43)–(13.46), is constructed by using the maximum entropy principle. Let $\mathscr{C}_{\text{ad}}$ be the admissible set of all the functions defined by

$$\mathscr{C}_{\text{ad}} = \{p_{[\mathbf{G}]} \colon \mathbb{M}_\mu^+(\mathbb{R}) \to \mathbb{R}^+;\, h_0(p_{[\mathbf{G}]}) = 0, [h_1(p_{[\mathbf{G}]})] = [0], h_2(p_{[\mathbf{G}]}) = 0\}. \tag{13.47}$$

Let $S(p_{[\mathbf{G}]})$ be the entropy of the probability density function $p_{[\mathbf{G}]}$ defined by Shannon in the context of the information theory as

$$S(p_{[\mathbf{G}]}) = -\int_{\mathbb{M}_\mu^+(\mathbb{R})} p_{[\mathbf{G}]}([G]) \log\big(p_{[\mathbf{G}]}([G])\big)\, \tilde{d}G, \tag{13.48}$$

which is a measure of uncertainties. The maximum entropy principle consists of identifying $p_{[\mathbf{G}]}$ as the probability density function of the random matrix $[\mathbf{G}]$, which maximizes the uncertainties under the constraints defined by Equations (13.43)–(13.46). We then have to solve the following optimization problem:

$$p_{[\mathbf{G}]} = \arg\{\max_{p_{[\mathbf{G}]} \in \mathscr{C}_{\text{ad}}} S(p_{[\mathbf{G}]})\}. \tag{13.49}$$

In order to solve this optimization problem, the method of Lagrange's multipliers is used. Let $(\lambda_0 - 1) \in \mathbb{R}$, $[\lambda_1] \in \mathbb{M}_\mu^+(\mathbb{R})$, and $(1 - \lambda_2) \in \mathbb{R}$ be the Lagrange multipliers corresponding to the constraints defined $h_0(p_{[\mathbf{G}]}) = 0$, $[h_1(p_{[\mathbf{G}]})] = [0]$, and $h_2(p_{[\mathbf{G}]}) = 0$. We then introduce the Lagrangian

$$\mathscr{L}(p_{[\mathbf{G}]}) = S(p_{[\mathbf{G}]}) - (\lambda_0 - 1) h_0(p_{[\mathbf{G}]}) - \sum_{j,k=1}^{\mu} [\lambda_1]_{jk} [h_1(p_{[\mathbf{G}]})]_{jk} - (1 - \lambda_2) h_2(p_{[\mathbf{G}]}). \tag{13.50}$$

The calculus of variations applied to Lagrangian $\mathscr{L}$ allows the maximum of $\mathscr{L}$ to be calculated and yields

$$p_{[\mathbf{G}]}([G]) = \mathbb{1}_{\mathbb{M}_\mu^+(\mathbb{R})}([G]) \times c_0 \times (\det[G])^{\lambda_2 - 1} \times e^{-\operatorname{tr}\{[\lambda_1][G]\}}, \tag{13.51}$$

in which $c_0 = \exp\{-\lambda_0\}$ and $\mathbb{1}_{\mathbb{M}_\mu^+(\mathbb{R})}([G])$ is equal to 1 if $[G] \in \mathbb{M}_\mu^+(\mathbb{R})$ and is equal to zero if $[G] \notin \mathbb{M}_\mu^+(\mathbb{R})$. From Equation (13.51), it can easily be deduced that the random matrix $[\mathbf{G}]$ is a second-order random variable, that is to say, $E\{\|[\mathbf{G}]\|_F^2\} < +\infty$. The constant of normalization c_0 and the Lagrange multipliers $[\lambda_1]$ and λ_2 must be calculated using Equations (13.43)–(13.46). However, the constraint defined by the third equation in (13.43) depends on an arbitrary real parameter v that has no physical meaning. We will then proceed to a change of parameter in replacing v by the coefficient of variation $\delta > 0$ of random matrix $[\mathbf{G}]$, which is defined by

$$\delta = \left\{ \frac{E\{\|[\mathbf{G}] - E\{[\mathbf{G}]\}\|_F^2\}}{\|E\{[\mathbf{G}]\}\|_F^2} \right\}^{1/2} = \left\{ \frac{1}{\mu} E\{\|[\mathbf{G}] - [I_\mu]\|_F^2\} \right\}^{1/2} \tag{13.52}$$

and which will allow the dispersion of the probability model of random matrix $[\mathbf{G}]$ to be fixed. Then from an algebraic calculation, it is deduced that for δ such that $0 < \delta < (\mu+1)^{1/2}(\mu+5)^{-1/2}$, the probability density function of random matrix $[\mathbf{G}]$ is written as

$$p_{[\mathbf{G}]}([G]) = \mathbf{1}_{\mathbb{M}_\mu^+(\mathbb{R})}([G]) \times C_G \times \left(\det[G]\right)^{(\mu+1)\frac{(1-\delta^2)}{2\delta^2}} \times \mathrm{e}^{-\frac{(\mu+1)}{2\delta^2}\,\mathrm{tr}[G]}, \tag{13.53}$$

in which the positive constant C_G is such that

$$C_G = (2\pi)^{-\mu(\mu-1)/4} \left(\frac{\mu+1}{2\delta^2} \right)^{\mu(\mu+1)(2\delta^2)^{-1}} \left\{ \prod_{j=1}^{\mu} \Gamma\left(\frac{\mu+1}{2\delta^2} + \frac{1-j}{2} \right) \right\}^{-1}, \tag{13.54}$$

where $\Gamma(z)$ is the gamma function defined for $z > 0$ by $\Gamma(z) = \int_0^{+\infty} t^{z-1}\mathrm{e}^{-t}\,dt$. Note that Equation (13.53) shows that $\{[\mathbf{G}]_{jk}, 1 \le j \le k \le n\}$ are dependent random variables.

13.4.3.3 Characteristic Function of a Random Matrix in Ensemble SG^+

For all $[Z]$ in $\mathbb{M}_\mu^S(\mathbb{R})$, the characteristic function of random matrix $[\mathbf{G}]$ with values in $\mathbb{M}_\mu^+(\mathbb{R}) \subset \mathbb{M}_\mu^S(\mathbb{R})$ is defined by $\Psi_{[\mathbf{G}]}([Z]) = E\left\{\exp(\mathrm{i}\langle\!\langle[Z],[\mathbf{G}]\rangle\!\rangle)\right\}$, which yields $\Psi_{[\mathbf{G}]}([Z]) = \left\{\det\left([I_\mu] - \mathrm{i}2\delta^2(\mu+1)^{-1}[Z]\right)\right\}^{-(\mu+1)(2\delta^2)^{-1}}$. If $(\mu+1)/\delta^2$ is an integer, then this last equation shows that the probability distribution defined by Equations (13.53) and (13.54) is a usual Wishart distribution. In general, $(n+1)/\delta^2$ is not an integer, and consequently, the probability distribution defined by Equations (13.53) and (13.54) is not a usual Wishart distribution.

13.4.3.4 Second-Order Moments of a Random Matrix in Ensemble SG^+

Because $E\{[\mathbf{G}]\} = [I_\mu]$, the covariance $C^G_{jk,j'k'}$ of the real-valued random variables $[\mathbf{G}]_{jk}$ and $[\mathbf{G}]_{j'k'}$ that is defined by $C^G_{jk,j'k'} = E\{([\mathbf{G}]_{jk} - [I_\mu]_{jk}) \times ([\mathbf{G}]_{j'k'} - [I_\mu]_{j'k'})\}$ can be written as $C^G_{jk,j'k'} = \delta^2(\mu+1)^{-1}\{[I_\mu]_{j'k}[I_\mu]_{jk'} + [I_\mu]_{jj'}[I_\mu]_{kk'}\}$. The variance of $[\mathbf{G}]_{jk}$ is then given by $V^G_{jk} = \delta^2(\mu+1)^{-1}(1 + [I_\mu]_{jk})$.

13.4.3.5 Invariance of Ensemble SG^+ under Real Orthogonal Transformations

Let $[R]$ be any real orthogonal matrix belonging to $\mathbb{M}_\mu(\mathbb{R})$ such that $[R]^T[R] = [R][R]^T = [I_\mu]$. Let $[\mathbf{G}']$ be the random matrix with values in $\mathbb{M}_\mu^+(\mathbb{R})$ defined by $[\mathbf{G}'] = [R]^T[\mathbf{G}][R]$. Let $p_{[\mathbf{G}']}([G'])$ be the probability density function of the random matrix $[\mathbf{G}']$ with respect to the volume element $\tilde{d}G'$ (see Equation (13.40)). It can easily be verified that $p_{[\mathbf{G}']}([G'])\,\tilde{d}G' = p_{[\mathbf{G}]}([G'])\,\tilde{d}G'$, which proves the invariance of the random matrix $[\mathbf{G}]$ under real orthogonal transformations.

13.4.3.6 Algebraic Representation of a Random Matrix in Ensemble SG^+

The following algebraic representation of the random matrix $[\mathbf{G}]$ is derived from Equation (13.53) and is a random generator of independent realizations for the random matrix $[\mathbf{G}]$. Such a generator is necessary to solve the random equations with the Monte Carlo method. Because $[\mathbf{G}]$ is a positive definite random matrix, it can be written (Cholesky decomposition) as $[\mathbf{G}] = [\mathbf{L}]^T[\mathbf{L}]$ in which $[\mathbf{L}]$ is an upper triangular random matrix with values in $\mathbb{M}_\mu(\mathbb{R})$ such that

(i) random variables $\{[\mathbf{L}]_{jj'},\, j \leq j'\}$ are independent;
(ii) for $j < j'$, the real-valued random variable $[\mathbf{L}]_{jj'}$ can be written as $[\mathbf{L}]_{jj'} = \sigma_\mu U_{jj'}$ in which $\sigma_\mu = \delta(\mu+1)^{-1/2}$ and $U_{jj'}$ is a real-valued Gaussian random variable with zero mean and variance equal to 1;
(iii) for $j = j'$, the positive valued random variable $[\mathbf{L}]_{jj}$ can be written as $[\mathbf{L}]_{jj} = \sigma_\mu\sqrt{2V_j}$ in which σ_μ is defined earlier and V_j is a positive-valued gamma random variable whose probability density function $p_{V_j}(v)$ with respect to dv is written as

$$p_{V_j}(v) = \mathbb{1}_{\mathbb{R}^+}(v)\frac{1}{\Gamma\left(\frac{\mu+1}{2\delta^2} + \frac{1-j}{2}\right)} v^{\frac{\mu+1}{2\delta^2} - \frac{1+j}{2}}\, e^{-v}. \tag{13.55}$$

It should be noted that the probability density function of each diagonal element $[\mathbf{L}]_{jj}$ of the random matrix $[\mathbf{L}]$ depends on the rank j of the element.

13.4.3.7 Invertibility and Convergence Property When Dimension Goes to Infinity

Because $[\mathbf{G}]$ is a positive definite random matrix, this random matrix is invertible almost surely, which means that for $\mathscr{P}$-almost θ in Θ, the inverse $[\mathbf{G}(\theta)]^{-1}$ of the matrix $[\mathbf{G}(\theta)]$ exists. However, in general, this last property does not guarantee that $[\mathbf{G}]^{-1}$ is a second-order random variable, that is to say, $E\{\|[\mathbf{G}]^{-1}\|_F^2\} = \int_\Theta \|[\mathbf{G}(\theta)]^{-1}\|_F^2\, d\mathscr{P}(\theta)$ is finite. In fact, from Equation (13.53), it can be deduced that this last property holds, and because $\|A\| \leq \|A\|_F$ for any matrix $[A]$, one has

$$E\{\|[\mathbf{G}]^{-1}\|^2\} \leq E\{\|[\mathbf{G}]^{-1}\|_F^2\} < +\infty. \tag{13.56}$$

Note that in general, Equation (13.56) does not imply that $\mu \mapsto E\{\|[\mathbf{G}]^{-1}\|^2\}$ is a bounded function with respect to μ. As we have explained in Section 13.3.3.2, one must have $E\{\|[\mathbf{G}]^{-1}\|^2\} \leq C < +\infty$ in which C is a positive constant independent of μ, in order that Equation (13.30) be verified (in infinite dimension). From Equation (13.53), it can be deduced that this last property holds, and one has the following fundamental property:

$$\forall \mu \geq 2, \quad E\{\|[\mathbf{G}]^{-1}\|^2\} \leq C_\delta < +\infty, \tag{13.57}$$

in which C_δ is a positive finite constant that is independent of μ but that depends on δ. Equation (13.57) means that $\mu \mapsto E\{\|[\mathbf{G}]^{-1}\|^2\}$ is a bounded function from $\{\mu \geq 2\}$ into $\mathbb{R}^+$. Figure 13.6(a) shows the graph of the function $\mu \mapsto E\{\|[\mathbf{G}]^{-1}\|^2\}$ for $\delta = 0.1$, 0.3, and 0.5, constructed by using Section 13.4.3.6 and the Monte Carlo numerical simulation with 300 realizations. For instance, these graphs show that

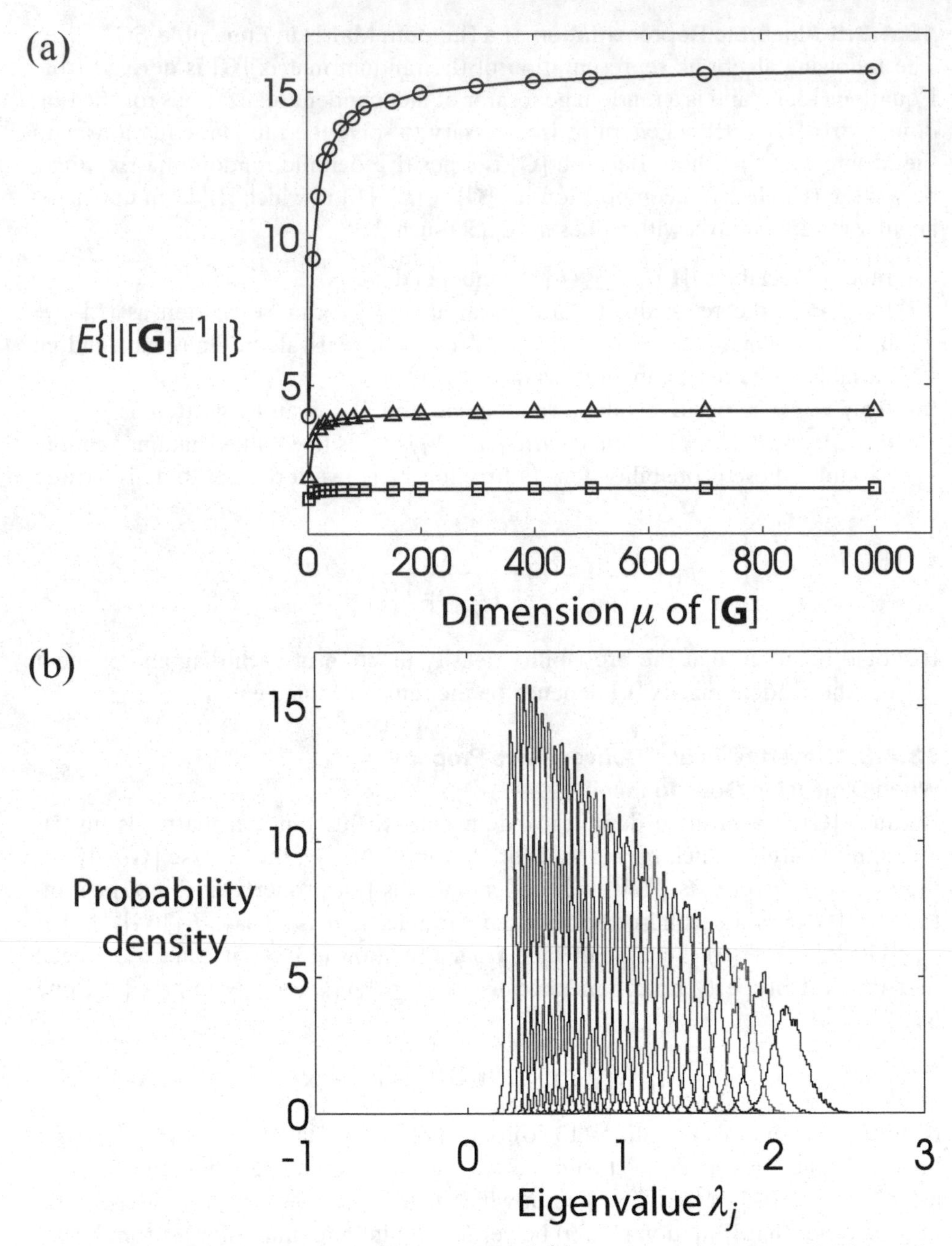

Figure 13.6. (a) Graph of function $\mu \mapsto E\{\|[\mathbf{G}]^{-1}\|^2\}$ as a function of δ: $\delta = 0.1$ (square), $\delta = 0.3$ (triangle), and $\delta = 0.5$ (circle). (b) Graphs of the probability density functions of the order-statistics random eigenvalues for the random matrix [**G**] with $\delta = 0.5$.

convergence is reached with a rate larger than 95% when $\mu \geq 20$ for $\delta = 0.1$, when $\mu \geq 80$ for $\delta = 0.3$, and when $\mu \geq 200$ for $\delta = 0.5$.

13.4.3.8 Probability Density Functions of the Random Eigenvalues

Let $\boldsymbol{\Lambda} = (\Lambda_1, \ldots, \Lambda_\mu)$ be the positive-valued random eigenvalues of the random matrix [**G**]. The joint probability density function $p_{\boldsymbol{\Lambda}}(\boldsymbol{\lambda}) = p_{\Lambda_1,\ldots,\Lambda_\mu}(\lambda_1, \ldots, \lambda_\mu)$ with respect to $d\boldsymbol{\lambda} = d\lambda_1 \cdots d\lambda_\mu$ of $\boldsymbol{\Lambda} = (\Lambda_1, \ldots, \Lambda_\mu)$ is given by

$$p_{\boldsymbol{\Lambda}}(\boldsymbol{\lambda}) = \mathscr{D}(\boldsymbol{\lambda}) \left\{ \prod_{j=1}^{\mu} \lambda_j^{(\mu+1)\frac{(1-\delta^2)}{2\delta^2}} \right\} \left\{ \prod_{\alpha<\beta} |\lambda_\beta - \lambda_\alpha| \right\} e^{-\frac{(\mu+1)}{2\delta^2} \sum_{k=1}^{\mu} \lambda_k}, \tag{13.58}$$

in which $\mathscr{D}(\boldsymbol{\lambda}) = c \times \mathbf{1}_{[0,+\infty[}(\lambda_1) \times \cdots \times \mathbf{1}_{[0,+\infty[}(\lambda_\mu)$, where c is a constant of normalization defined by the equation $\int_0^{+\infty} \cdots \int_0^{+\infty} p_{\boldsymbol{\Lambda}}(\boldsymbol{\lambda})\, d\boldsymbol{\lambda} = 1$. Let $p_{\widetilde{\Lambda}_j}(\lambda_j)$ be the probability density function with respect to $d\lambda_j$ of the random eigenvalue $\widetilde{\Lambda}_j$ for the order statistics $\widetilde{\Lambda}_1 \leq \widetilde{\Lambda}_2 \leq \cdots \leq \widetilde{\Lambda}_\mu$ of the random eigenvalues $\Lambda_1, \ldots, \Lambda_\mu$. The density $p_{\widetilde{\Lambda}_j}(\lambda_j)$ cannot be explicitly calculated. An estimation can be constructed by using the Monte Carlo numerical simulation. Figure 13.6(b) shows the graphs of the probability density functions $p_{\widetilde{\Lambda}_j}$ for $j = 1, \ldots, \mu$ with $\delta = 0.5$, $\mu = 30$, and 10,000 realizations for the Monte Carlo numerical simulation. This figure shows that the dispersion of the random eigenvalue $\widetilde{\Lambda}_j$ increases with its rank j and that the probability distributions are not Gaussian.

13.4.3.9 Probability Model of a Set of Random Matrices in SG^+

As explained in Sections 13.3.2 and 13.3.3, the random matrices $[\mathbf{H}]$, $[\mathbf{M}^s]$, $[\mathbf{D}^s(\omega)]$, $[\mathbf{K}^s]$, $[\mathbf{M}^a]$, $[\mathbf{D}^a(\omega)]$, and $[\mathbf{K}^a]$ are expressed in terms of the random matrices $[\mathbf{G}_H]$, $[\mathbf{G}_{M^s}]$, $[\mathbf{G}_{D^s}]$, $[\mathbf{G}_{K^s}]$, $[\mathbf{G}_{M^a}]$, $[\mathbf{G}_{D^a}]$, and $[\mathbf{G}_{K^a}]$. One has seen that the available information was only related to each one of these random matrices and that no information concerning statistical dependence between these random matrices was available. Then applying the maximum entropy principle with such an available information, it is deduced that these random matrices are independent.

13.4.4 Model Parameters, Stochastic Solver, and Convergence Analysis

13.4.4.1 Dispersion Parameters Controlling the Level of Uncertainties

One has seen that the dispersion parameter δ defined by Equation (13.52) allowed the dispersion of the random matrix $[\mathbf{G}]$ to be controlled. The dispersion parameters of the random matrices $[\mathbf{G}_H]$, $[\mathbf{G}_{M^s}]$, $[\mathbf{G}_{D^s}]$, $[\mathbf{G}_{K^s}]$, $[\mathbf{G}_{M^a}]$, $[\mathbf{G}_{D^a}]$, and $[\mathbf{G}_{K^a}]$ will then be denoted by

$$\delta_H, \delta_{M^s}, \delta_{D^s}, \delta_{K^s}, \delta_{M^a}, \delta_{D^a}, \delta_{K^a}. \tag{13.59}$$

Taking into account Equations (13.31) and (13.35)–(13.38), it can then be deduced that for all ω fixed in $\mathscr{F}$, the level of uncertainties in the computational structural–acoustic model will completely be defined by the dispersions Δ_H, Δ_{M^s}, Δ_{D^s}, Δ_{K^s}, Δ_{M^a}, Δ_{D^a}, and Δ_{K^a} of the random matrices $[\mathbf{H}]$, $[\mathbf{M}^s]$, $[\mathbf{D}^s(\omega)]$, $[\mathbf{K}^s]$, $[\mathbf{M}^a]$, $[\mathbf{D}^a(\omega)]$, and $[\mathbf{K}^a]$ and, consequently, by the dispersions of the random matrices $[\mathbf{G}_H]$, $[\mathbf{G}_{M^s}]$, $[\mathbf{G}_{D^s}]$, $[\mathbf{G}_{K^s}]$, $[\mathbf{G}_{M^a}]$, $[\mathbf{G}_{D^a}]$, and $[\mathbf{G}_{K^a}]$, that is to say, will completely be defined by the dispersion parameters δ_H, δ_{M^s}, δ_{D^s}, δ_{K^s}, δ_{M^a}, δ_{D^a}, and δ_{K^a}. It should be noted that there would be no interest to define the level of uncertainties in terms of Δ instead of δ because Δ depends on δ but also depends on μ and on the mean value of this random matrix and, consequently, can depend on ω whereas δ is independent of μ, of the mean value, and of the frequency.

13.4.4.2 Construction and Convergence of the Stochastic Solution

As explained in Sections 13.3 and 13.4, the probability model has been constructed to assure that for all ω fixed in $\mathcal{F}$, the system of random equations defined by Equations (13.20)–(13.23) has a unique second-order solution (see Equation (13.30)). The random equations are solved by the Monte Carlo method, which consists of generating ν independent realizations $[\mathbf{G}_H(\theta_\ell)]$, $[\mathbf{G}_{M^s}(\theta_\ell)]]$, $[\mathbf{G}_{D^s}(\theta_\ell)]]$, $[\mathbf{G}_{K^s}(\theta_\ell)]]$, $[\mathbf{G}_{M^a}(\theta_\ell)]]$, $[\mathbf{G}_{D^a}(\theta_\ell)]]$, and $[\mathbf{G}_{K^a}(\theta_\ell)]]$ for $\ell = 1, \ldots, \nu$ of the random matrices $[\mathbf{G}_H]$, $[\mathbf{G}_{M^s}]$, $[\mathbf{G}_{D^s}]$, $[\mathbf{G}_{K^s}]$, $[\mathbf{G}_{M^a}]$, $[\mathbf{G}_{D^a}]$, and $[\mathbf{G}_{K^a}]$, using the random generator defined in Section 13.4.3.6. For each realization θ_ℓ and for all ω in $\mathcal{F}$, the equations defined in (13.20)–(13.23) are solved in using Equations (13.31) and (13.35)–(13.38) (deterministic calculations). From these independent realizations of the random responses, ν independent realizations $\mathbf{U}^n(\omega, \theta_1), \ldots, \mathbf{U}^n(\omega, \theta_\nu)$ and $\mathbf{P}^m(\omega, \theta_1), \ldots, \mathbf{P}^m(\omega, \theta_\nu)$ of the random responses $\mathbf{U}^n(\omega)$ and $\mathbf{P}^m(\omega)$ are calculated. The probabilistic quantities of interest are then estimated using mathematical statistics. Using the usual statistical estimator of the mathematical expectation E, the mean square convergence of the stochastic solution with respect to n, m, and ν can be analyzed in studying the function $(n, m, \nu) \mapsto \mathcal{E}(n, m, \nu) = \nu^{-1/2}\{\sum_{\ell=1}^{\nu} \int_{\mathcal{F}} (\|\mathbf{U}^n(\omega, \theta_\ell)\|^2 + \|\mathbf{P}^m(\omega, \theta_\ell)\|^2)\, d\omega\}^{1/2}$.

13.4.4.3 Confidence Regions of the Random Responses

A probability quantity of interest is the confidence region of random frequency responses. Let $\omega \mapsto Y(\omega)$ be such a random frequency response function from $\mathcal{F}$ into $\mathbb{R}$. For instance, Y will be the modulus in logarithm scale of a component of the acceleration at a given point of the structure or the root mean square in decibel scale of the acoustic pressure inside the acoustic cavity. As explained in Section 13.4.4.2, the Monte Carlo simulation allows ν independent realizations $\omega \mapsto Y(\omega, \theta_1), \ldots, \omega \mapsto Y(\omega, \theta_\nu)$ to be constructed with the stochastic reduced computational structural–acoustic model. For the random function $\{Y(\omega), \omega \in \mathcal{F}\}$, the confidence region associated with the probability level P_c (for instance, $P_c = 0.95$ or 0.98) is limited by a lower and an upper envelope denoted by $\omega \mapsto y^-(\omega)$ and $\omega \mapsto y^+(\omega)$, such that

$$\mathcal{P}\{y^-(\omega) < Y(\omega) \leq y^+(\omega)\} = P_c, \quad \forall \omega \in \mathcal{F}. \tag{13.60}$$

For fixed ω in $\mathcal{F}$, let $F_{Y(\omega)}$ be the cumulative distribution function (continuous from the right) of the random variable $Y(\omega)$, which is such that $F_{Y(\omega)}(y) = \mathcal{P}\{Y(\omega) \leq y\}$. For $0 < p < 1$, the pth quantile (or fractile) of $F_{Y(\omega)}$ is defined as $\zeta(p) = \inf\{y\colon F_{Y(\omega)}(y) \geq p\}$. Then the upper envelope $y^+(\omega)$ and the lower envelope $y^-(\omega)$ for the symmetric interval are defined by $y^+(\omega) = \zeta((1 + P_c)/2)$ and $y^-(\omega) = \zeta((1 - P_c)/2)$. The estimations of $y^+(\omega)$ and $y^-(\omega)$ is performed by using the sample quantiles. Let $y_1(\omega) = Y(\omega, \theta_1), \ldots, y_\nu(\omega) = Y(\omega, \theta_\nu)$ be the ν independent realizations of the random variable $Y(\omega)$. Let $\tilde{y}_1(\omega) < \cdots < \tilde{y}_\nu(\omega)$ be the order statistics associated with $y_1(\omega), \ldots, y_\nu(\omega)$. One then has

$$y^\pm(\omega) \simeq \tilde{y}_{j^\pm}(\omega) \quad \text{with} \quad j^\pm = \text{fix}(\nu(1 \pm P_c)/2), \tag{13.61}$$

in which fix(z) is the integer part of the real number z.

13.5 Experimental Identification of the Probabilistic Model

There are several possible methodologies to identify the dispersion parameters δ_{M^s}, δ_{D^s}, δ_{K^s} for the structure, δ_{M^a}, δ_{D^a}, δ_{K^a} for the internal acoustic cavity, and δ_H for the structural–acoustic coupling interface. Next we present a methodology that has been used with success:

(i) The dispersion parameters δ_{M^a}, δ_{D^a}, and δ_{K^a} of the internal acoustic cavity are identified using an experimental acoustic database constructed (possibly for different configurations of the acoustic cavity analyzed with the same computational acoustic model) by measuring the experimental acoustic frequency response functions for the acoustic pressures at several given points inside the internal acoustic cavity and for an excitation that is an acoustic source located inside the cavity. For each configuration, the observation can be chosen as the root mean square in decibel scale of the acoustic pressures averaged on all the microphones as a function of the frequency f in hertz. The method used to identify the dispersion parameters can then be the maximum likelihood method.

(ii) For the structure, the dispersion parameters δ_{M^s}, δ_{D^s}, and δ_{K^s} are identified using an experimental structural database constructed by measuring the structural frequency response functions (and not the structural–acoustic frequency response functions) between excitation forces applied to the structure and accelerations at several given points of the structure. The stochastic reduced computational structural–acoustic model is used to identify these dispersions' parameters, and thus the dispersion parameters δ_{M^a}, δ_{D^a}, and δ_{K^a} are fixed to the experimental values that have been identified, and the value of δ_H is fixed to the nominal value. This last hypothesis is justified when the confidence regions of the random structural responses are not sensitive to the acoustic coupling (note that the structural–acoustic responses are sensitive to δ_H, but we are presently speaking about the identification of the dispersion parameters of the structure using only the structural responses). The identification of the dispersion parameters δ_{M^s}, δ_{D^s}, and δ_{K^s} can then be performed by using the maximum likelihood method, but for computational cost reasons such a method can eventually be substituted by the mean square method with a differentiable or a non-differentiable objective function.

(iii) For the identification of the dispersion parameter δ_H of the structural–acoustic coupling interface, a structural–acoustic database is constructed by measuring the structural–acoustic frequency responce functions between excitation forces applied to the structure and acoustic pressures at several given points inside the internal acoustic cavity. Therefore, fixing the dispersion parameters δ_{M^a}, δ_{D^a}, δ_{K^a}, δ_{M^s}, δ_{D^s}, and δ_{K^s} to their identified values, the stochastic reduced computational structural–acoustic model is used to identify δ_H with the maximum likelihood method, but as previously, for computational cost reasons, such a method can be substituted by the mean square method with a differentiable or a non-differentiable objective function.

13.6 Experimental Validation

In this section, an experimental validation of the approach developed in this chapter is presented for an application relative to the automotive industry. The experimental database has been constructed using 20 cars of the same type with different optional extras. We are interested in the booming noise that is defined as the modulus of the acoustic pressure in A-weighted decibels at the driver's ears for the structural excitation induced by the forces that are delivered by the engine on its four supports. The frequency of rotation varies in the frequency band of analysis 50–183 Hz corresponding to 1,500–5,500 rpm. Figure 13.7 is related to the booming noise. The 20 thin gray lines are the measurements of the booming noise for the 20 cars. It can be seen a very large experimental dispersion induced by the variability (optional extras) and by the manufacturing process. The mean computational structural–acoustic model of the car is a finite element model with 9,78,733 degrees of freedom for the structural displacements and 8,139 degrees of freedom for the acoustic pressures. The reduced mean computational model is constructed using $n = 1{,}723$ elastic modes of the structure in vacuo and $m = 57$ acoustic modes of the internal acoustic cavity with rigid walls. These values of n and m have been deduced from a mean square convergence analysis of the random response. In Figure 13.7, the dashed line corresponds to the reduced mean computational model, and the thick gray line corresponds to the mean value of the experiments over the 20 thin gray lines. Concerning the predictions performed by the stochastic reduced computational structural–acoustic model, the dispersion parameters δ_{M^a}, δ_{D^a}, δ_{K^a}, δ_{M^s}, δ_{D^s}, δ_{K^s}, and δ_H have been fixed to their optimal values that have experimentally be identified as explained in Section 13.5. The Monte Carlo stochastic solver is used with 1,500 realizations. A mean square convergence analysis of the stochastic response has been carried out, and convergence is reached for this number of realizations. In Figure 13.7, the upper and lower thick lines represent the upper and lower envelopes of the confidence region calculated as explained in Section 13.4.4 for a probability level of $P_c = 0.96$, and the middle thick line represents the mean value of the random response. Taking into account the complexity of the structural–acoustic system, the obtained results validate the stochastic reduced computational structural–acoustic model and demonstrate its capability to predict experimental measurements.

13.7 Bibliographical Comments

The equations of the mean boundary value problem given in Section 13.1.1 can be found in Ohayon and Soize (1998) and have been constructed using Pierce (1989) and Lighthill (1978) for the acoustic equations. The finite element method used in Section 13.1.2 is a powerful method to derive a computational model from a boundary value problem. For general considerations, we refer the reader to Zienkiewicz and Taylor (2000), and for the computational model in structural acoustics and for the construction of the reduced model introduced in Section 13.1.3, we refer the reader to Ohayon and Soize (1998).

The parametric probabilistic approach of system parameter uncertainties in the predictive model introduced in Section 13.2 is well understood today, and many papers have been published on this subject (see for instance Mace et al. 2005 for uncertainty in structural dynamics, Schueller 2005 for a recent overview on computational methods in stochastic mechanics and reliability analysis, and Ghanem and Spanos 2003 for stochastic finite element methods). Concerning mathematical statistics, such as parametric statistics and nonparametric statistics, we refer the

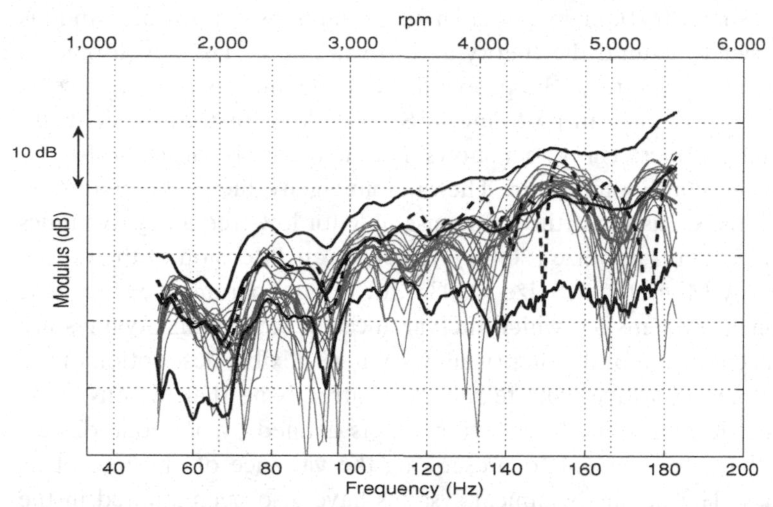

Figure 13.7. Comparisons of the mean and stochastic computational models with the experiments for the booming noise. Experiments (20 thin gray lines), mean value of the experiments (thick gray line), mean computational model (dashed line), mean value of the random response (middle thick line), confidence region: upper and lower envelopes are the upper and lower thick lines. From Durand (2007) and reprinted with permission from Durand et al. (2008). Copyright 2008, Acoustical Society of America.

reader to Serfling (1980), Spall (2003), and Walter and Pronzato (1997). Concerning the maximum entropy principle, we refer the reader to the fundamental papers by Shannon (1948) and Jaynes (1957).

The nonparametric probabilistic approach of both model uncertainties and system parameter uncertainties introduced in Section 13.3 has been proposed by Soize (2000) for dynamical systems by introducing a new concept to model uncertainties and in developing a new ensemble of positive definite random matrices in the context of the random matrix theory.

It should be noted that the random matrix theory was introduced and developed in mathematical statistics by Wishart and others in the 1930s and was intensively studied by physicists and mathematicians in the context of nuclear physics. These works began with Wigner in the 1950s and received an important effort in the 1960s by Wigner, Dyson, Mehta, and others. In 1965, Porter published a volume of important papers in this field, followed in 1967 by the first edition of the book by Mehta, whose second edition (Mehta 1991) is an excellent synthesis of the random matrix theory. We refer the reader to Weaver (Chapter 3 of this book) for an introduction to random matrix theory. Concerning multivariate statistical analysis and statistics of random matrices, the reader will find additional developments in Fougeaud and Fuchs (1967) and Anderson (1958). Concerning the generalities on the Monte Carlo method introduced in Section 13.4.4, we refer the reader to Kalos and Whitlock (1986) and MacKeown (1997). For the mathematical statistics, the statistical estimators, and the statistical methods such as the mean square method or the maximum likelihood method, we refer again the reader to Fougeaud and Fuchs (1967), Serfling (1980), Walter and Pronzato (1997), and Spall (2003)

We give herein a summary of the evolution of the concepts for the nonparametric probabilistic approach of uncertainties presented in this chapter. Since the first paper in 2000 (Soize 2000), many papers have been published by Soize and his collaborators in order to extend the theory and to validate it. The developments concerning the algebraic closure of the probabilistic model and its convergence as the dimension goes to infinity can be found in Soize (2001) for the transient linear stochastic elastodynamics. This paper shows that the theory is consistent for the continuous systems in infinite dimension. The random eigenvalue problem and the non-applicability of the Gaussian orthogonal ensemble for low-frequency dynamics has been analyzed in details in Soize (2003a). Recently, an extension of the theory has been proposed by Mignolet & Soize (2008a) for the nonparametric stochastic modeling of the linear systems for which the variances of several eigenvalues are prescribed. Such an extension of the theory allows a more flexible description of the dispersion levels of each positive definite random matrix. Note that the ensemble SG^{+} of random matrices presented in Section 13.4 is defined by only one dispersion parameter δ that is equivalent to prescribing the variance of any one of its random eigenvalues. The linear dynamical systems have also been studied in the medium-frequency range in taking into account the model uncertainties and the system parameter uncertainties (see Soize 2003b). The theory has been extended to dynamical systems for which model uncertainties are not homogeneous through the system. The dynamic substructuring techniques have thus been introduced (see Soize & Chebli 2003, Chebli & Soize 2004, Duchereau & Soize 2006). The model uncertainties in dynamical systems with cyclic symmetry have been studied in Capiez-Lernout et al. (2005, 2006). The construction of a probabilistic model for impedance matrices can be found in Cottereau et al. (2006, 2007). Recently, nonparametric stochastic modeling has been used to propose a methodology to analyze structural dynamic systems with uncertain boundary conditions (see Mignolet & Soize 2008b). The capability of the nonparametric probabilistic approach to take into account both the model uncertainties and the system parameter uncertainties (although the parametric probabilistic approach can take into account only the system parameter uncertainties) has been analyzed for several simple and complex dynamic systems (Soize 2005a, 2005b, Capiez-Lernout et al. 2006). Some new ensembles of random matrices for model uncertainties in coupled dynamical systems such as structural–acoustic systems have been introduced by Soize (2005c). Significant efforts have been performed to develop experimental identifications of the nonparametric probabilistic approach and to obtain experimental validations of the theory in structural dynamics (Chebli & Soize 2004, Soize 2005a, Arnst et al. 2006, Chen & Soize 2006, Duchereau & Soize 2006, Capiez-Lernout & Soize 2008b, Soize et al. 2008) and in structural acoustics (Durand 2007, Soize et al. 2007, 2008, Capiez-Lernout & Soize 2008a, Durand et al. 2008), and in particular, for the methodology presented in Section 13.5 and the experimental validation presented in Section 13.6, see Durand (2007), Durand et al. (2008), and Soize et al. (2008). Finally, the nonparametric probabilistic approach has been extended to analyze nonlinear elastodynamics with local nonlinearities (Desceliers et al. 2004, Sampaio & Soize 2007) and with distributed nonlinearities (Mignolet & Soize 2008c).

14 The Analysis of Random Built-Up Engineering Systems

Robin Langley

Department of Engineering, University of Cambridge, Cambridge, UK

14.1 Introduction

There has been much recent interest in predicting the response of structures that have uncertain or random properties, and this is particularly true in the field of vibroacoustics. In predicting interior or radiated noise a broad frequency range is of interest, and the response of an engineering structure such as a car or an airplane can be highly sensitive to small manufacturing variations, particularly at higher frequencies. Ideally this sensitivity should be identified and quantified at the design stage, so that the performance of the complete set of produced articles can be predicted, rather than just the performance of a nominally perfect design. Perhaps the most direct way to consider the effect of system uncertainty on the response is to adopt a *parametric model*. This approach consists of (i) identifying the uncertainties in the key physical properties (or "parameters") of the system, and (ii) propagating this uncertainty through the governing equations of motion to predict the uncertainty in the response. However both parts of this procedure can present severe practical difficulties. With regard to issue (i), the parametric model would ideally consist of the detailed statistical distribution of the system properties (material properties, dimensions, rigidity of joints, etc.), but it can be extremely difficult to acquire this data or even to identify which of the system properties are statistically significant. This problem can be overcome to some extent by adopting a less precise parametric model of the system uncertainties, consisting of perhaps bounds (Elishakoff 1995) or fuzzy-number descriptions (Moens & Vandepitte 2005), but the difficulty of propagating this uncertainty through the equations of motion, issue (ii), still remains. Typical industrial structures are described by finite element models that may have thousands or millions of degrees of freedom, and so the propagation of parametric uncertainty using Monte Carlo simulations, or more analytical means, can be highly computationally intensive. It is clear that parametric modeling is a very challenging research area, and progress is being made by the introduction of innovate Monte Carlo methods (Schuëller 2006) or alternative propagation methods (Moens & Vandepitte 2005).

Despite the foregoing discussion, the fact that engineering systems can be large and complex is not necessarily to the disadvantage of the analyst. In fact, very complex systems can sometimes be easier to analyze than less complex systems.

Of course, care is needed in the use of the term "complex" in this context; here, by "more complex" the intention is to say "more random." To consider the response of a single plate, for example, the introduction of a small degree of randomness will cause the natural frequencies, the mode shapes, and the plate response to be random, and it can be challenging to work out precisely the statistics of these quantities. The results obtained will be dependent on the statistics of the random input parameters, and thus a very detailed analytical model will be required. If, on the other hand, the plate is subjected to a high degree of randomness, then it is known that the natural frequencies (excluding the lower frequencies) approach a standard distribution described by the Gaussian orthogonal ensemble (GOE) (Weaver 1989a, Mehta 1991, Lobkis et al. 2000), regardless of the statistical details of the input parameters. In this case the prediction of the natural frequency statistics is unnecessary (because they are known), and the statistics of the response can be calculated in a relatively straightforward way (Langley & Brown 2004b). This opens up the possibility of *nonparametric* models of uncertainty for complex systems, that is, models that avoid the need to describe the detailed statistical distributions of the underlying physical parameters. One of the earliest nonparametric modeling approaches developed in engineering is known as statistical energy analysis (SEA) (Lyon 1975), in which the governing equations are based on the flow of vibrational energy between regions of the system known as "subsystems." By adopting a single (energy) degree of freedom for each subsystem, relatively efficient models of the system can be derived, and the mean energy response can be computed without the need for parametric modeling, large finite element models, or Monte Carlo simulations. The statistical assumptions employed in SEA are valid only at high frequencies, and significant current research effort is being directed at the development alternative, more general, nonparametric approaches. Much of this work is based on novel applications of random matrix theory (for example, Soize 2005a) or, as described below, the combination of SEA with other approaches.

Over the past decade, much research into the dynamic analysis of uncertain engineering systems has been motivated by the "mid-frequency" problem, where neither the finite element (FE) method nor SEA provide a satisfactory solution. At mid-frequencies the system is too complex for FE method to apply (too random, with too many degrees of freedom) and not complex enough for SEA to apply (not random enough for the averaging required by SEA to apply, so that very high response variances arise, which cannot be predicted by standard SEA). The obvious approach is therefore to try and combine FE method and SEA into a single model, so that advantage can be taken of the strengths of each method. This type of hybrid model has been proposed by many researchers, and the pioneering work of Soize (1993) on fuzzy structure theory highlighted many general underlying principles. An early attempt to produce a hybrid model consisted of partitioning the response of a system component into a "long-wavelength" contribution and a "short-wavelength" contribution (Langley & Bremner 1999). The long-wavelength part was then included in an FE model and the short-wavelength part in an SEA model. A computational difficulty with this approach for general systems is that the coupling between the FE and SEA parts of the model require the evaluation of some rather difficult integrals. An alternative method was then developed in which the response of the component is partitioned not into different wavelength components but into a "direct field" and a "reverberant field" (Shorter & Langley 2005a). One advantage of this approach is that the coupling between the SEA and FE parts of the model can be affected simply and efficiently through a "diffuse field reciprocity relation" (Shorter & Langley 2005a, Langley 2007). Also, the method can be combined with aspects of random matrix theory to predict not only the mean but also the variance of the system response (Langley & Cotoni 2007). This method is presented in detail in the present chapter by considering a succession of problems of increasing difficulty: (i) the statistics of the natural frequencies of a single random component (Section 14.2); (ii) the statistics of the response of a single random component (Section 14.3); (iii) the statistics of the dynamic stiffness matrix of a single random component (Section 14.4); (iv) the ensemble average of the energies and/or the cross-spectrum of the response of a built-up system (Section 14.5); and (v) the variance of the response of a built-up system (Section 14.6).

14.2 Natural Frequency Statistics

Various functions that are used to describe the statistics of the natural frequencies of a random system are reviewed in this section. The focus is general initially, but then the specific case of a single random component is considered. The literature in this area tends to be divided between two approaches, depending on whether random matrix theory (Mehta 1991) or random point process theory (Stratonovich 1963) is adopted. However, as will be discussed in what follows, the two approaches are identical other than through the use of different notation and terminology to describe the various functions that appear. It will be seen that the cumulant functions employed in random point process theory (or equivalently the cluster functions employed in random matrix theory) provide a very convenient vehicle for discussing the transition from parametric to nonparametric uncertainty models with increasing excitation frequency and/or system uncertainty.

Perhaps the most direct and complete way to describe the statistics of a set of natural frequencies $\boldsymbol{\omega} = (\omega_1, \omega_2, \ldots, \omega_N)$ is via the joint probability density function (JPDF) $p(\boldsymbol{\omega})$. Within this description, care is needed to distinguish whether the natural frequencies are considered to be *ordered*, that is, $\omega_1 < \omega_2 \cdots < \omega_N$, or *unordered*. By convention, the natural frequencies of an engineering system are almost always considered to be ordered, so that ω_{35} for example is generally understood to be the 35th natural frequency in an ascending list. However, from a mathematical point of view, there are advantages to be gained by considering an unordered set of natural frequencies; this unordered set can be constructed from the ordered set by randomly reassigning the ordered natural frequencies to the labels ω_j, so that ω_{35} for example is equally likely to be any member of the ordered set. The JPDF of the unordered set, $p(\boldsymbol{\omega})$, say, is then related to the JPDF of the ordered set, $p_{\text{ord}}(\boldsymbol{\omega})$, via

$$p(\boldsymbol{\omega}) = \frac{1}{N!} \sum_{\text{perm } \boldsymbol{\omega}} p_{\text{ord}}(\boldsymbol{\omega}), \tag{14.1}$$

where the summation is over all the possible permutations of the entries of $\boldsymbol{\omega}$ and N is the total number of natural frequencies.

Although the JPDF $p(\boldsymbol{\omega})$ provides a complete description of the system natural frequencies, it is not always the most convenient function to employ in calculations.

For this reason several other functions have been introduced, which can all be derived from $p(\boldsymbol{\omega})$. Random matrix theory (Mehta 1991) makes extensive use of the *n-point correlation function* R_n, which is defined as

$$\begin{aligned} R_n(\omega_1, \omega_2, \ldots, \omega_n) &= \frac{N!}{(N-n)!} \int_{-\infty}^{\infty} \cdots \int_{-\infty}^{\infty} p(\omega_1, \omega_2, \ldots, \omega_N)\, \mathrm{d}\omega_{n+1} \cdots \mathrm{d}\omega_N \\ &= f_n(\omega_1, \omega_2, \ldots, \omega_n). \end{aligned} \tag{14.2}$$

As indicated in the preceding equation, R_n is also sometimes denoted by f_n: this is the notation used in random point process theory (Stratonovich 1963), where the function is known as the nth *distribution function*. In physical terms, the probability of finding a natural frequency in each of the n regions $\omega_j < \omega < \omega_j + \mathrm{d}\omega_j$ (where $\mathrm{d}\omega_j$ is a small interval), without regard to the location of the other $N - n$ natural frequencies, is $f_n(\omega_1, \omega_2, \ldots, \omega_n)\, \mathrm{d}\omega_1 \cdots \mathrm{d}\omega_n$.

An insight into the usefulness of the distribution (or correlation) functions can be obtained by considering the statistics of a sum in the form

$$T = \sum_n a_n h(\omega_n), \tag{14.3}$$

where the coefficients a_n form a set of identically distributed statistically independent random variables. As will be seen in the following section, this type of expression arises frequently in structural dynamics. The average value of T can be written as

$$\begin{aligned} \mathrm{E}[T] &= \sum_n \mathrm{E}[a_n] \int_{-\infty}^{\infty} \cdots \int_{-\infty}^{\infty} h(\omega_n) p(\omega_1, \omega_2, \ldots, \omega_N)\, \mathrm{d}\omega_1\, \mathrm{d}\omega_2 \cdots \mathrm{d}\omega_N \\ &= \mathrm{E}[a_n] \int_{-\infty}^{\infty} h(\omega_n) f_1(\omega_n)\, \mathrm{d}\omega_n. \end{aligned} \tag{14.4}$$

Similarly, it can readily be shown that the mean squared value of T is given by

$$\begin{aligned} \mathrm{E}[T^2] &= \sum_n \sum_m \mathrm{E}[a_n a_m]\, \mathrm{E}[h(\omega_n) h(\omega_m)] \\ &= \mathrm{E}[a_n^2] \int_{-\infty}^{\infty} h^2(\omega_n) f_1(\omega_n)\, \mathrm{d}\omega_n \\ &\quad + \mathrm{E}[a_n]^2 \iint_{-\infty}^{\infty} h(\omega_n) h(\omega_m) f_2(\omega_n, \omega_m)\, \mathrm{d}\omega_n\, \mathrm{d}\omega_m, \end{aligned} \tag{14.5}$$

where the first term arises from contributions with $n = m$ in the summations, and the second term arises from contributions with $n \neq m$. It follows from Equations (14.4) and (14.5) that the variance of T has the form

$$\begin{aligned} \mathrm{Var}[T] &= \mathrm{E}[a_n^2] \int_{-\infty}^{\infty} h^2(\omega_n) g_1(\omega_n)\, \mathrm{d}\omega_n \\ &\quad + \mathrm{E}[a_n]^2 \int_{-\infty}^{\infty} \int_{-\infty}^{\infty} h(\omega_n) h(\omega_m) g_2(\omega_n, \omega_m)\, \mathrm{d}\omega_n\, \mathrm{d}\omega_m, \end{aligned} \tag{14.6}$$

$$g_1(\omega_n) = f_1(\omega_n) = T_1(\omega_n), \tag{14.7}$$

$$g_2(\omega_n, \omega_m) = f_2(\omega_n, \omega_m) - f_1(\omega_n) f_1(\omega_m) = -T_2(\omega_n, \omega_m). \tag{14.8}$$

The function g_n is known as the nth *cumulant function* in random point process theory (Stratonovich 1963), whereas an identical function, apart from a $(-1)^n$ sign change, is known as the *n-level cluster function* T_n in random matrix theory (Mehta 1991).

It is clear from Equations (14.4) and (14.6) that the second-order statistics of a quantity in the form of T are completely determined by the first two cumulant functions g_1 and g_2. The nature of these functions will depend upon the degree of randomness of the system and the frequency range of interest. In the extreme case of a deterministic system, the functions will be composed of delta functions located at the natural frequencies. At low frequencies and/or low levels of randomness, where random perturbations in the natural frequencies are small compared with the natural frequency spacing, it can be expected that g_1 and g_2 will vary strongly with frequency; in this case the functions will depend on the detailed properties of the system randomness, and in principle a parametric model of this randomness can be used to yield $p(\boldsymbol{\omega})$ and thus g_1 and g_2. In practice this approach faces two difficulties: (i) the data required to develop a realistic parametric model of the system uncertainty can be very difficult to obtain; (ii) for a complex system the probabilistic transformation from the statistics of the system parameters to $p(\boldsymbol{\omega})$ is analytically and computationally extremely challenging. As discussed in Section 14.1, the parametric modeling of uncertainty is a major current research area and it will not be considered in detail here, where the focus is on the development of nonparametric models.

At higher frequencies and/or higher levels of randomness, the random perturbations in the natural frequency positions will be relatively large, and it can be expected that a natural frequency is likely to occur at any point on the frequency axis, regardless of the location of the natural frequencies of the "perfect" system. The extreme limit of this process would be "stationary" natural frequency statistics, meaning that the natural frequencies of the perfect system are of little significance; in this case g_1 will be independent of frequency and $g_2(\omega_n, \omega_m) = g_2(\omega_n - \omega_m)$. Up to this point in the discussion the random system under consideration has not been described in any detail, and no distinction has been made between a complete built-up system and a subcomponent, such as a plate, shell, or acoustic volume. However, in considering the applicability or otherwise of a stationary natural frequency model, it is important to address this issue. The dynamic properties of a system are dependent upon both the natural frequencies and the mode shapes, and the statistical description of the natural frequencies must be accompanied by a statistical description of the mode shapes. The natural accompaniment to a stationary model of the natural frequencies would be the assumption that each mode shape has the same statistical properties. However, in a built-up system, modes are very often "localized" to restricted regions, and the assumption that each mode is statistically identical would miss the distinction between modes that are localized to different regions. Thus the stationary natural frequency model is most reasonably applied to systems that are relatively homogeneous, to the extent that the modes are "global." This does *not* mean that this type of model cannot be applied to built-up systems, but rather that it must first be applied separately to relatively homogeneous subcomponents of the system, which are then coupled to form a description of the full system. A coupling procedure of this type is described in Sections 14.5 and 14.6.

If the natural frequencies are taken to be statistically stationary, then the cumulant function g_1 has a very straightforward physical interpretation: the probability of finding a natural frequency in an interval $d\omega$ is $g_1\, d\omega$, and so the average number of modes in a frequency band Δ is $g_1\Delta$, which means that g_1 is equal to the *asymptotic modal density* (i.e., the average number of natural frequencies in a unit frequency band). This value is available is closed form for a wide range of structural and acoustic components (Lyon & DeJong 1995): for example, the modal density of a plate of surface area A, mass per unit area m_a, and flexural rigidity D is given by $g_1 = (A/4\pi)\sqrt{m_a/D}$. Some components, such as a beam or an acoustic volume, have a modal density that varies with frequency, and in this case the stationary model can be applied by assuming that the natural frequencies are *locally* stationary. The response of a system is normally dominated by those modes that are resonant in the vicinity of the excitation frequency, and providing the modal density is approximately constant in this local vicinity, longer range variations in the modal density can be reasonably neglected.

It follows from the preceding discussion, and from Equation (14.4), that under the assumption of stationary natural frequency statistics, the mean value of a variable in the form of Tdepends only on the modal density and not on any more complex features of the natural frequency statistics. This is a very simple and powerful result (known as Campbell's theorem in the random vibration literature; Lin 1967). From Equation (14.6) it can be seen that the variance of T depends on both the modal density and the second cumulant function $g_2(\omega_n - \omega_m)$. The second cumulant function is not a simple universal physical measure (like the modal density) but rather depends on more detailed features of the natural frequency statistics. The simplest possible case is that of a Poisson natural frequency process, for which $g_2 = 0$. However, there is a substantial body of literature that suggests that (in the absence of symmetries) whatever the underlying cause of the system randomness, the system natural frequencies are not Poisson but rather conform to the GOE. Numerical and experimental evidence for the occurrence of the GOE in structural systems that are sufficiently random has been given (for example) by Mehta (1991), Weaver (1989a), Lobkis et al. (2000), and Langley and Brown (2004a, 2004b). The GOE result for $g_2(\omega_n - \omega_m)$ is rather complicated, and it is given in Equation (6.4.14) of Mehta (1991) in terms of a function $Y_2 = -g_2$ for a system in which the modal density has been scaled to unity.

On the basis of the above discussion, it can be seen that if the system is random to the extent that the natural frequency statistics can be considered to be locally stationary, then g_1 and g_2 can be found without reference to the underlying cause or nature of the system uncertainty, and a relatively simple nonparametric model of the system natural frequencies can be developed. The implications of this for response prediction are discussed in the following sections.

14.3 The Response of a Single Component

This section will consider the statistics of the response of a single random component, such as a plate or an acoustic volume. It is assumed that the component properties are reasonably homogeneous, so that the mode shapes can be considered to be global (i.e., extended throughout the whole component) and the damping can be taken to be "proportional," which implies that the equations of motion of the component are uncoupled when expressed in modal coordinates. Under these conditions, the response of the component at spatial position $\mathbf{x}$ to applied harmonic loading of frequency ω can be written as

$$u(\mathbf{x}) = \sum_n \frac{F_n \phi_n(\mathbf{x})}{\omega_n^2 - \omega^2 + i\eta\omega\omega_n}, \tag{14.9}$$

where ω_n and ϕ_n are the nth natural frequency and mode shape (scaled to unit generalized mass), F_n is the generalized force in the nth mode, and η is the loss factor. It follows that the vibrational energy of the component (defined here as twice the time-averaged kinetic energy) has the form

$$E = (1/2)m_a\omega^2 \int_A |u(\mathbf{x})|^2 \, d\mathbf{x} = \sum_n \frac{c_n}{(\omega_n^2 - \omega^2)^2 + (\eta\omega\omega_n)^2}, \quad c_n = (1/2)\omega^2 |F_n|^2, \tag{14.10}$$

where A represents the domain of the component and m_a is the mass per unit area (or equivalent). Equations (14.9) and (14.10) each have the form of Equation (14.3), providing the numerator expressions have the properties that were assigned to a_n: that is, they are statistically independent and identically distributed for each n. To consider the case of Equation (14.10), this condition requires that the generalized forces are statistically independent and identically distributed, and certain types of forcing automatically meet this requirement – for example, rain-on-the-roof loading. For other types of loading, the requirement will place a demand on the properties of the mode shapes; for example, for single point forcing of amplitude F applied at $\mathbf{x}_0$ the generalized forces are given by $F_n = F\phi_n(\mathbf{x}_0)$, and the numerators c_n will be statistically independent and identically distributed only if the mode shapes $\phi_n(\mathbf{x}_0)$ also have this property. The statistical properties of the mode shapes of a random system have been considered by Brody et al. (1981), and it was shown from invariance arguments that if the system has a large number of degrees of freedom, then the mode shapes are statistically independent and Gaussian. This ultimately implies that the numerators of both Equations (14.9) and (14.10) can be considered to have the required statistical properties, and Equations (14.4)–(14.6) can thus be used to find the mean and variance of the response.

The statistics of the energy, Equation (14.10), has been considered by Langley and Brown (2004a) who used Equations (14.4)–(14.6) to show that

$$\mathrm{E}[E] = \frac{\mathrm{E}[c_n]\pi\nu}{2\eta\omega^3}, \quad \mathrm{RelVar}[E] = \frac{1}{\pi m}\{\alpha + q(m)\}, \quad \alpha = \frac{\mathrm{E}[c_n^2]}{\mathrm{E}[c_n]^2}, \quad m = \omega\eta\nu, \tag{14.11a–d}$$

where RelVar represents the relative variance (the variance divided by the squares of the mean), ν is the modal density, m is termed the modal overlap factor, and the function $q(m)$ depends on the model adopted for the system natural frequencies. For Poisson natural frequencies $q(m) = 0$, whereas for GOE natural frequencies (the case considered in the present chapter unless otherwise stated) the result is

$$q(m) = -1 + \frac{1}{2\pi m}[1 - \exp(-2\pi m)] + \mathrm{E}_1(\pi m)\left[\cosh(\pi m) - \frac{1}{\pi m}\sinh(\pi m)\right], \tag{14.12}$$

where E_1 is the exponential integral (Abramowitz & Stegun 1972). If the energy given by Equation (14.10) is frequency averaged over a bandwidth Δ centered on ω, then Equation (14.11) remains valid for the mean energy, but the relative variance is modified to (Langley & Brown 2004b)

$$\mathrm{RelVar}[E] = \frac{\alpha - 1}{\pi m B^2}\left\{2B\tan^{-1} B - \ln(1 + B^2)\right\} + \frac{\ln(1 + B^2)}{(\pi m B)^2}, \quad B = \frac{\Delta}{\omega\eta}. \tag{14.13}$$

In principle Equation (14.13) should reduce to Equations (14.11b)–(14.12) when $\Delta = 0$; however, in practice the equation reduces to $\mathrm{RelVar}[E] = (\alpha - 1)/(\pi m) + 1/(\pi m)^2$, which is an approximate form of Equations (14.11b)–(14.12) that is valid for $m > 1$. This aspect of the equations arises from approximations that are employed in the derivation of Equation (14.13) (Langley & Brown 2004a). Equation (14.11a) for the mean energy is in agreement with a well-known result derived by Cremer et al. (1990), and it is also the result which would be obtained by applying SEA (Lyon & DeJong 1995) to the component – in the latter context, it can be noted that the result can be rephrased in the standard form $\mathrm{E}[E] = \mathrm{E}[P]/(\omega\eta)$, where P is the power input from the applied loading.

The variance equations, Equations (14.11b) and (14.13), have been verified numerically (Langley & Brown 2004a, 2004b) and experimentally (Cotoni et al. 2005) for a single plate that is randomized by the addition of a number of small masses. Results for the experimental plate reported by Cotoni et al. (2005) are reproduced in Figure 14.1; the plate was made from steel, with area 2 m^2, thickness 1 mm, and loss factor 0.014 (arising mainly from added damping treatment). The plate mass was 15.6 kg, and nine point masses with a total mass of 1.6 kg were added in 19 sets of random locations to produce an ensemble of 19 random plates. The two cases shown in Figure 14.1, (a) and (b), correspond to different loading conditions and hence to different values of α. In case (a) a single point load is applied, and for Gaussian mode shapes this would yield $\alpha = 3$ according to Equation (14.11c). In practice an empirical value of $\alpha = 2.7$ has been employed, and a qualitative justification for this lower figure based on Brody statistics has been given by Langley and Cotoni (2005). Loading case (b) in Figure 14.1 corresponds to averaging the energy over five locations of the point load, and in this case the value of α is $\alpha = 1 + (2.7 - 1)/5 = 1.34$ (Cotoni et al. 2005). Given the limited size of the experimental ensemble, the level of agreement displayed between theory and experiment in Figure 14.1 is considered to be good, giving strong evidence in favor of the nonparametric natural frequency model discussed in the previous section. The fact that the results are relatively poor at low frequencies is consistent with the fact that the system natural frequencies cannot be expected to form a stationary point process at low frequencies, unless the degree of randomization is extremely large.

The statistics of the response at location $\mathbf{x}$, as given by Equation (14.9), has been studied by Langley and Cotoni (2005) for the case of a point load applied at location $\mathbf{x}_0$, so that $F_n = F\phi_n(\mathbf{x}_0)$. The response can be considered to have two components: (i) the ensemble average value and (ii) a random fluctuation around the ensemble average value. These components have an alternative more physical interpretation, in that the ensemble average value is equal to the response that would be obtained were the system of infinite extent; this is referred to as the "direct field" response, indicating the response obtained in the absence of any reflections from the component boundaries. In contrast, the fluctuating component is the response produced

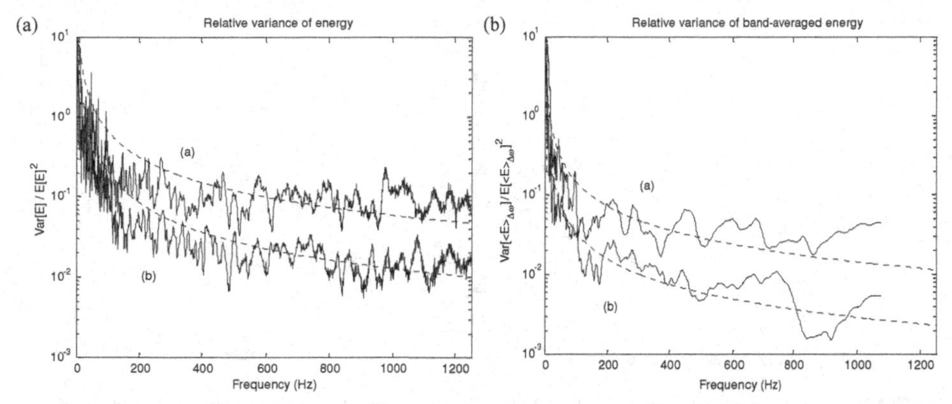

Figure 14.1. Experimental results for the variance of the energy of a point excited random plate. (a) Narrow band experimental results (irregular curves), compared with Equation (14.11b). (b) Band averaged experimental results (irregular curves) with $B = 10$, compared with Equation (14.13). In both figures, (a) results for a single point load and (b) results when the energy is averaged over five point load locations. Reprinted from Cotoni et al. (2005). Copyright 2005, with permission from Elevier.

by repeated reflections from the boundaries, and this is termed the "reverberant" response. It is shown by Langley and Cotoni (2005) that

$$\mathrm{E}\left[|u_{\mathrm{rev}}(\mathbf{x})|^2\right] = \frac{\pi \nu}{2M^2\omega^3\eta}\left\{1 + R^2(\mathbf{x}-\mathbf{x}_0)[2+q(m)]\right\}, \tag{14.14}$$

where the subscript rev represents the reverberant part of the response, M is the total mass of the component, and the function R is defined by

$$R(\mathbf{x}-\mathbf{x}_0) = M\,\mathrm{E}[\phi_n(\mathbf{x})\phi_n(\mathbf{x}_0)] = \begin{cases} \mathrm{J}_0(kr), & \text{two-dimensional systems,} \\ \sin(kr)/kr, & \text{three-dimensional systems.} \end{cases} \tag{14.15}$$

The function R is the correlation function of the mode shapes, which has been shown by Weaver and Burkhardt (1994) to be equal to the diffuse field correlation function of the component, as indicated on the right-hand side of Equation (14.15). Given that $R(\mathbf{x}-\mathbf{x}_0) \to 0$ as $|\mathbf{x}-\mathbf{x}_0| \to \infty$, Equation (14.14) suggests that the reverberant response is larger in the vicinity of the drive point than elsewhere in the component. In particular, for small values of the modal overlap factor m, $q(m) \to 0$, and the response *at* the forcing point is three times that at remote points; conversely for large m, $q(m) \to -1$, and the response at the forcing point is twice that at remote points. These results are initially surprising, because the reverberant response consists only of reflections from the component boundaries, and random scattering from the boundaries might be expected to produce a uniform energy density over the component, regardless of the location of the forcing point. Weaver and Burkhardt (1994) have used a time-reversal argument to explain the physics of this result, which arises because the reflected waves are correlated in the vicinity of the forcing point. Langley and Cotoni (2005) have investigated the validity of Equation (14.14) via numerical simulations of the response of a random plate, and the results obtained are shown in Figure 14.2.

The plate considered had a modal density of 0.104 modes/Hz, and an ensemble of 200 plates was created by adding small masses in random positions. The concentration of vibrational energy density at the forcing point is shown Figure 14.2(a)

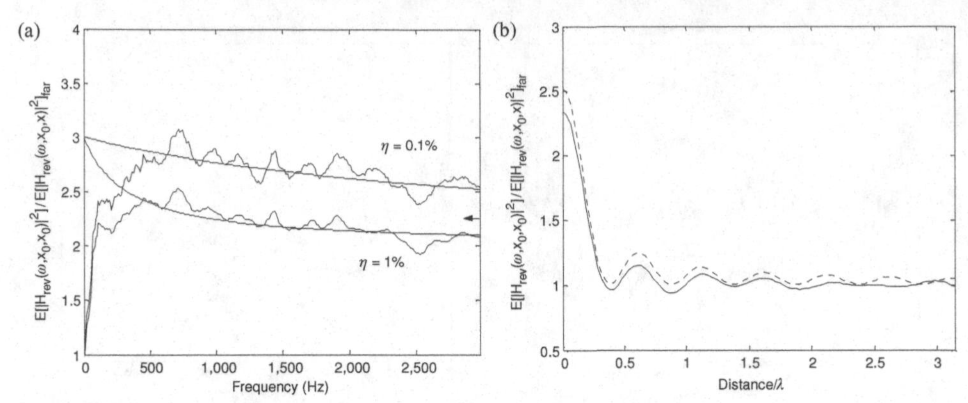

Figure 14.2. Validation of Equation (14.14) by simulations of a random plate. (a) Reverberant energy density at the forcing point divided by the energy density in the far field for two levels of damping (irregular curves from simulations, smooth curves from theory). (b) Reverberant, energy density as a function of the distance from the forcing point (dashed curve, theory; solid curve, simulations). Reprinted with permission from Langley and Cotoni (2005). Copyright 2005, Acoustical Society of America.

for two levels of damping – the lower level of damping gives a modal overlap of $m = 0.314$ at 3 kHz, whereas the higher level of damping has $m = 3.14$ at this frequency. Good agreement between the simulations and the theory can be seen, apart from at frequencies below 400 Hz, where it is unreasonable to assume that the natural frequencies form a stationary random process, because of the level of randomization employed. The arrow in the figure indicates the value obtained by assuming Poisson rather than GOE natural frequencies – this result is independent of frequency and damping and clearly does not agree well with the simulation data. The energy density of the plate is shown as a function of the distance from the forcing point in Figure 14.2(b), for the case of excitation at 3 kHz and the lower value of the loss factor (so that $m = 0.314$). Again good agreement between theory and simulation can be seen, with the discrepancies being traced to the fact that the mode shapes are not entirely Gaussian, as has been discussed previously regarding the choice of the value of α for use in Equation (14.11b).

The relative variance of the modulus squared reverberant response has been considered by Lobkis et al. (2000) and Langley and Cotoni (2005) . The result for a point remote from the forcing point is

$$\mathrm{RelVar}\left[\left|u_{\mathrm{rev}}(\mathbf{x})\right|^{2}\right] = 1 + \left(\frac{1}{\pi m}\right)\left\{\frac{\mathrm{E}[\phi_n^4]^2}{\mathrm{E}[\phi_n^2]^4} - f(\pi m)\right\}, \tag{14.16}$$

where

$$f(x) = \left(\frac{x}{4} + 2 - \frac{5}{4x}\right) + \mathrm{e}^{-2x}\left(\frac{x}{4} + \frac{1}{2} + \frac{5}{4x}\right) - \mathrm{E}_1(x)\mathrm{e}^{-x}\left(\frac{x^2}{4} + \frac{3x}{4} + \frac{5}{2} + \frac{5}{2x}\right)$$
$$- \mathrm{E}_1(x)\mathrm{e}^{x}\left(\frac{x^2}{4} - \frac{3x}{4} + \frac{5}{2} - \frac{5}{2x}\right). \tag{14.17}$$

This result is compared with numerical simulations for a random plate in Figure 14.3, after Langley & Cotoni (2005). The plate is the same as that employed in Figure 14.2, and as before, two levels of damping are considered. For each level of damping, four

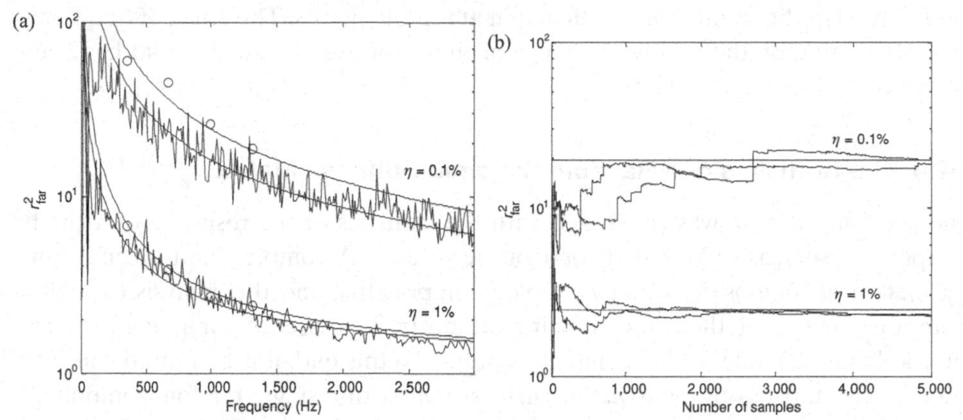

Figure 14.3. The relative variance of the modulus squared response at a point remote from the forcing point on a random plate. (a) The relative variance (irregular curve, 200 samples; circles, 10,000 samples; solid curve, GOE prediction; dashed curve, Poisson prediction). (b) Convergence of the relative variance at 1322 Hz – results for two different response points are shown for each level of damping. The horizontal lines are the GOE predictions. Reprinted with permission from Langley and Cotoni (2005). Copyright 2005, Acoustical Society of America.

sets of results are shown in Figure 14.3(a): (i) simulation results obtained using 200 samples, (ii) simulation results obtained using 10,000 samples, (iii) analytical results from Equation (14.16), and (iv) analytical results obtained by using Poisson (rather than GOE) natural frequencies, in which case $f(\pi m) = 0$ in Equation (14.16). The use of a limited number of samples (200) suggests that the Poisson assumption is more realistic, although this conclusion is reversed when more samples are considered. The convergence of the relative variance with the number of samples is shown in Figure 14.3(b) for a frequency of 1322 Hz, at which the modal overlap factors for the two damping values are 0.137 and 1.37. It is clear that at low damping a very large number of samples is required to obtain an accurate result for the relative variance. It can be seen that particular samples cause a step increase in the relative variance – such samples have a very large response, and this is caused by a mode with a natural frequency that is very close to the excitation frequency and a mode shape that is large at both the forcing point and the response point. This is a very rare occurrence, but it has a significant influence on the response variance, and enough samples must be considered to capture this effect. The behavior is less pronounced at higher levels of damping because the modal overlap is higher and the response is less sensitive to the properties of any single mode. A comparison of Figures 14.1 and 14.3 highlights the fact that the relative variance of the energy density at a single point, as given by Equation (14.16), is higher than the relative variance of the total energy of the system, as given by Equation (14.11b). At very high modal overlap the relative variance of the total energy tends to zero, whereas that of the energy density at a point tends to unity. The latter behaviour is consistent with a result owing to Schroeder (1987), who showed that the response becomes a complex Gaussian process at high modal overlap, which means that the modulus squared response has an exponential distribution and a relative variance of unity.

The relative variance of the modulus squared response *at* the forcing point has been considered by Langley and Cotoni (2005) , and it has been found that at low modal overlap the result is higher than that at remote points. The effect of frequency band averaging on the relative variance of the response at a single point has been discussed by Cotoni et al. (2005).

14.4 The Statistics of a Random Dynamic Stiffness Matrix

The previous section was concerned with the statistics of the response of a single component, such as a plate, shell, or acoustic volume. A complex built-up engineering system is composed of many coupled components, and the analysis of such a system requires a method of combining the properties of each component to produce a dynamic model of the complete system. To this end, it is helpful to consider initially the statistics of the *dynamic stiffness matrix* of a single random component because the dynamic properties of the whole system can be assembled from the dynamic stiffness matrices of the individual components. The dynamic stiffness matrix $\mathbf{D}$ of a component is defined by the relation

$$\mathbf{Dq} = \mathbf{f}, \tag{14.18}$$

where $\mathbf{q}$ is the set of degrees of freedom of the component and $\mathbf{f}$ is the vector of applied forces. The degrees of freedom employed in Equation (14.18) can be specified in many ways, and the details of $\mathbf{D}$ will depend on the precise definition adopted. If, for example, the component is modeled by the finite element method and $\mathbf{q}$ is taken to be complete set of finite element degrees of freedom, then the dynamic stiffness matrix will have the form $\mathbf{D} = -\omega^2\mathbf{M} + \mathrm{i}\omega\mathbf{C} + \mathbf{K}$, where ω is the excitation frequency and $\mathbf{M}$, $\mathbf{C}$, and $\mathbf{K}$ are respectively the component mass, damping, and stiffness matrices. On the other hand, $\mathbf{q}$ might be restricted to the degrees of freedom at which the component is coupled with other components (for example, the ends of a beam or the boundary of a plate), in which case $\mathbf{D}$ will be a complex function of ω, which fully accounts for the dynamics of all the degrees of freedom that are not explicitly included in $\mathbf{q}$. The following analysis is applicable to any definition of $\mathbf{q}$. Given that the component has random properties, the dynamic stiffness matrix will be random, and Equation (14.18) can be written in the form

$$\left\{\mathrm{E}[\mathbf{D}] + \mathbf{D}_{\mathrm{ran}}\right\}\mathbf{q} = \mathbf{f}, \tag{14.19}$$

where $\mathbf{D}_{\mathrm{ran}}$ represents the zero-mean random variation of $\mathbf{D}$ around the average value. Langley (2007) has shown that $\mathrm{E}[\mathbf{D}]$ can be interpreted physically as the dynamic stiffness matrix arising from the direct field response, that is, the response in the absence of any reflections from the system boundaries, and the notation $\mathrm{E}[\mathbf{D}] = \mathbf{D}_{\mathrm{dir}}$ is adopted to highlight this fact. Similarly, the term $\mathbf{D}_{\mathrm{ran}}\mathbf{q}$ can be interpreted as the set of forces arising from the reverberant field, that is, repeated reflections from the boundary, with the notation $\mathbf{D}_{\mathrm{ran}}\mathbf{q} = -\mathbf{f}_{\mathrm{rev}}$. Equation (14.19) then becomes

$$\mathbf{D}_{\mathrm{dir}}\mathbf{q} = \mathbf{f} + \mathbf{f}_{\mathrm{rev}}. \tag{14.20}$$

The matrix $\mathbf{D}_{\mathrm{dir}}$ can be computed by a variety of methods (Shorter & Langley 2005b) and thus the only remaining uncertainty in Equation (14.20) lies in $\mathbf{f}_{\mathrm{rev}}$; the major advantage of this rearrangement of Equation (14.19) is that the component randomness is transferred to the right-hand side of the equation and can be considered to act through an applied load, but to make further progress the relevant statistical properties of $\mathbf{f}_{\text{rev}}$ need to be established. To this end, Equations (14.19) and (14.20) can be combined to yield

$$\mathbf{f}_{\text{rev}} = \mathbf{D}_{\text{dir}}(\mathbf{H} - \mathbf{H}_{\text{dir}})\mathbf{f}, \tag{14.21}$$

where $\mathbf{H} = \mathbf{D}^{-1}$is the receptance matrix and $\mathbf{H}_{\text{dir}} = \text{E}[\mathbf{H}]$. In deriving Equation (14.21) the relationship $\mathbf{H}_{\text{dir}} = \mathbf{D}_{\text{dir}}^{-1}$ has been employed; this relationship does *not* follow directly from the definition of the two matrices – in fact the result is initially rather surprising because it states that the average value of the inverse of the dynamic stiffness matrix is equal to the inverse of the average value of the matrix. As discussed by Langley (2007), the result arises from the fact that $\mathbf{D}_{\text{dir}}$ and $\mathbf{H}_{\text{dir}}$ correspond to the same *physical* system, that is, a system in which there are no reflections from the boundary, or, equivalently, an infinitely extended system. Equation (14.21) can be used to derive the statistical properties of $\mathbf{f}_{\text{rev}}$ by noting that the entries of the receptance matrix have the form of Equation (14.9), and thus random point process theory can be applied. It is shown by Langley (2007) that $\mathbf{E}[\mathbf{f}_{\text{rev}}] = 0$ and that

$$\text{E}[\mathbf{f}_{\text{rev}}\mathbf{f}_{\text{rev}}^{*T}] = \left(\frac{4E}{\pi\omega\nu}\right)\Im\{(\mathbf{D}_{\text{dir}})\} + \left(\frac{2}{\pi m}\right)\{2\Re\{\mathbf{S}_{ff}\} + q(m)\mathbf{S}_{ff}\}, \tag{14.22}$$

$$\mathbf{S}_{ff} = \mathbf{E}[\hat{\mathbf{f}}\hat{\mathbf{f}}^{*T}], \qquad \hat{\mathbf{f}} = \mathbf{i}\mathbf{D}_{\text{dir}}\Im\{\mathbf{H}_{\text{dir}}\}\mathbf{f}, \tag{14.23}$$

where E is the (ensemble averaged) vibrational energy of the component, which is defined as twice the time-averaged kinetic energy. In the majority of cases of interest the first term dominates Equation (14.22), so that

$$\text{E}[\mathbf{f}_{\text{rev}}\mathbf{f}_{\text{rev}}^{*T}] = \left(\frac{4E}{\pi\omega\nu}\right)\Im\{\mathbf{D}_{\text{dir}}\}. \tag{14.24}$$

This result was first derived by Shorter and Langley (2005a) by using wave scattering arguments, and it represents a generalization of earlier results obtained for a single degree of freedom system by Smith (1962). The equation has been termed the "diffuse field reciprocity relation," in that it relates the force exerted on the coordinates $\mathbf{q}$ by a diffuse wavefield (corresponding to the reverberant response of the component) to the direct field radiation properties of the coordinates, as described by $\Im\{\mathbf{D}_{\text{dir}}\}$. Equations (14.20) and (14.24) allow the analysis of complex built-up random systems to be performed in a relatively straightforward way: rather than consider the properties of the system via the assembly of a set of random dynamic stiffness matrices, a set of deterministic matrices ($\mathbf{D}_{\text{dir}}$) can be considered together with a set of random forces $\mathbf{f}_{\text{rev}}$ whose properties can be expressed in terms of the component energies via Equation (14.24). The component energies E are not known a priori but rather must be computed as part of the analysis procedure, as explained in the following section.

It can be noted that the form of the second term in Equation (14.22) depends on the model adopted for the natural frequencies of the system (for example, Poisson and GOE, but the leading term, that is, Equation (14.24), is obtained for all types of stationary natural frequency model and is therefore robust against modeling assumptions.

14.5 The Mean Response of a Built-Up System

In this section a complex built-up system is considered to be an assembly of (i) deterministic components and (ii) random components, and the random components are taken to have the properties described in the previous section. Clearly this is an idealization, in that, first, no component is ever truly deterministic, and second, some random components may not be sufficiently random to display the stationary natural frequency statistics assumed in the previous section. Nonetheless, most practical systems display a clear demarcation between highly random components and near-deterministic components – those components that have a short vibrational wavelength, such as thin plates and shells, are very sensitive to imperfections, whereas those that have a longer vibrational wavelength, such as beams and columns, are fairly robust against imperfections. Thus in an automotive structure, the roof and door panels are very sensitive to imperfection, whereas the side rails, pillars, and other stiff parts of the structure are not. A generalization of the present approach would be to consider a built-up system to be an assembly of (i) components that are not highly random and (ii) components that are sufficiently random to display stationary natural frequency statistics and conform to nonparametric statistical models. The analysis would then proceed by applying the present approach with the components in category (i) being assumed to be deterministic. The results obtained could then be randomized by using a parametric model of the uncertainties in the category (i) components, assuming that the appropriate statistical data is available. In reality, statistical data of this type is very difficult to obtain, and furthermore, the uncertainty in the system response can be expected to be dominated by the highly random components (providing enough components of this type are present), thus implying that the effort required to perform the parametric analysis might not be rewarded by any significant change in the results. This argument suggests that the assumptions employed in the present approach should in most cases be adequate, except at very low frequencies where few if any of the components fall into category (ii), thus making some type of parametric model essential.

The complete set of degrees of freedom of the deterministic components is labeled $\mathbf{q}$, and the assembled dynamic stiffness matrix of these components is labeled $\mathbf{D}_d$; this matrix can be obtained by using the finite element method or any other appropriate deterministic analysis method. The random components are taken to be coupled with the deterministic system, and the dynamic stiffness matrix of the kth component, when expressed in the coordinates $\mathbf{q}$, is written as $\mathbf{D}^{(k)}$. It can be noted that

(i) typically the kth random component will be connected to only a subset of $\mathbf{q}$ but nonetheless by adding appropriate zeros $\mathbf{D}^{(k)}$ can be expanded to the full dimension of $\mathbf{q}$;

(ii) in some situations two random components may be directly coupled with each other with no obvious deterministic component lying between them – in this case the set $\mathbf{q}$ is taken to include a set of degrees of freedom that describes the motion of the junction between the two random components, and the appropriate partition of $\mathbf{D}_d$ is set to zero (i.e., a zero stiffness deterministic component is considered to be present);

(iii) the complete set of degrees of freedom of a random component includes freedoms other than $\mathbf{q}$ – the matrix $\mathbf{D}^{(k)}$ should be interpreted as the *reduced* dynamic stiffness matrix that fully accounts for the presence of these freedoms (for example, the reduced dynamic stiffness matrix for the boundary of a plate takes due account of all of the interior dynamics of the plate).

If the dynamic stiffness matrix of each of the random components is taken to have the properties outlined in Equations (14.19) and (14.20), then the equations of motion of the whole system can be written in the form

$$\mathbf{D}_{\text{tot}}\mathbf{q} = \mathbf{f} + \sum_k \mathbf{f}_{\text{rev}}^{(k)}, \qquad \mathbf{D}_{\text{tot}} = \mathbf{D}_d + \sum_k \mathbf{D}_{\text{dir}}^{(k)}, \tag{14.25}$$

where the superscript denotes the kth component. It follows immediately from Equations (14.24) and (14.25) that

$$\mathbf{S}_{qq} = \text{E}\left[\mathbf{q}\mathbf{q}^{*\mathrm{T}}\right] = \mathbf{D}_{\text{tot}}^{-1}\left[\mathbf{S}_{ff} + \sum_k \left(\frac{4E_k}{\omega\pi\nu_k}\right)\Im\left\{\mathbf{D}_{\text{dir}}^{(k)}\right\}\right]\mathbf{D}_{\text{tot}}^{-1*T}, \tag{14.26}$$

where $\mathbf{S}_{ff}$ is the cross-spectrum of the forces applied to the deterministic system, and E_k and ν_k are respectively the (ensemble-averaged) vibrational energy and the modal density of the kth random component. Now the net energy flow into the kth random component must be dissipated through damping and therefore be equal to $\omega\eta_k E_k$, where η_k is the internal loss factor. A detailed consideration of this condition for each random component leads to the following set of equations (Shorter and Langley 2005b):

$$\omega(\eta_j + \eta_{\text{d},j})E_j + \sum_k \omega\eta_{jk}\nu_j(E_j/\nu_j - E_k/\nu_k) = P_j + P_{\text{in},j}^{\text{ext}}, \qquad j = 1, 2, \ldots, N_r, \tag{14.27}$$

where N_r is the number of random components and

$$\omega\eta_{\text{d},j} = \left(\frac{2}{\pi\nu_j}\right)\sum_{r,s}\Im\left\{D_{d,rs}\right\}\left(\mathbf{D}_{\text{tot}}^{-1}\Im\left\{\mathbf{D}_{\text{dir}}^{(j)}\right\}\mathbf{D}_{\text{tot}}^{-1*T}\right)_{rs}, \tag{14.28}$$

$$\omega\eta_{jk}\nu_j = (2/\pi)\sum_{r,s}\Im\left\{D_{\text{dir},rs}^{(j)}\right\}\left(\mathbf{D}_{\text{tot}}^{-1}\Im\left\{\mathbf{D}_{\text{dir}}^{(k)}\right\}\mathbf{D}_{\text{tot}}^{-1*T}\right)_{rs}, \tag{14.29}$$

$$P_{\text{in},j}^{\text{ext}} = (\omega/2)\sum_{r,s}\Im\left\{D_{\text{dir},rs}^{(j)}\right\}\left(\mathbf{D}_{\text{tot}}^{-1}\mathbf{S}_{ff}\mathbf{D}_{\text{tot}}^{-1*T}\right)_{rs}. \tag{14.30}$$

In Equation (14.27), $P_{\text{in},j}^{\text{ext}}$ is the power input to component j arising from the external forces on the deterministic system, whereas P_j is the power input from external forces applied directly to the component, which can be estimated by standard methods (Lyon and DeJong 1995). Equation (14.27) has exactly the form of the SEA equations; in the present approach all of the coefficients that appear in the equation are given in closed form by Equations (14.28)–(14.30), and thus Equation (14.27) can be solved to yield the component energies E_j, which can then be substituted into Equation (14.26) to yield the response of the deterministic system. It should be emphasized that E_j represents the ensemble-averaged energy of component j, and thus Equations (14.26)–(14.30) predict the mean response of the system in the form of (i) the ensemble-averaged energy of the random components and (ii) the

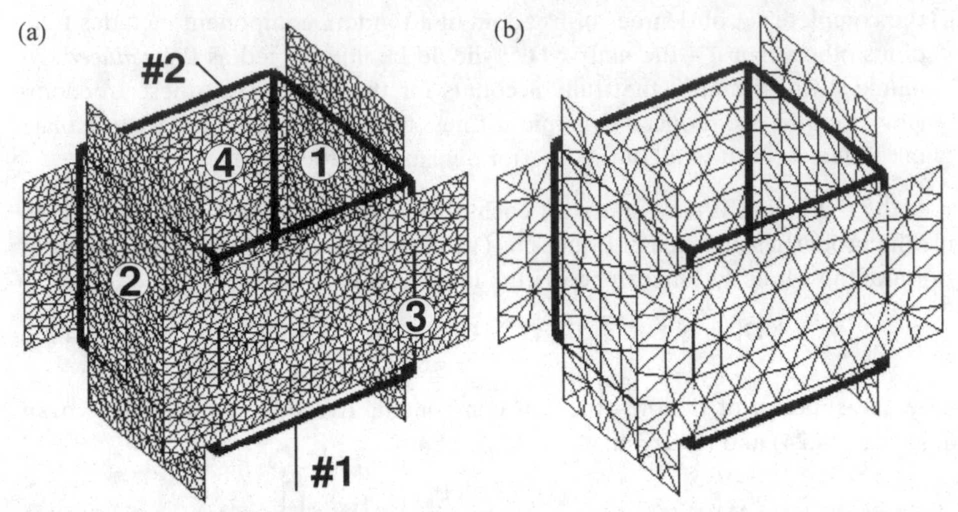

Figure 14.4. The plate/beam structure (Cotoni et al. 2008). (a) The benchmark FE model. (b) The deterministic part of the hybrid FE–SEA model (beam elements and in-plane plate elements). Reprinted with permission from Cotoni et al. (2008). Copyright 2007, Acoustical Society of America.

ensemble-averaged cross-spectrum of the response of the deterministic system. In line with standard SEA terminology, the random components can also be referred to as "subsystems."

Equations (14.26)–(14.30) enable a complex system to be analysed by using a combination of the FE method and SEA – the system is modeled as an assembly of FE components and SEA subsystems, and the coupling between these very different types of description is affected by using Equation (14.24). In addition, other deterministic modeling approaches, such as the boundary element method, can readily be included within the method by simply adding the appropriate degrees of freedom to $\mathbf{q}$. Some features of the method are as follows: (i) short-wavelength random components are modeled by using a single energy variable, rather than the hundreds or thousands of degrees of freedom that would be required to model the component using a standard approach, and hence the method is very computationally efficient; (ii) the method automatically computes the ensemble average response, and thus there is no need to perform Monte Carlo simulations and then average the computed response, and furthermore, the extension of the method to response variance prediction is discussed in the following section; (iii) the method does not require input regarding the physical uncertainties in the system because the uncertainty is captured using a nonparametric model; (iv) if required, the method could readily be combined with parametric models of the uncertainty in the "deterministic" parts of the system.

Equations (14.28)–(14.30) have been applied to a wide range of example systems (for example, Charpentier et al. 2006, Cotoni et al. 2008). The example considered here is taken from Cotoni et al. (2008) and concerns a simple plate/beam structure, which is shown in Figure 14.4. The structure is made from aluminum, and the beams are all 0.6-m long, with a square hollow section of external width 25.4 mm and thickness 3.2 mm. The plates are of dimension 0.6 m $\times$ 1.1 m and thickness 1 mm.

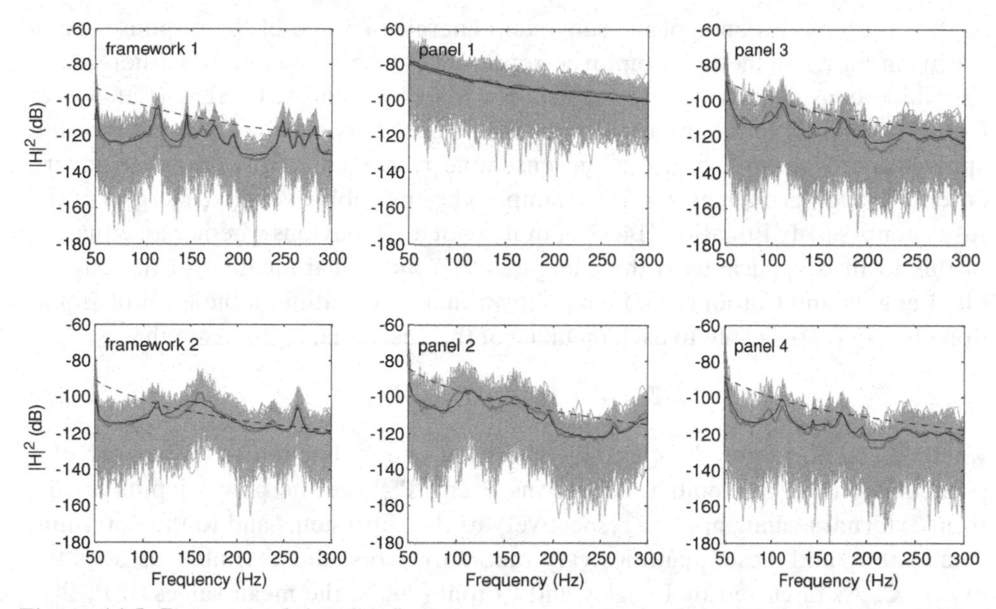

Figure 14.5. Response of the plate/beam structure to a point force applied to plate 1. The light gray line denotes Monte Carlo results from each of the 200 detailed FE models. The bold gray line denotes ensemble average of the Monte Carlo results. The solid black line denotes prediction from the hybrid FE–SEA method. The dotted line denotes result obtained using a pure SEA model. Reprinted with permission from Cotoni et al. (2008). Copyright 2007, Acoustical Society of America.

Each plate is connected to the beam framework via four bolts, and a benchmark ensemble of 200 FE models has been generated by adding 10 small masses (each having 2% of the plate mass) to each of the plates in random positions.

The detailed FE model employed 10,481 degrees of freedom, whereas the hybrid FE–SEA model (i.e., the model used in the present approach) employed 1,184 degrees of freedom to model the deterministic beams and the in-plane motion of the plates, together with four SEA subsystems, to model the plate bending motion. The response of the system to a force applied to plate 1 (as indicated in Figure 14.4) is shown in Figure 14.5. The results are shown in terms of the modulus squared response – for the plate subsystems this has been deduced from the computed plate energy by assuming that the energy density is constant over the plate. Good agreement between the hybrid calculations and the detailed FE Monte Carlo results can be seen in Figure 14.5. Also shown in the figure are the results obtained by treating each component of the system as an SEA subsystem, which leads to poor results for this structure.

The hybrid equations, Equations (14.26)–(14.30), lead to a prediction of the ensemble mean response. Also of interest from a practical engineering point of view is the spread in the results, as represented by the gray cloud of individual sample results shown in Figure 14.5. This issue is considered in the following section.

14.6 The Variance of the Response of a Built-Up System

The hybrid FE–SEA method described in the previous section has recently been extended to the prediction of the response variance (Langley & Cotoni 2007), that is, the ensemble variance of the subsystem energies E_j and of the response cross-spectrum $\mathbf{S}_{qq}$. The method employed represents an extension of an earlier analysis, which considered variance prediction within the context of SEA (Langley & Cotoni 2004). To consider variance issues it is necessary to develop equations that apply to *each member* of a random ensemble rather than equations that govern ensemble-averaged quantities. For example, the ensemble-averaged energies of the subsystems satisfy Equation (14.27), but it is not at all obvious whether an equation of this form is applicable to the energies of an individual member of the ensemble. Langley and Cotoni (2007) have shown that an equation in the form of Equation (14.27) *is* applicable to each member of the ensemble, in the sense that

$$\mathbf{C}\bar{\mathbf{E}} = \mathbf{P} + \mathbf{P}_{\text{in}}^{\text{ext}}, \quad \bar{E}_j = E_j/\nu_j, \tag{14.31}$$

where E_j is no longer the ensemble-averaged energy but rather the energy of a particular ensemble member. The terms $\mathbf{P}$ and $\mathbf{P}_{\text{in}}^{\text{ext}}$ are the power inputs arising from external loading applied respectively to the subsystems and to the deterministic system, and these quantities vary randomly across the ensemble, as does the matrix $\mathbf{C}$. As discussed by Langley and Cotoni (2007), the mean values of $\mathbf{P}$, $\mathbf{P}_{\text{in}}^{\text{ext}}$, and $\mathbf{C}$ correspond to those values employed in Equations (14.27)–(14.30), and this implies that the mean value of $\mathbf{C}$ is symmetric, although this is not necessarily the case for a particular realization of $\mathbf{C}$. By considering a perturbation expansion of Equation (14.31) and enforcing power conservation requirements that apply to $\mathbf{C}$, it can be shown that

$$\begin{aligned}\text{Cov}[\bar{E}_i, \bar{E}_j] = {} & \sum_k \sum_s C_{0,ik}^{-1} C_{0,js}^{-1} \,\text{Cov}[P_k + P_{\text{in},k}^{\text{ext}}, P_s + P_{\text{in},s}^{\text{ext}}] \\ & + \sum_k \sum_s \sum_{r \neq k} \left[(C_{0,ik}^{-1} - C_{0,ir}^{-1}) C_{0,js}^{-1} + (C_{0,jk}^{-1} - C_{0,jr}^{-1}) C_{0,is}^{-1} \right] \hat{E}_r \,\text{Cov}[C_{kr}, P_s + P_{\text{in},s}^{\text{ext}}] \\ & + \sum_k \sum_p \sum_{s \neq k} \sum_{r \neq p} (C_{0,ik}^{-1} - C_{0,is}^{-1})(C_{0,jp}^{-1} - C_{0,jr}^{-1}) \hat{E}_s \hat{E}_r \,\text{Cov}[C_{ks}, C_{pr}],\end{aligned} \tag{14.32}$$

where $C_{0,jk}^{-1}$ represents the jkth entry of E[$\mathbf{C}$] and $\hat{E}_j$ is the ensemble average of $\bar{E}_j$. Thus everything on the right-hand side of Equation (14.32) is known in terms of the mean hybrid equations and their solution, apart from the various Cov terms, which involve the power inputs and the entries of the matrix $\mathbf{C}$. These terms can be derived by using random matrix theory and the diffuse field reciprocity relation, and full details have been given by Langley and Cotoni (2007). Similarly, the variance of the response of the deterministic system (because of the randomness in the subsystems) can be shown to be

$$\text{Var}[(\mathbf{S}_{qq})_{ij}] = 2(\mathbf{D}_{\text{tot}}^{-1}\mathbf{S}_{ff}\mathbf{D}_{\text{tot}}^{-\mathbf{T}*})_{ij} \sum_k \hat{E}_k G_{ij}^{(k)} + \sum_{k,s} \{2\,\text{Cov}[\bar{E}_k, \bar{E}_s] + \hat{E}_k \hat{E}_s\} G_{ij}^{(k)} G_{ij}^{(s)}, \tag{14.33}$$

$$\mathbf{G}^{(k)} = \left(\frac{4a_k}{\omega\pi}\right) \mathbf{D}_{\text{tot}}^{-1} \Im\{\mathbf{D}_{\text{dir}}^{(k)}\} \mathbf{D}_{\text{tot}}^{-\mathbf{T}*}, \tag{14.34}$$

where the parameter a_k has been detailed by Langley and Cotoni (2007).

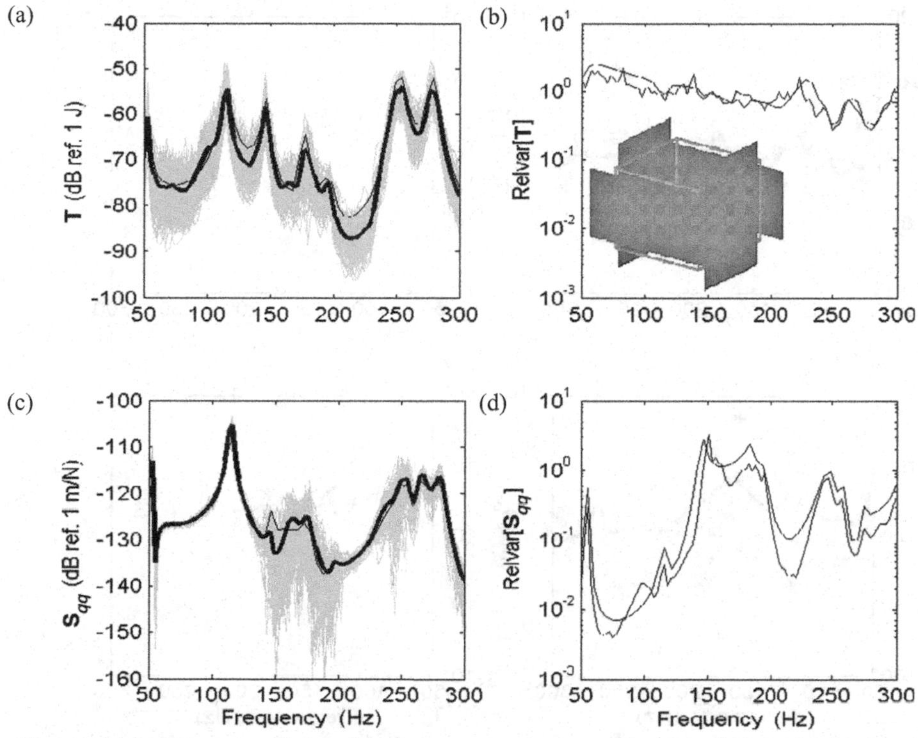

Figure 14.6. Response of the plate beam structure to a force applied to the framework, as calculated by the Monte Carlo method (irregular curves) and the hybrid method (smooth curves). (a), (b) The response of a plate. (c), (d) The response at a point on the framework. (a), (c) The mean response, together with the Monte Carlo simulations. (b), (d) The relative variance. Reprinted with permission from Langley and Cotoni (2007). Copyright 2007, Acoustical Society of America.

The variance equations have been applied to the beam/plate structure considered previously in Figures 14.4 and 14.5, and the results obtained are shown in Figures 14.6 and 14.7. It can be seen that the predictions are generally in good agreement with the detailed FE Monte Carlo simulations, whether the forcing is applied to the framework (Figure 14.6) or to a plate (Figure 14.7). Note that although the framework itself is deterministic, the fact that it is coupled with a set of random plates produces a random response, and this can be seen clearly in both the Monte Carlo results and the hybrid predictions.

14.7 Concluding Remarks

This chapter has reviewed some of the physical aspects of the response of a complex random system, and has summarized recent work on the development of a hybrid FE–SEA prediction method. This is a nonparametric approach that has the following main features relative to the standard FE method: (i) random subsystems that have a short wavelength of vibration are modeled as SEA subsystems, leading to a very large reduction in the number of degrees of freedom in the model; (ii) the effect of uncertainties is modeled by using a nonparametric method, and so there is no requirement for detailed data regarding the statistics of the system uncertainties;

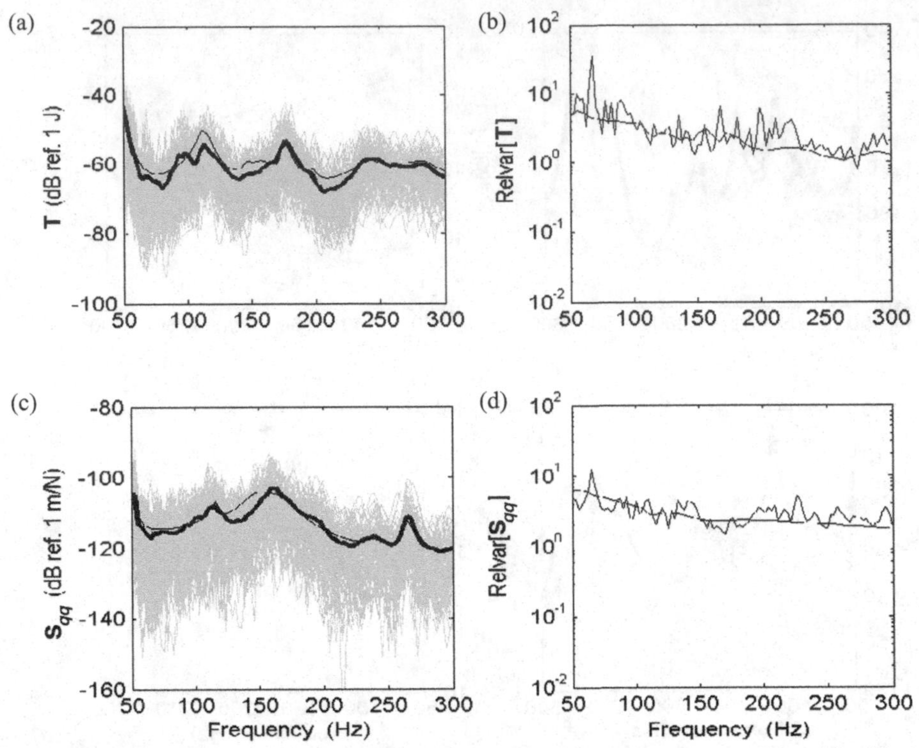

Figure 14.7. As for Figure 14.6 except the force is now applied to a plate. (a), (b) A non-driven plate (i.e., the force is applied to a different plate). Reprinted with permission from Langley and Cotoni (2007). Copyright 2007, Acoustical Society of America.

(iii) the method yields the mean and variance of the system response, without any requirement for Monte Carlo simulations or any other numerically intensive uncertainty propagation scheme. When compared with a standard SEA analysis, the main distinguishing features are as follows: (i) the method employs the FE method to model low-modal-density regions of the structure that cannot be represented by SEA subsystems; (ii) the coupling loss factors required for the SEA part of the analysis are all expressed in terms of the subsystem direct-field dynamic stiffness matrices; (iii) the method yields the ensemble variance of the response, in addition to the mean value. The key enabling result behind the method is the diffuse field reciprocity relation (Shorter & Langley 2005a, Langley 2007), which allows SEA and FE system descriptions to be employed and coupled within the same model. As discussed in Section 14.5, one possible enhancement of the method would be to combine the approach with a parametric model of uncertainties in the FE part of the model.

The chapter has illustrated one way in which recent advances in random matrix theory can be exploited in engineering. Other approaches and viewpoints are presented in the other chapters, and it is hoped that together these would form a wide-ranging overview of developments in this exciting field.

References

Abramowitz, M. & Stegun, I. (1972), *Handbook of Mathematical Functions*, Dover.

Achenbach, J. D., Gautesen, A. K. & McMaken, H. (1982), *Ray Methods for Waves in Elastic Solids*, Pitman, Boston.

Adler, R. J. (1981), *The Geometry of Random Fields*, Wiley, New York.

Aki, K. (1992), "Scattering conversions P to S versus S to P," *Bull. Seism. Soc. Am.* **82**(4), 1969–1972.

Aki, K. & Chouet, B. (1975), "Origin of coda waves: Source, attenuation, and scattering effects," *J. Geophys. Res.* **80**(23), 3322–3342.

Aki, K. & Richards, P. G. (2002), *Quantitative Seismology, Theory and Methods*, University Science Books, New York.

Akolzin, A. & Weaver, R. L. (2004), "Generalized Berry conjecture and mode correlation in chaotic plates," *Phys. Rev. E* **70**, 046212.

Albeverio, S. & Šeba, P. (1991), "Wave chaos in quantum systems with point interactions," *J. Stat. Phys.* **64**(1–2), 369–383.

Andersen, A., Ellegaard, C., Jackson, A. D. & Schaadt, K. (2001), "Random matrix theory and acoustic resonances in plates with an approximate symmetry," *Phys. Rev. E* **63**, 066204.

Anderson, T. W. (1958), *Introduction to Multivariate Statistical Analysis*, John Wiley, New York.

Apresyan, L. A. & Kravstov, Y. A. (1996), *Radiation Transfer: Statistical and Wave Aspects*, Gordon and Breach, Amsterdam.

Argaman, N., Imry, Y. & Smilansky, U. (1993), "Semiclassical analysis of spectral correlations in mesoscopic systems," *Phys. Rev. B* **47**(8), 4440–4447.

Arnol'd, V. I. (1978), *Mathematical Methods of Classical Mechanics*, Springer, Berlin.

Arnst, M., Clouteau, D., Chebli, H., Othman, R. & Degrande, G. (2006), "A non-parametric probabilistic model for ground-borne vibrations in buildings," *Probab. Eng. Mech.* **21**(1), 18–34.

Auld, B. A. (1973), *Acoustic Fields and Waves in Solids I, II*, Wiley, New York.

Aurich, R., Bäcker, A., Schubert, R. & Taglieber, M. (1999), "Maximum norms of chaotic quantum eigenstates and random waves," *Physica D* **129**, 1–14.

Babic, V. M. & Buldyrev, V. S. (1991), *Short Wavelength Diffraction Theory*, Springer, Berlin.

Balazs, N. L. & Voros, A. (1989), "The quantized Baker's transformation," *Ann. Phys.* **190**(1), 1–31.

Balian, R. (1968), "Random matrices and information theory," *Il Nuovo Cimento B* **57**, 183–193.

Balian, R. & Bloch, C. (1972), "Distribution of eigenfrequencies for the wave equation in a finite domain: III. Eigenfrequency density oscillations," *Ann. Phys.* **69**, 76–160.

Balian, R. & Bloch, C. (1974), "Solution of the Schrödinger equation in terms of classical paths," *Ann. Phys.* **85**(2), 514–545.

Baltes, H. P. & Hilf, E. R. (1976), *Spectra of Finite Systems*, Bibliographisches Insititut Wissenschaftsverlag, Mannheim.

Baranger, H. U., Jalabert, R. A. & Stone, A. D. (1993), "Quantum-chaotic scattering effects in semiconductor microstructures," *Chaos* **3**(4), 665–682.

Baranger, H. U. & Mello, P. A. (1994), "Mesoscopic transport through chaotic cavities: A random s-matrix theory approach," *Phys. Rev. Lett.* **73**(1), 142–145.

Barthélemy, J., Legrand, O. & Mortessagne, F. (2005), "Inhomogeneous resonance broadening and statistics of complex wave functions in a chaotic microwave cavity," *Europhys. Lett.* **70**, 162–168.

Beenakker, C. W. J. (1993), "Universality in the random-matrix theory of quantum transport," *Phys. Rev. Lett.* **70**(8), 1155–1158.

Beenakker, C. W. J. (1997), "Random matrix theory of quantum transport," *Rev. Mod. Phys.* **69**(3), 731–808.

Berkolaiko, G., Harrison, J. M. & Novaes, M. (2008), "Full counting statistics of chaotic cavities from classical action correlations," *J. Phys. A* **41**(36), 365102.

Beron-Vera, F. J. & Brown, M. G. (2003), "Ray stability in weakly range-dependent sound channels," *J. Acoust. Soc. Am.* **114**, 122–130.

Beron-Vera, F. J. & Brown, M. G. (2004), "Travel time stability in weakly range-dependent sound channels," *J. Acoust. Soc. Am.* **115**, 1068–1077.

Beron-Vera, F. J. & Brown, M. G. (2009), "Underwater acoustic beam dynamics," *J. Acoust. Soc. Am.* **126**(1), 80–91.

Beron-Vera, F. J., Brown, M. G., Colosi, J. A., Tomsovic, S., Virovlyansky, A. L., Wolfson, M. A. & Zaslavsky, G. M. (2003), "Ray dynamics in a long-range acoustic propagation experiment," *J. Acoust. Soc. Am.* **114**, 1226–1242.

Berry, M. V. (1977), "Regular and irregular semiclassical wave functions," *J. Phys. A* **10**, 2083–2091.

Berry, M. V. (1981a), "Quantizing a classically ergodic system: Sinai's billiard and the KKR method," *Ann. Phys.* **131**, 163–216.

Berry, M. V. (1981b), "Regularity and chaos in classical mechanics, illustrated by three deformations of a circular billiard," *Eur. J. Phys.* **2**, 91–102.

Berry, M. V. (1983), "Semiclassical mechanics of regular and irregular motion," *in* G. Iooss, R. H. G. Helleman & R. Stora, eds, *Chaotic Behavior of Deterministic Systems Les Houches Lecture Series*, Vol. 36, North Holland, Amsterdam, pp. 171–271.

Berry, M. V. (1985), "Semiclassical theory of spectral rigidity," *Proc. R. Soc. A* **400**, 229–251.

Berry, M. V. (1987), "Quantum chaology (the Bakerian lecture)," *Proc. R. Soc. A* **413**, 183–198.

Berry, M. V. (1989), "Quantum scars of classical closed orbits in phase space," *Proc. R. Soc. A* **423**, 219–231.

Berry, M. V. (1991), "Some quantum-to-classical asymptotics," *in* M.-J. Giannoni, A. Voros, & J. Zinn-Justin, eds, *Chaos and Quantum Physics*, Vol. 52 of Les Houches Lecture Series, North Holland, Amsterdam, pp. 251–304.

Berry, M. V. (2002), "Statistics of nodal lines and points in chaotic quantum billiards: Perimeter corrections, fluctuations, curvature," *J. Phys. A* **35**, 3025–3038.

Berry, M. V. & Dennis, M. R. (2008), "Boundary-condiiton-varying circle billiards and gratings: The Dirichlet singularity," *J. Phys. A* **41**, 135203.

Berry, M. V. & Ishio, H. (2002), "Nodal densities of Gaussian random waves satisfying mixed boundary condition," *J. Phys. A* **35**, 5961–5972.

Berry, M. V. & Ishio, H. (2005), "Nodal-line densities of chaotic quantum billiard modes satisfying mixed boundary conditions," *J. Phys. A* **38**, L513–L518.

Berry, M. V. & Keating, J. P. (1990), "A rule for quantizing chaos?," *J. Phys. A* **23**, 4839–4849.

Berry, M. V. & Mount, K. E. (1972), "Semiclassical approximations in wave mechanics," *Rep. Prog. Phys.* **35**, 315–97.

Berry, M. V. & Robnik, M. (1986), "Statistics of energy levels without time-reversal symmetry: Aharonov–Bohm chaotic billiards," *J. Phys. A* **19**, 649–668.

Berry, M. V. & Tabor, M. (1976), "Closed orbits and the regular bound spectrum," *Proc. R. Soc. A* **349**, 101–23.

Berry, M. V. & Tabor, M. (1977), "Level clustering in the regular spectrum," *Proc. R. Soc. A* **356**, 375–94.

Bertelsen, P. (1997), Quantum chaos and vibration of elastic plates, MSc thesis, Niels Bohr Institute, University of Copenhagen.

Bertelsen, P., Ellegaard, C. & Hugues, E. (2000), "Distribution of eigenfrequencies for vibrating plates," *Eur. Phys. J. B* **15**(1), 87–96.

Bies, W. E. & Heller, E. J. (2002), "Nodal structure of chaotic eigenfunction," *J. Phys. A* **35**, 5673–5685.

Biham, O. & Kvale, M. (1992), "Unstable periodic orbits in the stadium billiard," *Phys. Rev. A* **46**(10), 6334–6339.

Birkhoff, G. D. (1927), "On the periodic motions of dynamical systems," *Acta Math.* **50**, 359–379.

Blomgren, P., Papanicolaou, G. & Zhao, H. (2002), "Super-resolution in time-reversal acoustics," *J. Acoust. Soc. Am.* **111**(1), 230–248.

Blum, G., Gzutzmann, S. & Smilansky, U. (2002), "Nodal domain statistics: A criterion for quantum chaos," *Phys. Rev. Lett.* **88**(11), 114101.

Blümel, R. & Smilansky, U. (1988), "Classical irregular scattering and its quantum-mechanical implications," *Phys. Rev. Lett.* **60**(6), 477–480.

Blümel, R. & Smilansky, U. (1990), "Random matrix description of chaotic scattering: Semiclassical approach," *Phys. Rev. Lett.* **64**(3), 241–244.

Boasman, P. A. (1994), "Semiclassical accuracy for billiards," *Nonlinearity* **7**, 485–537.

Bogomolny, E. B. (1992), "Semiclassical quantization of multidimensional systems," *Nonlinearity* **5**(4), 805–866.

Bogomolny, E. B. (1988), "Smoothed wave-functions of chaotic quantum systems," *Physica D* **31**(2), 169–189.

Bogomolny, E. B. & Keating, J. P. (1996), "Gutzwiller's trace formula and spectral statistics: Beyond the diagonal approximation," *Phys. Rev. Lett.* **77**(8), 1472–1475.

Bogomolny, E., Gerland, U. & Schmit, C. (2001), "Singular statistics," *Phys. Rev. E* **63**, 036206.

Bogomolny, E. & Hugues, E. (1998), "Semiclassical theory of flexural vibrations of plates," *Phys. Rev. E* **57**, 5404–5424.

Bogomolny, E., Lebœuf, P. & Schmit, C. (2000), "Spectral statistics of chaotic systems with a pointlike scatterer," *Phys. Rev. Lett.* **85**(12), 2486–2489.

Bogomolny, E. & Schmit, C. (2002), "Percolation model of nodal line of nodal domains of chaotic wave functions," *Phys. Rev. Lett.* **88**(11), 114102.

Bohigas, O. (1991), "Random matrix theories and chaotic dynamics," *in* M.-J. Giannoni, A. Voros & J. Zinn-Justin, eds, *Chaos and Quantum Physics*, Vol. 52 of Les Houches Lecture Series, North Holland, Amsterdam, pp. 89–199.

Bohigas, O., Giannoni, M. J. & Schmit, C. (1984), "Characterization of chaotic quantum spectra and universality of level fluctuation laws," *Phys. Rev. Lett.* **52**(1), 1–4.

Bohigas, O., Legrand, O., Schmit, C. & Sornette, D. (1991), "Comment on spectral statistics in elastodynamics," *J. Acoust. Soc. Am.* **89**(3), 1456–1458.

Bolt, R. H. (1947), "Normal frequency spacing statistics," *J. Acoust. Soc. Am.* **19**(1), 79–90.

Bolt, R. H. & Roop, R. W. (1950), "Frequency response fluctuations in rooms," *J. Acoust. Soc. Am.* **22**(2), 280–289.

Bonnet, M. (1995), *Boundary Integral Equation Methods for Solids and Fluids*, Wiley, Chichester.

Born, M. & Wolf, E. (1959), *Principles of Optics*, Pergamon, Oxford.

Borondo, F., Vergini, E., Wisniacki, D. A., Zembekov, A. A. & Benito, R. M. (2005), "Homoclinic motions in the vibrational spectra of floppy systems: The LiCN molecule," *J. Chem. Phys.* **122**(11), 111101.

Brack, M. & Bhaduri, R. K. (1997), *Semiclassical Physics*, Addison–Wesley, Reading, MA.

Brenguier, F., Shapiro, N. M., Campillo, M., Ferrazzini, V., Duputel, Z., Coutant, O. & Nercessian, A. (2008), "Towards forecasting volcanic eruptions using seismic noise," *Nature Geosci.* **1**(2), 126–130.

Brodier, O., Neicu, T. & Kudrolli, A. (2001), "Eigenvalues and eigenfunctions of a clover plate," *Eur. Phys. J. B* **23**(3), 365–372.

Brody, T. A., Flores, J., French, J. B., Mello, P. A., Pandey, A. & Wong, S. S. M. (1981), "Random-matrix physics – spectrum and strength fluctuations," *Rev. Mod. Phys.* **53**, 385–479.

Brouwer, P. W. (2003), "Wave function statistics in open chaotic billiards," *Phys. Rev. E* **68**(4), 046205.

Brouwer, P. W. & Rahav, S. (2006), "Semiclassical theory of the Ehrenfest time dependence of quantum transport in ballistic quantum dots," *Phys. Rev. B* **74**(7), 075322.

Brown, M. G., Colosi, J. A., Tomsovic, S., Virovlyansky, A. L., Wolfson, M. A. & Zaslavsky, G. M. (2003), "Ray dynamics in long-range deep ocean sound propagation," *J. Acoust. Soc. Am.* **113**, 2533. nlin.CD/0109027.

Bunimovich, L. A. (1974), "On the ergodic properties of some billiards," *Funct. Anal. App.* **8**, 73–74.

Burkhardt, J. (1997), "Experimental studies of diffuse decay curvature (a)," *J. Acoust. Soc. Am.* **101**(5), 3168.

Campillo, M. (2006), "Phase and correlation in 'random' seismic fields and the reconstruction of the Green function," *Pure Appl. Geophys.* **163**(2–3), 475–502.

Campillo, M. & Paul, A. (2003), "Long-range correlations in the diffuse seismic coda," *Science* **299**(5606), 547–549.

Cao, J. & Worsley, K. J. (2001), *Spatial Statistics: Methodological Aspects and Applications*, Springer Lecture Notes in Statistics, Springer, Berlin, pp. 169–182.

Capiez-Lernout, E., Pellissetti, M., Pradlwarter, H., Schueller, G. & Soize, C. (2006), "Data and model uncertainties in complex aerospace engineering systems," *J. Sound Vib.* **295**(3–5), 923–938.

Capiez-Lernout, E. & Soize, C. (2008a), "Design optimization with an uncertain vibroacoustic model," *ASME J. Vib. Acoust.* **130**(2), 021001.

Capiez-Lernout, E. & Soize, C. (2008b), "Robust design optimization in computational mechanics," *J. Appl. Mech.* **75**(2), 021001.

Capiez-Lernout, E., Soize, C., Lombard, J.-P., Dupont, C. & Seinturier, E. (2005), "Blade manufacturing tolerances definition for a mistuned industrial bladed disk," *J. Eng. Gas Turbines and Power* **127**(3), 621–628.

Carlo, G. G., Vergini, E. G. & Lustemberg, P. (2002), "Scar functions in the Bunimovich stadium billiard," *J. Phys. A* **35**, 7965–7982.

Casati, G., Maspero, G. & Shepelyansky, D. L. (1999), "Quantum fractal eigenstates," *Physica D* **131**(1–4), 311–316.

Casati, G., Valz-Gris, F. & Guarneri, L. (1980), "On the connection between quantization of nonintegrable systems and statistical theory of spectra," *Lett. Nuovo Cimenti* **28**(8), 279–282.

Cassereau, D. & Fink, M. (1992), "Time reversal of ultrasonic fields 3 theory of the closed time-reversal cavity," *IEEE Trans. Ultrason. Ferroelec. Freq. Contr.* **39**(5), 579–592.

Cerruti, N. R. & Tomsovic, S. (2002), "Sensitivity of wave field evolution and manifold stability in chaotic systems," *Phys. Rev. Lett.* **88**, 054103.

Chandrasekhar, S. (1960), *Radiative Transfer*, Dover, New York.

Charpentier, A., Cotoni, V. & Fukui, K. (2006), "Using the hybrid FE–SEA method to predict structure-borne noise in a car body-in-white," *in Proceedings of Inter-Noise 2006*.

Chebli, H. & Soize, C. (2004), "Experimental validation of a nonparametric probabilistic model of nonhomogeneous uncertainties for dynamical systems," *J. Acoust. Soc. Am.* **115**(2), 697–705.

Chen, C., Duhamel, D. & Soize, C. (2006), "Probabilistic approach for model and data uncertainties and its experimental identification in structural dynamics: Case of composite sandwich panels," *J. Sound Vib.* **294**, 64–81.

Cherroret, N. & Skipetrov, S. E. (2008), "Microscopic derivation of self-consistent equations of Anderson localization in a disordered medium of finite size," *Phys. Rev. E* **77**(4), 046608.

Ching, E. S. C., Leung, P. T., van den Brink, A. M., Suen, W. M., Tong, S. S. & Young, K. (1998), "Quasinormal-mode expansion for waves in open systems," *Rev. Mod. Phys.* **70**(4), 1545–1554.

Chirikov, B. V. (1959), "Resonance processes in magnetic traps," *Atomic Energy* **6**, 630. Russian; English *J. Nucl. Energy Part C: Plasma Phys.* **1**, 253 (1960).

Chirikov, B. V. (1979), "A universal instability of many-dimensional oscillator systems," *Phys. Rep.* **52**, 263–379.

Colin de Verdière, Y. (1985), "Ergodicité et fonctions propres du laplacien," *Commun. Math. Phys.* **102**(3), 497–502.

Colin de Verdière, Y. (2006), "Mathematical models for passive imaging I: General background," Mathematical Physics, e-prints ArXiv: math-ph/0610043.

Colosi, J. A. & Brown, M. G. (1998), "Efficient numerical simulation of stochastic internal-wave induced sound-speed perturbation fields," *J. Acoust. Soc. Am.* **103**, 2232–2235.

Colosi, J. A., Scheer, E. K., Flatté, S. M., Cornuelle, B. D., Dzieciuch, M. A., Munk, W. H., Worcester, P. F., Howe, B. M., A.Mercer, J., Spindel, R. C., Metzger, K., Birdsall, T. & Baggeroer, A. B. (1999), "Comparisons of measured and predicted acoustic fluctuations for a 3250-km propagation experiment in the eastern north Pacific Ocean," *J. Acoust. Soc. Am.* **105**, 3202–3218.

Cotoni, V., Langley, R. S. & Kidner, M. R. F. (2005), "Numerical and experimental validation of variance prediction in the statistical energy analysis of built-up systems," *J. Sound Vib.* **288**(3), 701–728.

Cotoni, V., Shorter, P. J. & Langley, R. S. (2008), "Numerical and experimental validation of a hybrid finite element – Statistical energy analysis method," *J. Acoust. Soc. Am.* **122**(1), 259–270.

Cottereau, R., Clouteau, D. & Soize, C. (2006), "Probabilistic nonparametric model of impedance matrices: Application to the seismic design of a structure," *Eur. J. Comput. Mech.* **15**(1–3), 131–142.

Cottereau, R., Clouteau, D. & Soize, C. (2007), "Construction of a probabilistic model for impedance matrices," *Comput. Meth. Appl. Mech. Eng.* **196**, 17–20.

Couchman, L., Ott, E. & Antonsen, T. M. (1992), "Quantum chaos in systems with ray splitting," *Phys. Rev. A* **47**(10), 6193–6210.

Creagh, S. C. (1996), "Trace formula for broken symmetry," *Ann. Phys.* **248**, 60–94.

Creagh, S. C. & Littlejohn, R. G. (1991), "Semiclassical trace formulas in the presence of continuous symmetries," *Phys. Rev. A* **44**, 836–850.

Creagh, S. C. & Littlejohn, R. G. (1992), "Semiclassical trace formulas for systems with non-Abelian symmetry," *J. Phys. A* **25**, 1643–1669.

Cremer, L., Heckl, M. & Ungar, E. E. (1990), *Structure Borne Sound: Structural Vibrations and Sound Radiation at Audio Frequencies*, 2nd edn, Springer, Berlin.

Cvitanović, P., Artuso, R., Mainieri, R., Tanner, G. & Vattay, G. (2005), *Chaos: Classical and Quantum*, Niels Bohr Institute, Copenhagen. http://ChaosBook.org.

Cvitanovic, P. & Eckhardt, B. (1989), "Periodic-orbit quantization of chaotic systems," *Phys. Rev. Lett.* **63**(8), 823–826.

Cvitanovic, P. & Eckhardt, B. (1993), "Symmetry decomposition of chaotic dynamics," *Nonlinearity* **6**(2), 277–311.

Dainty, A. & Toksöz, M. N. (1990), "Array analysis of seismic scattering," *Bull. Seism. Soc. Am.* **80**, 2242–2260.

Davy, J. L. (1981), "The relative variance of the transmission function of a reverberation room," *J. Sound Vib.* **77**(4), 455–479.

Davy, J. L. (1986), "The ensemble variance of random noise in a reverberation room," *J. Sound Vib.* **107**(3), 361–373.

Davy, J. L. (1987), "Improvements to formulae for the ensemble relative variance of random noise in a reverberation room," *J. Sound Vib.* **115**(1), 145–161.

Davy, J. L. (1990), "The distribution of modal frequencies in a reverberation room," *in Proceedings of Internoise '90*, Vol. 1, Noise Control Foundation, Poughkeepsie, NY, pp. 159–164.

de Rosny, J. & Fink, M. (2002), "Overcoming the diffraction limit in wave physics using a time-reversal mirror and a novel acoustic sink," *Phys. Rev. Lett.* **89**(12), 124301.

de Rosny, J., Tourin, A. & Fink, M. (2000), "Coherent backscattering of an elastic wave in a chaotic cavity," *Phys. Rev. Lett.* **84**(8), 1693–1695.

Delande, D. (2001), "The semiclassical approach for chaotic systems," *in* P. Sebbah, ed., *Waves and Imaging through Complex Media*, Kluwer Academic, Amsterdam. International Physics School, Cargése, 1999.

Delande, D. & Sornette, D. (1997), "Acoustic radiation from mebranes at high frequencies: The quantum chaos regime," *J. Acoust. Soc. Am.* **101**(4), 1793–1807.

Dennis, M. R. (2007), "Nodal densities of planar gaussian random waves," *Eur. Phys. J. Special Topics* **145**, 191–210.

Derode, A., Larose, E., Campillo, M. & Fink, M. (2003), "How to estimate the green's function of a heterogeneous medium between two passive sensors? Application to acoustic waves," *Appl. Phys. Lett.* **83**(15), 3054–3056.

Derode, A., Mamou, V. & Tourin, A. (2006), "Influence of correlations between scatterers on the attenuation of the coherent wave in a random medium," *Phys. Rev. E* **74**(3), 036606.

Derode, A., Roux, P. & Fink, M. (1995), "Robust acosutic time reversal with high-order multiple scattering," *Phys. Rev. Lett.* **75**(23), 4206–4209.

Derode, A., Tourin, A. & Fink, M. (2001), "Random multiple scattering of ultrasound II. Is time reversal a self-averaging process?," *Phys. Rev. E* **64**(3), 036606.

Desceliers, C., Soize, C. & Cambier, S. (2004), "Non-parametric – Parametric model for random uncertainties in nonlinear structural dynamics – Application to earthquake engineering," *Earthquake Eng. Struct. Dyn.* **33**(3), 315–327.

Doya, V., Legrand, O., Mortessagne, F. & Miniatura, C. (2002), "Speckle statistic in a chaotic multimode fiber," *Phys. Rev. E* **65**, 056223.

Draeger, C., Aime, J.-C. & Fink, M. (1999), "One-channel time-reversal in chaotic cavities: Experimental results," *J. Acoust. Soc. Am.* **105**(2), 618–625.

Draeger, C. & Fink, M. (1997), "One-channel time reversal of elastic waves in a chaotic 2D silicon cavity," *Phys. Rev. Lett.* **79**(3), 407–410.

Draeger, C. & Fink, M. (1999), "One-channel time-reversal in chaotic cavities: Theoretical limits," *J. Acoust. Soc. Am.* **105**(2), 611–617.

Duchereau, J. & Soize, C. (2006), "Transient dynamics in structures with nonhomogeneous uncertainties induced by complex joints," *Mech. Syst. Sig. Proc.* **20**, 854–867.

Durand, J.-F. (2007), Modélisation de véhicules automobiles en vibroacoustique numérique avec incertitudes de modélisation et validation expérimentale, PhD thesis, Université de Marne-la-Vallée.

Durand, J.-F., Soize, C. & Gagliardini, L. (2008), "Structural–acoustic modeling of automotive vehicles in presence of uncertainties and experimental identification and validation," *J. Acoust. Soc. Am.* **124**(3), 1513–1525.

Duvall, Jr., T. L., Jefferies, S. M., Harvey, J. W. & Pomerantz, M. A. (1993), "Time–distance helioseismology," *Nature* **362**(6419), 430–432.

Dyson, F. J. (1962), "Statistical theory of the energy levels of complex systems. I–III," *J. Math. Phys.* **3**(1), 140–175.

Ebeling, K. J. (1984), "Statistical properties of random wave fields," *in* W. P. Mason & R. N. Thurston, eds, *Physical Acoustics*, Vol. 17, Academic, New York, chapter 4, pp. 233–310.

Edelmann, G. F., Akal, T., Hodgkiss, W. S., Kim, S., Kuperman, W. A. & Song, H. C. (2002), "An initial demonstration of underwater acoustic communication using time reversal," *IEEE J. Ocean. Eng.* **27**(3), 602–609.

Efetov, K. B. (1983), "Supersymmetry and the theory of disordered metals," *Adv. Phys.* **32**(1), 53–127.

Efetov, K. B. (1999), *Supersymmetry in Disorder and Chaos*, Cambridge University Press, Cambridge, UK.

Elishakoff, I. (1995), "Essay on uncertainties in elastic and viscoelastic structures: From A. M. Freudenthal"s criticisms to modern convex modelling," *Comput. Struct.* **56**(6), 871–1079.

Ellegaard, C., Guhr, T. & Lindemann, K. (1995), "Spectral statistics of acoustic resonances in aluminum blocks," *Phys. Rev. Lett.* **75**, 1546–1549.

Ellegaard, C., Guhr, T., Lindemann, K., Lorensen, H. Q., Nygård, J. & Oxborrow, M. (1995), "Spectral statistics of acoustic resonances in aluminum blocks," *Phys. Rev. Lett.* **75**(8), 1546–1549.

Ellegaard, C., Schaadt, K. & Bertelsen, P. (2001), "Acoustic chaos," *Physica Scripta* **T90**, 223–230.

Exner, P. & Šeba, P. (1996), "Point interactions in two and three dimensions as models of small scatterers," *Phys. Lett. A* **222**(1–2), 1–4.

Filippi, P., Habault, D., Lefebvre, J.-P. & Bergassoli, A. (1989), *Acoustics: Basic Physics, Theory and Methods*, Academic, San Diego, CA.

Fink, M. (1992), "Time-reversal of ultrasonic fields 1 basic principles," *IEEE Trans. Ultrason. Ferroelec. Freq. Contr.* **39**(5), 555–566.

Fink, M. (1997), "Time reversed acoustics," *Phys. Today* **50**(3), 34–40.

Fink, M., Cassereau, D., Derode, A., Prada, C., Roux, P., Tanter, M., Thomas, J. L. & Wu, F. (2000), "Timer-reversed acoustics," *Rep. Prog. Phys.* **63**(12), 1933–1995.

Flatté, S. M., Dashen, R., Munk, W. H. & Zachariasen, F. (1979), *Sound Transmission through a Fluctuating Ocean*, Cambridge University Press, Cambridge, UK.

Fletcher, J. B., Fumal, T., Liu, H. & Porcella, R. (1990), "Near-surface velocities and attenuation at two bore-holes near Anza," *Bull. Seism. Soc. Am.* **80**(4), 807–831.

Foldy, L. L. (1945), "The multiple scattering of waves: I. General theory of isotropic scattering by randomly distributed scatterers," *Phys. Rev.* **67**, 107–119.

Forrester, P. J. (2006), "Quantum conductance problems and the Jacobi ensemble," *J. Phys. A* **39**(22), 6861–6870.

Fougeaud, C. & Fuchs, A. (1967), *Statistique*, 2nd edn, Dunod, Paris. (1972).

Frisch, U. (1968), *Probabilistic Methods in Applied Mathematics*, Vol. 1, Academic, New York, chapter "Wave propagation in random media," pp. 75–198.

Fulling, S. A. (2002), "Spectral oscillations, periodic orbits, and scaling," *J. Phys. A* **35**, 4049–4066.

Fyodorov, Y. V. (1995), "Basic features of Efetov's supersymmetry approach," *in Mesoscopic Quantum Physics*, Vol. 61 of Les Houches Lecture Series, Elsevier, Amsterdam.

Fyodorov, Y. V. & Sommers, H.-J. (2003), "Random matrices close to Hermitian or unitary: Overview of methods and results," *J. Phys. A* **36**(12), 3303–3347.

Galdi, V., Pinto, I. M. & Felsen, L. B. (2005), "Wave propagation in ray-chaotic enclosures: Paradigms, oddities and examples," *IEEE Antennas Propag. Mag.* **47**(1), 62–81.

Gaspard, P. (1998), *Chaos, Scattering and Statistical Mechanics*, Cambridge University Press, Cambridge, UK.

Gaspard, P., Alonso, D., Okuda, T. & Nakamura, K. (1994), "Chaotic scattering on C_{4v} four-disk billiards: Semiclassical and exact quantum theories," *Phys. Rev. E* **50**(4), 2591–2596.

Gaspard, P. & Baras, F. (1995), "Chaotic scattering and diffusion in the Lorentz gas," *Phys. Rev. E* **51**(6), 5332–5352.

Gaspard, P. & Ramirez, D. A. (1992), "Ruelle classical resonances and dynamical chaos: The three- and four-disk scatterers," *Phys. Rev. A* **45**(12), 8383–8397.

Gaspard, P. & Rice, S. A. (1989), "Semiclassical quantization of the scattering from a classically chaotic repellor," *J. Chem. Phys.* **90**(4), 2242–2254.

Gaudin, M. (1961), "Sur la loi limite de l'espacement des valeurs propres d'une matrice aléatoire," *Nucl. Phys.* **25**, 447–458.

Georgeot, B. & Prange, R. E. (1995), "Exact and quasiclassical fredholm solutions of quantum billiards," *Phys. Rev. Lett.* **74**(15), 2851–2854.

Ghanem, R. & Spanos, P. D. (2003), *Stochastic Finite Elements: A Spectral Approach*, revised edn, Dover, New York.

Giannoni, M. J., Voros, A. & Jinn-Justin, J., eds (1991), *Chaos and Quantum Physics*, North Holland, Amsterdam.

Gizon, L. & Birch, A. (2004), "Time-distance helioseismology: Noise estimation," *Astrophys. J.* **614**(1), 472–489.

Gizon, L. & Birch, A. C. (2005), "Local helioseismology," *Living Rev. in Solar Phys.* **2**(6).

Gmachl, C., Capasso, F., Narimanov, E. E., Nockel, J. U., Stone, A. D., Faist, J., Sivco, D. L. & Cho, A. Y. (1998), "High-power directional emission from microlasers with chaotic resonators," *Science* **280**(5369), 1556–1564.

Goldstein, H. (1980), *Classical Mechanics*, Addison-Wesley, Reading, MA.

Goodman, J. W. (1985), *Statistical Optics*, John Wiley, New York.

Goodman, J. W. (2007), *Speckle Phenomena in Optics*, Ben Roberts, New York.

Gorin, T., Seligman, T. H. & Weaver, R. L. (2006a), "Scattering fidelity in elastodynamics," *Phys. Rev. E* **73**, 015202.

Gorin, T., Seligman, T. H. & Weaver, R. L. (2006b), "Scattering fidelity in elastodynamics," *Phys. Rev. E* **73**(1), 015202.

Gouédard, P., Stehly, L., Brenguier, F., Campillo, M., Colin de Verdière, Y., Larose, E., Margerin, L., Roux, P., Sánchez-Sesma, F. J., Shapiro, N. M. & Weaver, R. L. (2008), "Cross-correlation of random fields: Mathematical approach and applications," *Geophys. Prospect.* **56**(3), 375–393.

Grassberger, P., Badii, R. & Politi, A. (1988), "Scaling laws for invariant measures on hyperbolic and nonhyperbolic attractors," *J. Stat. Phys.* **51**, 135–178.

Grêt, A., Snieder, R., Aster, R. C. & Kyle, P. R. (2005), "Monitoring rapid temporal change in a volcano with coda wave interferometry," *Geophys. Res. Lett.* **32**(6), L06304.

Guhr, T., Müller-Groeling, A. & Weidenmüller, H. A. (1998), "Random-matrix theories in quantum physics: Common concepts," *Phys. Rep.* **299**, 189–425.

Gutkin, E. (2003), "Billiard dynamics: A survey with the emphasis on open problems," *Regular Chaotic Dyn.* **8**(1), 1–13.

Gutzwiller, M. C. (1970), "Energy spectrum according to classical mechanics," *J. Math. Phys.* **11**, 1791–1806.

Gutzwiller, M. C. (1971), "Periodic orbits and classical quantization conditions," *J. Math. Phys.* **12**, 343–358.

Gutzwiller, M. C. (1980), "The quantum mechanical Toda lattice," *Ann. Phys.* **124**, 347–381.

Gutzwiller, M. C. (1990), *Chaos in Classical and Quantum Mechanics*, Springer, Berlin.

Haake, F. (2001), *Quantum Signatures of Chaos*, 2nd edn, Springer, Berlin.

Ham, C. J. (2008), Periodic orbit analysis of the Helmholtz equation in two-dimensional enclosures, PhD thesis, University of Southampton.

Hannay, J. H. & Berry, M. V. (1980), "Quantization of linear maps on a torus – Fresnel diffraction by a periodic grating," *Physica* **1D**, 267–290.

Hannay, J. H., Keating, J. P. & de Almeida, A. M. O. (1994), "Optical realization of the baker's transformation," *Nonlinearity* **7**(5), 1327–1342.

Hansen, K. T. (1993a), "Symbolic dynamics: I. Finite dispersive billiards," *Nonlinearity* **6**(5), 753–769.

Hansen, K. T. (1993b), "Symbolic dynamics: II. Bifurcations in billiards and smooth potentials," *Nonlinearity* **6**(5), 771–778.

Hansen, K. T. (1993c), Symbolic Dynamics in Chaotic Systems, PhD thesis, Physics Department, University of Oslo.

Hansen, K. T. (1995), "Alternative method to find orbits in chaotic systems," *Phys. Rev. E* **52**(3), 2388–2391.

Hansen, K. T. & Cvitanović, P. (1995), Symbolic dynamics and Markov partitions for the stadium billiard. *ArXiv*: chao-dyn/9502005.

Hegewisch, K. C., Cerruti, N. R. & Tomsovic, S. (2001), "Ocean acoustic wave propagation and ray method correspondence: Internal waves," *J. Acoust. Soc. Am.* **117**, 1582–.

Heller, E. J. (1984), "Bound-state eigenfunctions of classically chaotic Hamiltonian systems: Scars of periodic orbits," *Phys. Rev. Lett.* **53**(16), 1515–1518.

Heller, E. J. (1991), "Wave packet dynamics and quantum chaology," *in* M.-J. Giannoni, A. Voros & J. Zinn-Justin eds, *Chaos and Quantum Physics*, Vol. 52 of Les Houches Lecture Series, North Holland, Amsterdam, pp. 547–663.

Heller, E. & Tomsovic, S. (1993), "Postmodern quantum mechanics," *Phys. Today* **46**(7), 38–46.

Hennino, R., Trégourès, N., Shapiro, N. M., Margerin, L., Campillo, M., Van Tiggelen, B. A. & Weaver, R. L. (2001), "Observation of equipartition of seismic waves," *Phys. Rev. Lett.* **86**(15), 3447–3450.

Heusler, S., Muller, S., Altland, A., Braun, P. & Haake, F. (2007), "Periodic-orbit theory of level correlations," *Phys. Rev. Lett.* **98**, 044103.

Heusler, S., Müller, S., Braun, P. & Haake, F. (2006), "Semiclassical theory of chaotic conductors," *Phys. Rev. Lett.* **96**(6), 066804.

Hirsekorn, S. (1988), "The scattering of ultrasonic waves by multiphase polycrystals," *J. Acoust. Soc. Am.* **83**, 1231–1242.

Hodges, C. H. (1982), "Confinement of vibration by structural irregularity," *J. Sound Vib.* **82**(3), 411–424.

Höhmann, R., Kuhl, U., Stöckmann, H.-J., Urbina, J. D. & Dennis, M. R. (2009), "Density and correlation functions of vortex and saddle points in open billiard systems," *Phys. Rev. E* **79**(1), 016203.

Hortikar & Srednicki, M. (1998), "Correlations in chaotic eigenfunctions at large separation," *Phys. Rev. Lett.* **80**, 1646–1649.

Hu, H., Strybulevych, A., Page, J. H., Skipetrov, S. E. & van Tiggelen, B. A. (2008), "Localization of ultrasound in a three-dimensional elastic network," *Nature Phys.* **4**(12), 945–948.

Ing, R. K., Quieffen, N., Catheline, S. & Fink, M. (2005), "In solid localization of finger impacts using acoustic time-reversal process," *App. Phys. Lett.* **87**(20), 204104.

Insana, M. F. & Hall, T. J. (1990), "Parametric ultrasound imaging from backscatter coefficient measurements: Image formation and interpretation," *Ultrason. Imaging* **12**, 245–267.

Ishimaru, A. (1977), "Theory and application of wave propagation and scattering in random media," *Proc. IEEE* **65**, 1030–1061.

Ishimaru, A. (1978), *Wave propagation and scattering in radom media*, Academic, New York.

Ishio, H. & Keating, J. P. (2004), "Semiclassical wavefunctions in chaotic scattering systems," *J. Phys. A* **37**(22), L217–L223.

Ishio, H., Saichev, A. I., Sadreev, A. F. & Berggren, K.-F. (2001), "Wave function statistics for ballistic quantum transport through chaotic open billiards: Statistical crossover and coexistence of regular and chaotic waves," *Phys. Rev. E* **64**(5), 156208.

Izrailev, F. M. (1990), "Simple models of quantum chaos – Spectrum and eigenfunctions," *Phys. Rep.* **196**, 299–392.

Jackson, D. R. & Dowling, D. R. (1991), "Phase conjugation in underwater acoustics," *J. Acoust. Soc. Am.* **89**(1), 171–181.

Jacquod, P. & Sukhorukov, E. V. (2004), "Breakdown of universality in quantum chaotic transport: The two-phase dynamical model," *Phys. Rev. Lett.* **92**(11), 116801.

Jalabert, R. A., Baranger, H. U. & Stone, A. D. (1990), "Conductance fluctuations in the ballistic regime: A probe of quantum chaos?," *Phys. Rev. Lett.* **65**(19), 2442–2445.

Jalabert, R. A., Pichard, J.-L. & Beenakker, C. W. J. (1994), "Universal quantum signatures of chaos in ballistic transport," *Europhys. Lett.* **27**(4), 255–260.

Jarzynsky, C. (1997), "Berry's conjecture and information theory," *Phys. Rev. E* **56**, 2254–2256.

Jaynes, E. T. (1957), "Information theory and statistical mechanics," *Phys. Rev.* **106**(4), 620–630.

Jorba, A. & Simo, C. (1996), "On quasiperiodic perturbations of elliptic equilibrium points," *SIAM J. Math. Anal.* **27**, 1704–1737.

Joyce, W. B. (1975), "Sabine's reverberation time and ergodic auditoriums," *J. Acoust. Soc. Am.* **58**(3), 643–655.

Kalos, M. H. & Whitlock, P. A. (1986), *Monte Carlo Methods, Volume 1: Basics*, John Wiley, New York.

Kang, T.-S. & Shin, J. S. (2006), "Surface-wave tomography from ambient seismic noise of accelerograph networks in southern Korea," *Geophys. Res. Lett.* **33**(17), L17303.

Karal, F. C. & Keller, J. B. (1964), "Elastic, electromagnetic and other waves in random media," *J. Math. Phys.* **5**, 537–547.

Keating, J., Novaes, M., Prado, S. & Sieber, M. (2006), "Semiclassical structure of chaotic resonance eigenfunctions," *Phys. Rev. Lett.* **97**(15), 150406.

Keating, J. P. & Prado, S. D. (2001), "Orbit bifurcations and the scarring of wavefunctions," *Proc. R. Soc. A* **457**(2012), 1855–1872.

Kitahara, M. (1985), *Boundary Integral Equation Methods in Eigenvalue Problems of Elastodynamics and Thin Plates*, Elsevier, Amsterdam.

Kozlov, V. V. & Treshchëv, D. V. (1980), *Billiards: A Genetic Introduction to the Dynamics of Systems with Impacts*, Vol. 89 of Translations of Mathematical Monographs, American Mathematical Society, Providence, RI.

Kubota, Y. & Dowell, E. H. (1992), "Asymptotic modal analysis for sound fields of a reverberant chamber," *J. Acoust. Soc. Am.* **92**(2), 1106–1112.

Kuhl, U., Stöckmann, H. J. & Weaver, R. (2005), "Classical wave experiments on chaotic scattering," *J. Phys. A* **38**(49), 10433–10463.

Kuperman, W. A., Hodgkiss, W. S., Song, H. C., Akal, T., Ferla, C. & Jackson, D. R. (1998), "Phase conjugation in the ocean: Experimental demonstration of an acoustic time-reversal mirror," *J. Acoust. Soc. Am.* **103**(1), 25–40.

Landau, L. D. & Lifshitz, E. M. (1959), *Theory of Elasticity*, Pergamon, Oxford, UK.

Langley, R. S. (2007), "On the diffuse field reciprocity relationship and vibrational energy variance in a random subsystem at high frequencies," *J. Acoust. Soc. Am.* **121**(2), 913–921.

Langley, R. S. & Bremner, P. (1999), "A hybrid method for the vibration analysis of complex structural–acoustic systems," *J. Acoust. Soc. Am.* **105**(3), 1657–1671.

Langley, R. S. & Brown, A. W. M. (2004*a*), "The ensemble statistics of the band-averaged energy of a random system," *J. Sound Vib.* **275**, 847–857.

Langley, R. S. & Brown, A. W. M. (2004*b*), "The ensemble statistics of the energy of a random system subjected to harmonic excitation," *J. Sound Vib.* **275**, 823–846.

Langley, R. S. & Cotoni, V. (2004), "Response variance prediciton in the statistical energy analysis of built-up systems," *J. Acoust. Soc. Am.* **115**(2), 706–718.

Langley, R. S. & Cotoni, V. (2005), "The ensemble statistics of the vibration energy density of a random system subjected to single point harmonic excitation," *J. Acoust. Soc. Am.* **118**(5), 3064–3076.

Langley, R. S. & Cotoni, V. (2007), "Response variance prediction for uncertain vibro-acoustic systems using a hybrid deterministic-statistical method," *J. Acoust. Soc. Am.* **122**(6), 3445–3463.

Larose, E., Derode, A., Campillo, M. & Fink, M. (2004), "Imaging from one-bit correlations of wideband diffuse wave fields," *J. Appl. Phys.* **95**(12), 8393–8399.

Larose, E., Lobkis, O. I. & Weaver, R. L. (2006), "Coherent backscattering of ultrasound without a source," *Europhys. Lett.* **76**(3), 422–428.

Larose, E., Margerin, L., Derode, A., van Tiggelen, B., Campillo, M., M. Shapiro, N., Paul, A., Stehly, L. & Tanter, M. (2006), "Correlation of random wavefields: An interdisciplinary review," *Geophysics* **71**(4), SI11–SI21.

Larose, E., Margerin, L., van Tiggelen, B. A. & Campillo, M. (2004), "Weak localization of seismic waves," *Phys. Rev. Lett.* **93**(4), 048501.

Laurent, D., Legrand, O. & Mortessagne, F. (2006), "Diffractive orbits in the length spectrum of a two-dimensional microwave cavity with a small scatterer," *Phys. Rev. E* **74**, 046219.

Lax, M. (1951), "Multiple scattering of waves," *Rev. Mod. Phys.* **23**, 287–310.

Lax, M. (1952), "Multiple scattering of waves: II. The effective field in dense systems," *Phys. Rev.* **85**, 621–629.

Legrand, O. & Mortessagne, F. (1996), "On spectral correlations in chaotic reverberant rooms," *Acustica* **82**, S150–S150.

Legrand, O., Mortessagne, F. & Sornette, D. (1995), "Spectral rigidity in the large modal overlap regime – Beyond the Ericson–Schroeder hypothesis," *J. de Physique I* **5**(8), 1003–1010.

Legrand, O., Mortessagne, F. & Weaver, R. L. (1997), "Semiclassical analysis of spectral correlations in regular billiards with point scatterers," *Phys. Rev. E* **55**(6), 7741–7744.

Legrand, O. & Sornette, D. (1990), "Test of Sabine"s reverberation time in ergodic auditoriums within geometrical acoustics," *J. Acoust. Soc. Am.* **88**(2), 865–870.

Legrand, O. & Sornette, D. (1991), "Quantum chaos and sabine's law of reverberation in ergodic rooms," *Lect. Notes Phys.* **392**, 267–274.

Lerosey, G., de Rosny, J., Tourin, A. & Fink, M. (2007), "Focusing beyond the diffraction limit with far-field time reversal," *Science* **315**(5815), 1120–1122.

Li, B. & Robnik, M. (1994), "Statistical properties of high-lying chaotic eigenstates," *J. Phys. A* **27**, 5509–5523.

Liddle, A. R. & Lyth, D. H. (2000), *Cosmological Inflation and Large-Scale Structure*, Cambridge University Press, Cambridge, UK.

Lighthill, J. (1978), *Waves in Fluids*, Cambridge University Press, Cambridge, UK.

Lin, F.-C., Ritzwoller, M. H., Townend, J., Bannister, S. & Savage, M. K. (2007), "Ambient noise Rayleigh wave tomography of New Zealand," *Geophys. J. Int.* **170**(2), 649–666.

Lin, K. K. & Zworski, M. (2002), "Quantum resonances in chaotic scattering," *Chem. Phys. Lett.* **355**(1–2), 201–205.

Lin, Y. K. (1967), *Probabilistic Theory of Structural Dynamics*, McGraw-Hill, New York.

Lobkis, O. I. & Weaver, R. L. (2000), "Complex modes in a reverberant dissipative body," *J. Acoust. Soc. Am.* **108**(4), 1480–1485.

Lobkis, O. I. & Weaver, R. L. (2001), "On the emergence of the Green's function in the correlations of a diffuse field," *J. Acoust. Soc. Am.* **110**, 3011–3017.

Lobkis, O. I. & Weaver, R. L. (2003), "Coda-wave interferometry in finite solids: Recovery of *p*-to-*s* conversion rates in an elastodynamic billiard," *Phys. Rev. Lett.* **90**(25), 254302.

Lobkis, O. I., Weaver, R. L. & Rozhkov, I. (2000), "Power variances and decay curvature in a reverberant system," *J. Sound Vib.* **237**(2), 281–302.

Lobkis, O., Rozhkov, I. & Weaver, R. (2003), "Non-exponential dissipation in a lossy elastodynamic billiard, comparison with Porter–Thomas and random matrix predictions," *Phys. Rev. Lett.* **91**, 194101.

Longuet-Higgins, M. S. (1957a), "The statistical analysis of a random, moving surface," *Phil. Trans. R. Soc.* **249**, 321–387.

Longuet-Higgins, M. S. (1957b), "Statistical properties of an isotropic random surface," *Phil. Trans. R. Soc.* **250**, 157–174.

Lu, W., Rose, M., Pance, K. & Sridhar, S. (1999), "Quantum resonances and decay of a chaotic fractal repeller observed using microwaves," *Phys. Rev. Lett.* **82**(26), 5233–5236.

Lu, W. T., Sridhar, S. & Zworski, M. (2003), "Fractal Weyl laws for chaotic open systems," *Phys. Rev. Lett.* **91**(15), 154101.

Lyon, R. H. (1969), "Statistical analysis of power injection and response in structures and rooms," *J. Acoust. Soc. Am.* **45**(3), 545–565.

Lyon, R. H. (1975), *Statistical Energy Analysis of Dynamical Systems*, MIT press, Boston.

Lyon, R. H. & DeJong, R. G. (1995), *Theory and Application of Statistical Energy Analysis*, Butterworth-Heinemann, Boston.

Mace, B. R., Worden, K. & Manson, G. (2005), "Special issue on 'uncertainty in structural dynamics," *J. Sound Vib.* **288**(3), 431–790.

MacKeown, P. K. (1997), *Stochastic Simulation in Physics*, Springer, Singapore.

Mandel, L. & Wolf, E. (1995), *Optical Coherence and Quantum Optics*, Cambridge University Press, Cambridge, UK.

Margerin, L., Campillo, M. & van Tiggelen, B. (2000), "Monte carlo simulation of multiple scattering of elastic waves," *J. Geophys. Res.* **105**(B4), 7873–7892.

Margerin, L., Campillo, M. & van Tiggelen, B. A. (2001), "Coherent backscattering of acoustic waves in the near field," *Geophys. J. Int.* **145**(3), 593–603.

Margerin, L., Campillo, M., van Tiggelen, B. & Hennino, R. (2009), "Energy partition of seismic coda waves in layered media: Theory and application to pinyon flats observatory," *Geophys. J. Int.* **177**(2), 571–585.

Margerin, L., van Tiggelen, B. & Campillo, M. (2001), "Effect of absorption on energy partition of elastic waves in the seismic coda," *Bull. Seism. Soc. Am.* **91**(3), 624–627.

McCoy, J. J. (1981), *Macroscopic Response of Continua with Random Microsctructure*, Vol. 6, Pergamon, New York.

McDonald, S. W. & Kaufman, A. N. (1979), "Spectrum and eigenfunctions for a Hamiltonian with stochastic trajectories," *Phys. Rev. Lett.* **42**(18–30), 1189–1191.

Mehta, M. L. (1960), "On the statistical properties of the level-spacings in nuclear spectra," *Nucl. Phys.* **18**, 395–419.

Mehta, M. L. (1991), *Random Matrices*, 2nd edn, Academic, New York.

Mehta, M. L. & Gaudin, M. (1960), "On the density of eigenvalues of a random matrix," *Nucl. Phys.* **18**, 420–427.

Mignolet, M. P. & Soize, C. (2008a), "Nonparametric stochastic modeling of linear systems with prescribed variance of several natural frequencies," *Probabilistic Eng. Mech.* **32**(2–3), 267–278.

Mignolet, M. P. & Soize, C. (2008b), "Nonparametric stochastic modeling of structural dynamic systems with uncertain boundary conditions," *in* AIAA Conference, Schaumburg, Chicago.

Mignolet, M. P. & Soize, C. (2008c), "Stochastic reduced order models for uncertain nonlinear dynamical systems," *Comput. Meth. Appl. Mech. Eng.* **197**(45–48), 3951–3963.

Mirlin, A. D. (2000), "Statistics of energy levels and eigenfunctions in disordered systems," *Phys. Rep.* **326**, 259–382.

Mishchenko, M. I., Travis, L. D. & Lacis, A. A. (2006), *Multiple Scattering of Light by Particles: Radiative Transfer and Coherent Backscattering*, Cambridge University Press, Cambridge, New York.

Moens, D. & Vandepitte, D. (2005), "A fuzzy finite element procedure for the calculation of uncertain frequency-response functions of damped structures: Part 1 – Procedure," *J. Sound Vib.* **288**(3), 431–462.

Morse, P. & Bolt, R. H. (1944), "Sound waves in rooms," *Rev. Mod. Phys.* **16**, 69–150.

Morse, P. M. & Feshbach, H. (1953), *Methods of Theoretical Physics*, McGraw-Hill, New York.

Morse, P. M. & Ingard, K. U. (1968), *Theoretical Acoustics*, Princeton University Press, Princeton, NJ.

Mortessagne, F. & Legrand, O. (1996), "Semi-classical acoustic time response in 2-d rooms," *Acustica* **82**, S152–S152.

Mortessagne, F., Legrand, O. & Sornette, D. (1993a), "Role of the absorption distribution and generalization of exponential reverberation law in chaotic rooms," *J. Acoust. Soc. Am.* **94**(1), 154–161.

Mortessagne, F., Legrand, O. & Sornette, D. (1993b), "Transient chaos in room acoustics," *Chaos* **3**(4), 529–541.

Moschetti, M. P., Ritzwoller, M. H. & Shapiro, N. M. (2007), "Surface wave tomography of the western United States from ambient seismic noise: Rayleigh wave group velocity maps," *Geochem. Geophys. Geosyst.* **8**, Q08010.

Munk, W. H. (1981), Internal waves and small scale processes, *in* B. A. Warren & C. Wunsch, eds, "Evolution of Physical Oceanography," MIT Press, Cambridge, MA, pp. 264–291.

Munk, W. H., Worcester, P. & Wuncsh, C. (1995), *Ocean Acoustic Tomography*, Cambridge University Press, Cambridge, UK.

Nakamura, K. & Harayama, T. (2004), *Quantum Chaos and Quantum Dots*, Oxford University Press, Oxford, UK.

Neicu, T. & Kudrolli, A. (2002), "Periodic orbit analysis of an elastodynamic resonator using shape deformation," *Europhys. Lett.* **57**(3), 341–347.

Neicu, T., Schaadt, K. & Kudrolli, A. (2001), "Spectral properties of a mixed system using an acoustical resonator," *Phys. Rev. E* **63**, 026206.

Nishida, K., Kawakatsu, H. & Obara, K. (2008), "Three-dimensional crustal S-wave velocity structure in japan using microseismic data recorded by Hi-net tiltmeters," *J. Geophys. Res.*, **113**(B10), B10302.

Nockel, J. U. & Stone, A. D. (1997), "Ray and wave chaos in asymmetric resonant optical cavities," *Nature* **385**(6611), 45–47.

Nonnenmacher, S. & Rubin, M. (2007), "Resonant eigenstates for a quantized chaotic system," *Nonlinearity* **20**(6), 1387–1420.

Nonnenmacher, S. & Zworski, M. (2005), "Fractal Weyl laws in discrete models of chaotic scattering," *J. Phys. A* **38**(49), 10683–10702.

Nonnenmacher, S. & Zworski, M. (2007a), "Distribution of resonances for open quantum maps," *Commun. Math. Phys.* **269**(2), 311–365.

Nonnenmacher, S. & Zworski, M. (2007b), "Quantum decay rates in chaotic scattering." ArXiv: 0706.3242v1.

O'Connor, P., Gehlen, J. & Heller, E. J. (1987), "Properties of random superpositions of plane waves," *Phys. Rev. Lett.* **58**, 1296–1299.

Ohayon, R. & Soize, C. (1998), *Structural Acoustics and Vibration*, Academic, San Diego, CA.

O'Holleran, K., Dennis, M. R. & Padgett, M. J. (2009), "Topology of light's darkness," *Phys. Rev. Lett.* **102**(14), 143902.

Oleze, M. L. & Zachary, J. F. (2006), "Examination of cancer in mouse models using high-frequency quantitative ultrasound," *Ultrasound in Med. and Biol.* **32**, 1639–1648.

Ott, E. (1993), *Chaos in Dynamical Systems*, Cambridge University Press, Cambridge, UK.

Ozorio de Almeida, A. M. & Vallejos, R. O. (2000), "Poincaré's recurrence theorem and the unitarity of the s-matrix," *Chaos Solitons Fractals* **11**(7), 1015–1020.

Ozorio de Almeida, A. M. & Vallejos, R. O. (2001), "Poincaré section decomposition for quantum scatterers," *Physica E* **9**(3), 488–493.

Palmer, D. R., Brown, M. G., Tappert, F. D. & Bezdek, H. F. (1988), "Classical chaos in nonseparable wave propagation problems," *Geophys. Res. Lett.* **15**, 569–572.

Palmer, D. R., Georges, T. M. & Jones, R. M. (1991), "Classical chaos and the sensitivity of the acoustic field to small-scale ocean structure," *Comput. Phys. Commun.* **65**, 219–223.

Pandolfi, D., Bean, C. J. & Saccorotti, G. (2006), "Coda wave interferometric detection of seismic velocity changes associated with the 1999 $m = 3.6$ event at Mt. Vesuvius," *Geophys. Res. Lett.* **33**(6), L06306.

Papoulis, A. (1991), *Probability, Random Variables, and Stochastic Processes*, 3rd edn, McGraw-Hill, New York.

Pato, M. P., Schaadt, K., Tufaile, A. P. B., Ellegaard, C., Nogueira, T. N. & Sartorelli, J. C. (2005), "Universality of rescaled curvature distributions," *Phys. Rev. E* **71**(3), 037201.

Paul, A., Campillo, M., Margerin, L., Larose, E. & Derode, A. (2005), "Empirical synthesis of time-asymmetrical Green functions from the correlation of coda waves," *J. Geophys. Res.* **110**(B8), B09312.

Pavloff, N. & Schmit, C. (1995), "Diffractive orbits in quantum billiards," *Phys. Rev. Lett.* **75**(1), 61–64. Erratum **75**(20), 3779.

Photiadis, D. M. (1997), "Acoustics of a fluid-loaded plate with attached oscillators. Part I. Feynman rules," *J. Acoust. Soc. Am.* **102**, 348–357.

Pierce, A. D. (1989), *Acoustics: An Introduction to Its Physical Principles and Applications*, Acoustical Society of America, Woodbury, NY.

Pierre, C. (1988), "Rayleigh quotient and perturbation-theory for the eigenvalue problem – Comment," *J. App. Mech.* **55**(4), 986–988.

Pierre, C. (1990), "Weak and strong vibration localization in disordered structures: A statistical investigation," *J. Sound Vib.* **139**(1), 111–132.

Pnini, R. & Shapiro, B. (1996), "Intensity fluctuations in closed and open systems," *Phys. Rev. E* **54**(2), R1032–R1035.

Porter, C. E., ed. (1965), *Statistical Theories of Spectra: Fluctuations*, Academic, New York.

Poupinet, G., Ellsworth, V. L. & Frechet, J. (1984), "Monitoring velocity variations in the crust using earthquake doublets: An application to the Calaveras fault, California," *J. Geophys. Res.* **89**(NB7), 5719–5732.

Prigodin, V. N., Altshuler, B. L., Efetov, K. B., & Iida, S. (1994), "Mesoscopic dynamical echo in quantum dots," *Phys. Rev. Lett.* **72**, 546–549.

Primack, H. & Smilansky, U. (1998), "On the accuracy of the semiclassical trace formula," *J. Phys. A* **31**, 6253–6277.

Rahav, S. & Brouwer, P. W. (2006), "Ehrenfest time and the coherent backscattering off ballistic cavities," *Phys. Rev. Lett.* **96**(19), 196804.

Rahav, S. & Fishman, S. (2002), "Spectral statistics of rectangular billiards with localized perturbations," *Nonlinearity* **15**, 1541–1594.

Rautian, T. G. & Khalturin, V. I. (1978), "The use of the coda for determination of the earthquake source spectrum," *Bull. Seism. Soc. Am.* **68**, 923–948.

Rayleigh, L. (1892), "On theinfluence of obstacles arranged in rectangular order upon the properties of a medium," *Philos. Mag.* **34**, 481–502.

Rayleigh, L. (1945), *Theory of Sound*, 2nd edn, Dover, New York.

Reichl, L. E. (2004), *The Transition to Chaos: Conservative Classical Systems and Quantum Manifestations*, 2nd edn, Springer, Berlin.

Richens, P. J. & Berry, M. V. (1981), "Pseudo-integrable systems in classical and quantum mechanics," *Physica* **1D**, 495–512.

Richter, A. (1998), "Playing billiards with microwaves – Quantum manifestations of classical chaos, *in Emerging Applications of Number Theory,* Vol. 109 of The IMA Volumes in Mathematics and Its Applications, Springer, New York, pp. 479–523.

Richter, K. (2000), *Semiclassical Theory of Mesoscopic Quantum Physics*, Springer, Berlin.

Richter, K. & Sieber, M. (2002), "Semiclassical theory of chaotic quantum transport," *Phys. Rev. Lett.* **89**(20), 206801.

Richter, K., Ullmo, D. & Jalabert, R. A. (1996), "Orbital magnetism in the ballistic regime: Geometrical effects," *Phys. Rep.* **276**, 1–83.

Rosenzweig, N. & Porter, C. E. (1962), "'Repulsion of energy levels' in complex atomic spectra," *Phys. Rev.* **120**(5), 1698–1714.

Rouseff, D., Jackson, D. R., Fox, W. L. J., Ritcey, J. A. & Dowling, D. R. (2001), "Underwater acoustic communication by passive-phase conjugation: Theory and experimental results," *IEEE J. Ocean. Eng.* **26**(4), 821–831.

Roux, P. & Fink, M. (2000), "Time reversal in a waveguide: Study of the temporal and spatial focusing," *J. Acoust. Soc. Am.* **107**(5), 2418–2429.

Roux, P., Roman, B. & Fink, M. (1997), "Time-reversal in an ultrasonic waveguide," *App. Phys. Lett.* **70**(14), 1811–1813.

Roux, P., Sabra, K. G., Kuperman, W. A. & Roux, A. (2005), "Ambient noise cross correlation in free space: Theoretical approach," *J. Acoust. Soc. Am.* **117**(1), 79–84.

Rypina, I. I. & Brown, M. G. (2007), "On the width of a ray," *J. Acoust. Soc. Am.* **122**, 1440–1448.

Rypina, I. I., Brown, M. G., Beron-Vera, F. J., Koak, H., Olascoaga, M. J. & Udovydchenkov, I. A. (2007), "Robust transport barriers resulting from strong kolmogorov-arnold-moser stability," *Phys. Rev. Lett.* **98**, 104102.

Ryzhik, L. V., Papanicolaou, G. C. & Keller, J. B. (1996), "Transport equations for elastic and other waves in random media," *Wave Motion* **24**, 327–370.

Sabra, K. G., Gerstoft, P., Roux, P., Kuperman, W. A. & Fehler, M. C. (2005), "Surface wave tomography from microseisms in Southern California," *Geophys. Res. Lett.* **32**(14), L14311.

Sabra, K. G., Roux, P. & Kuperman, W. A. (2005), "Emergence rate of the time-domain Green's function from the ambient noise cross-correlation function," *J. Acoust. Soc. Am.* **118**(6), 3524–3531.

Saichev, A. I., Ishio, H., Sadreev, A. F. & Berggren, K.-F. (2002), "Statistics of interior current distributions in two-dimensional open chaotic billiards," *J. Phys. A* **35**, L87–L93.

Sampaio, R. & Soize, C. (2007), "On measures of non-linearity effects for uncertain dynamical systems – Application to a vibro-impact system," *J. Sound Vib.* **303**(3–5), 659–674.

Schaadt, K. (1997), The quantum chaology of acoustic resonators, MSc thesis, Niels Bohr Institute, University of Copenhagen.

Schaadt, K., Guhr, T., Ellegaard, C. & Oxborrow, M. (2003), "Experiments on elastomechanical wave functions in chaotic plates and their statistical features," *Phys. Rev. E* **68**(3), 036205.

Schaadt, K. & Kudrolli, A. (1999), "Experimental investigation of universal parametric correlators using a vibrating plate," *Phys. Rev. E* **60**(4), R3479–R3482.

Schaadt, K., Simon, G. & Ellegaard, C. (2001), "Ultrasound resonances in a rectangular plate described by random matrices," *Phys. Scr.* **T90**, 231–237.

Schnirelman, A. I. (1974), "Ergodic properties of eigenfunctions," *Usp. Mat. Nauk.* **29**(6), 181–182.

Schomerus, H. & Tworzydlo, J. (2004), "Quantum-to-classical crossover of quasibound states in open quantum systems," *Phys. Rev. Lett.* **93**(15), 154102.

Schroeder, M. (1954a), "Die statisischen Parameter der Frequenzkurvven von grosser Raumen," *Acustica* **4**, 594–600.

Schroeder, M. (1954b), "Eigenfrequenzstatistik und Anregungsstatistik in Raumen," *Acustica* **4**, 456–468.

Schroeder, M. R. (1959), "Measurement of sound diffusion in reverberation chambers," *J. Acoust. Soc. Am.* **31**(11), 1407–1414.

Schroeder, M. R. (1962), "Frequency-correlation functions of frequency responses in rooms," *J. Acoust. Soc. Am.* **34**(12), 1819–1823.

Schroeder, M. R. (1965), "Some new results in reverberation theory and measurement methods," *in* 5ème Congres International d'Acoustique (Liège, 7–14 Sept.), p. G31.

Schroeder, M. R. (1969a), "Effect of frequency and space averaging on the transmission responses of multimode media," *J. Acoust. Soc. Am.* **46**(2A), 277–283.

Schroeder, M. R. (1969b), "Spatial averaging in a diffuse sound field and the equivalent number of independent measurements," *J. Acoust. Soc. Am.* **46**(3A), 534.

Schroeder, M. R. (1987), "Statistical parameters of the frequency response curves of large rooms," *J. Audio Eng. Soc.* **35**(5), 299–306.

Schroeder, M. R. & Kuttruff, K. H. (1962), "On frequency response curves in rooms. comparison of experimental, theoretical and Monte Carlo results for the average frequency spacing between maxima," *J. Acoust. Soc. Am.* **34**(1), 76–80.

Schueller, G. I. (2005), "Special issue on 'computational methods in stochastic mechanics and reliability analysis," *Comput. Meth. Appl. Mech. Eng.* **194**(12–16), 1251–1795.

Schuëller, G. I. (2006), "Developments in stochastic structural mechanics," *Arch. Appl. Mech.* **75**(10–12), 755–773.

Schuster, A. (1905), "Radiation through a foggy atmosphere," *Atrophys. J.* **21**, 1–22.

Šeba, P. (1990), "Wave chaos in singular quantum billiard," *Phys. Rev. Lett.* **64**(16), 1855–1858.

Self, R. H. (2005), "Asymptotic expansion of integrals," *in* M. C. M. Wright, ed., *Lecture Notes on the Mathematics of Acoustics*, Imperial College Press, London, chapter 4, pp. 91–105.

Sens-Schönfelder, C. & Wegler, U. (2006), "Passive image interferometry and seasonal variations of seismic velocities at merapi volcano, indonesia," *Geophys. Res. Lett.* **33**(21), L21302.

Serfling, R. J. (1980), *Approximation Theorems of Mathematical Statistics*, John Wiley, New York.

Sevryuk, M. B. (2007), "Invariant tori in quasiperiodic nonautonomous dynamical systems via Herman's method," *Dis. Cont. Dyn. Sys.* **18**, 569–595.

Shannon, C. E. (1948), "A mathematical theory of communication," *Bell Syst. Tech. J.* **27**, 379–423, 623–659.

Shapiro, N. M. & Campillo, M. (2004), "Emergence of broadband Rayleigh waves from correlations of the ambient seismic noise," *Geophys. Res. Lett.* **31**(7), L07614.

Shapiro, N. M., Campillo, M., Stehly, L. & Ritzwoller, M. H. (2005), "High-resolution surface-wave tomography from ambient seismic noise," *Science* **307**(5715), 1615–1618.

Shnirelman, A. I. (1974), "Ergodic properties of eigenfunctions," *Usp. Mat. Nauk.* **29**, 181–182.

Shorter, P. J. & Langley, R. S. (2005a), "On the reciprocity relationship between direct field radiation and diffuse reverberant loading," *J. Acoust. Soc. Am.* **117**(1), 85–95.

Shorter, P. J. & Langley, R. S. (2005b), "Vibro-acoustic analysis of complex systems," *J. Sound Vib.* **288**(3), 669–699.

Sieber, M., Primack, H., Smilansky, U., Ussishkin, I. & Schanz, H. (1995), "Semiclassical quantization of billilards with mixed boundary conditions," *J. Phys. A* **28**, 5041–5078.

Sjöstrand, J. & Zworski, M. (2007), "Fractal upper bounds on the density of semiclassical resonances," *Duke Math. J.* **137**(3), 381–459.

Smilansky, U. (1991), "The classical and quantum theory of chaotic scattering," *in* M.-J. Giannoni, A. Voros & J. Zinn-Justin, eds, *Chaos and Quantum Physics*, Vol. 52 of Les Houches Lecture Series, North Holland, Amsterdam, pp. 371–441.

Smilansky, U. (1995), "Semiclassical quantization of chaotic billiards – A scattering approach," *in* G. M. E. Akkermans & J. L. Pichard, eds, *Mesoscopic Quantum Physics*, Les Houches Lecture Series, Elsevier, Amsterdam, pp. 373–434.

Smilansky, U. & Ussishkin, I. (1996), "The smooth spectral counting function and the total phase shift for quantum billiards," *J. Phys. A* **29**(10), 2587–2597.

Smith, K. B., Brown, M. G. & Tappert, F. D. (1992a), "Acoustic ray chaos induced by mesoscale ocean structure," *J. Acoust. Soc. Am.* **91**, 1950–1959.

Smith, K. B., Brown, M. G. & Tappert, F. D. (1992b), "Ray chaos in underwater acoustics," *J. Acoust. Soc. Am.* **91**, 1939–1949.

Smith, P. W., Jr. (1962), "Response and radiation of structural modes excited by sound," *J. Acoust. Soc. Am.* **34**(5), 640–647.

Snieder, R. (2002), "Coda wave interferometry and the equilibration of energy in elastic media," *Phys. Rev. E* **66**(4), 046615.

Snieder, R. (2004), "Extracting the Green's function from the correlation of coda waves: A derivation based on stationary phase," *Phys. Rev. E* **69**(4), 046610.

Snieder, R. & Page, J. (2007), "Multiple scattering in evolving media," *Phys. Today* **60**(5), 49–55.

Soize, C. (1993), "A model and numerical method in the medium frequency range for vibroacoustic predictions using the theory of structural fuzzy," *J. Acoust. Soc. Am.* **94**(2), 849–865.

Soize, C. (2000), "A nonparametric model of random uncertainties for reduced matrix models in structural dynamics," *Probab. Eng. Mech.* **15**(3), 277–294.

Soize, C. (2001), "Maximum entropy approach for modeling random uncertainties in transient elastodynamics," *J. Acoust. Soc. Am.* **109**(5), 1979–1996.

Soize, C. (2003a), "Random matrix theory and non-parametric model of random uncertainties in vibration analysis," *J. Sound Vib.* **263**, 893–916.

Soize, C. (2003b), "Uncertain dynamical systems in the medium-frequency range," *J. Eng. Mech.* **129**(6), 1017–1027.

Soize, C. (2005a), "A comprehensive overview of a non-parametric probablistic approach of model uncertainties for predictive models in structural dynamics," *J. Sound Vib.* **288**(3), 623–652.

Soize, C. (2005b), "Probabilistic models for computational stochastic mechanics and applications," *in* G. Augusti, G. I. Schueller & M. Ciampoli, eds, *Structural Safety and Reliability ICOSSAR '05*, Millpress, Rotterdam, pp. 23–42. ISBN 90 5966 040 4.

Soize, C. (2005c), "Random matrix theory for modeling random uncertainties in computational mechanics," *Comput. Meth. Appl. Mech. Eng.* **194**(12–16), 1333–1366.

Soize, C., Capiez-Lernout, E., Durand, J.-F., Fernandez, C. & Gagliardini, L. (2008), "Probabilistic model identification of uncertainties in computational models for dynamical systems and experimental validation," *Comput. Meth. Appl. Mech. Eng.* **198**(1), 150–163.

Soize, C. & Chebli, H. (2003), "Random uncertainties model in dynamic substructuring using a nonparametric probabilistic model," *J. Eng. Mech.* **129**(4), 449–457.

Soize, C., Chen, C., Durand, J.-F., Duhamel, D. & Gagliardini, L. (2007), "Computational elastoacoustics of uncertain complex systems and experimental validation," *in*

M. Papadrakakis, D. C. Charmpis, N. D. Lagaros & Y. Tsompanakis, eds, *Computational Methods in Structural Dynamics and Earthquake Engineering COMPDYN 2007*, National Technical University of Athens, Athens.

Solomon, H. (1978), *Geometric Probability*, SIAM.

Sondergaard, N. & Tanner, G. (2002), "Wave chaos in the elastic disk," *Phys. Rev. E* **66**, 066211.

Soven, P. (1967), "Coherent-potential model of substitutional disordered alloys," *Phys. Rev.* **156**(3), 809–813.

Spall, J. C. (2003), *Introduction to Stochastic Search and Optimization*, John Wiley, Hoboken, NJ.

Srednicki, M. & Stiernelof, F. (1996), "Gaussian fluctuations in chaotic eigenstates," *J. Phys. A* **29**(18), 5817–5826.

Stanke, F. E. & Kino, G. S. (1984), "A unified theory for elastic wave propagation in polycrystalline materials," *J. Acoust. Soc. Am.* **75**, 665–681.

Stehly, L., Campillo, M., Froment, B. & Weaver, R. L. (2008), "Reconstructing Green's function by correlation of the coda of the correlation (c^3) of ambient seismic noise," *J. Geophys. Res.* **113**, B11306.

Stehly, L., Campillo, M. & Shapiro, N. M. (2006), "A study of the seismic noise from its long-range correlation properties," *J. Geophys. Res.* **111**, B10306.

Stehly, L., Campillo, M. & Shapiro, N. M. (2007), "Traveltime measurements from noise correlation: Stability and detection of instrumental time-shifts," *Geophys. J. Int.* **171**(1), 223–230.

Stehly, L., Fry, M., Campillo, M., Shapiro, N. M., Guilbert, J., Boschi, L. & Giardini, D. (2009), "Tomography of the Alpine region from observations of seismic ambient noise," *Geophys. J. Int.* **178**(1), 338–350.

Stöckmann, H.-J. (1999), *Quantum Chaos*, Cambridge University Press, Cambridge, UK.

Stöckmann, H.-J. & Stein, J. (1990), "'Quantum' chaos in billiards studied by microwave absorption," *Phys. Rev. Lett.* **64**, 2215–2218.

Stratonovich, R. L. (1963), *Topics in the Theory of Random Noise*, Gordon & Breach, New York.

Tabachnikov, S. (2005), *Geometry and Billiards*, Vol. 30 of Student Mathematical Library, American Mathematical Society.

Tabor, M. (1989), *Chaos and Integrability in Nonlinear Dynamics: An Introduction*, Wiley, New York, pp. 20–31.

Tanner, G. & Søndergaard, N. (2007), "Short wavelength approximation of a boundary integral operator for homogeneous and isotropic elastic bodies," *Phys. Rev. E* **75**, 036607.

Tanner, G. & Søndergaard, N. (2007), "Wave chaos in acoustics and elasticity," *J. Phys. A* **40**(50), R443–R509.

Tanter, M., Aubry, J.-F., Gerber, J., Thomas, J.-L. & Fink, M. (2001), "Optimal focusing by spatio-temperal inverse filter. I Basic principles," *J. Acoust. Soc. Am.* **110**(1), 37–47.

Tanter, M., Thomas, J.-L. & Fink, M. (2000), "Time reversal and the inverse filter," *J. Acoust. Soc. Am.* **108**(1), 223–234.

Tappert, F. D. (1977), "The parabolic approximation method," *in* J. B. Keller & J. S. Papadakis, eds, *Lecture Notes in Physics*, Vol. 70, of Wave Propagation and Underwater Acoustics, Springer-Verlag, New York, pp. 224–287.

Tolstikhin, O. I., Ostrovsky, V. N. & Nakamura, H. (1997), "Siegert pseudo-states as a universal tool: Resonances, *s* matrix, Green function," *Phys. Rev. Lett.* **79**(11), 2026–2029.

Tolstikhin, O. I., Ostrovsky, V. N. & Nakamura, H. (1998), "Siegert pseudostate formulation of scattering theory: One-channel case," *Phys. Rev. A* **58**(3), 2077–2096.

Tomsovic, S., Grinberg, M. & Ullmo, D. (1995), "Semiclassical trace formulas of near-integrable systems – Resonances," *Phys. Rev. Lett.* **75**(24), 4346–4349.

Turner, J. A. & Weaver, R. L. (1994), "Radiative transfer and multiple scattering of diffuse ultrasound in polycrystalline media," *J. Acoust. Soc. Am.* **96**(6), 3675–3683.

Twersky, V. (1977), "Coherent scalar field in pair-correlated random distributions of aligned scatterers," *J. Math. Phys.* **18**, 2468–2486.

Udovydchenkov, I. A. & Brown, M. G. (2008), "Modal group time spreads in weakly range-dependent sound channels," *J. Acoust. Soc. Am.* **123**, 41–50.

Ullmo, D., Grinberg, M. & Tomsovic, S. (1996), "Near-integrable systems: Resonances and semiclassical trace formulas," *Phys. Rev. E* **54**(1), 136–152.

Urbina, J. D. & Richter, K. (2004), "Semiclassical construction of random wave functions for confined systems," *Phys. Rev. E* **70**(1), 015201.

Vallejos, R. O. & Ozorio de Almeida, A. M. (1999), "Decomposition of resonant scatterers by surfaces of section," *Ann. Phys.* **278**(1), 86–108.

van Manen, D.-J., Robertasson, J. O. A. & Curtis, A. (2005), "Modeling of wave propagation in inhomogeneous media," *Phys. Rev. Lett.* **94**, 164301.

van Tiggelen, B. (2005), "Analogies between quantum and classical waves: Deceiving, surprising and complementary," Presentation given at the Cuernavaca Gathering on Chaotic and Random Scattering of Quantum and Acoustic Waves. http://lpm2c.grenoble.cnrs.fr/Themes/tiggelen/Cuernavaca2005.pdf

van Tiggelen, B. A., Margerin, L. & Campillo, M. (2001), "Coherent backscattering of elastic waves: Specific role of source, polarization, and near field," *J. Acoust. Soc. Am.* **110**(3), 1291–1298.

Varadan, V. K., Varadan, V. V. & Pao, Y. H. (1978), "Multiple scattering of elastic waves by cylinders of arbitrary cross section. I. SH waves," *J. Acoust. Soc. Am.* **63**(5), 1310–1319.

Vasil'ev, D. G. (1987), "Asymptotics of the spectrum of a boundary value problem," *Trans. Moscow Math. Soc.* **49**, 173–245.

Verbaarschot, J. J. M., Weidenmuller, H. A. & Zirnbauer, M. R. (1985), "Grassman integration in stochastic quantum physics – The case of compound nucleus scattering," *Phys. Rep.* **129**(6), 367–438.

Vergini, E. G. (2000), "Semiclassical theory of short periodic orbits in quantum chaos," *J. Phys. A* **33**(25), 4709–4716.

Vergini, E. G. (2004), "Semiclassical limit of chaotic eigenfunctions," *J. Phys. A* **37**, 6507–6519.

Vergini, E. G. & Carlo, G. G. (2000), "Semiclassical quantization with short periodic orbits," *J. Phys. A* **33**(25), 4717–4724.

Vergini, E. G. & Carlo, G. G. (2001), "Semiclassical construction of resonances with hyperbolic structure: The scar function," *J. Phys. A* **34**, 4525–4552.

Vergini, E. G. & Schneider, D. (2005), "Asymptotic behaviour of matrix elements between scar functions," *J. Phys. A* **38**, 587–616.

Vergini, E. G. & Wisniacki, D. A. (1998), "Localized structures embedded in the eigenfunctions of chaotic Hamiltonian systems," *Phys. Rev. E* **58**(5), R5225–R5228.

Vergini, E. & Saraceno, M. (1995), "Calculation by scaling of highly excited states of billiards," *Phys. Rev. E* **52**(3), 2204–2207.

Vesperinas, M. N. & Wolf, E. (1985), "Phase conjugation and symmetries with wave fields in free space containing evanescent components," *JOSAA* **2**(9), 1429–1434.

Viktorov, I. A. (1967), *Rayleigh and Lamb Waves*, Plenum, New York.

Villasenor, A., Yang, Y., Ritzwoller, M. H. & Gallart, J. (2007), "Ambient noise surface wave tomography of the Iberian Peninsula: Implications for shallow seismic structure," *Geophys. Res. Lett.* **34**(11), L11304.

Virovlyansky, A. L., Kazarova, A. Y. & Lyubavin, L. Y. (2005), "Ray-based description of normal mode amplitudes in a range-dependent waveguide," *Wave Motion* **42**, 317–334.

Virovlyansky, A. L., Kazarova, A. Y. & Lyubavin, L. Y. (2007), "Statistical description of chaotic rays in a deep water acoustic waveguide," *J. Acoust. Soc. Am.* **121**, 2542–2552.

Voros, A. (1979), "Semi-classical ergodicity of quantum eigenstates in the Wigner representation," *Lect. Not. Phys.* **93**, 326–333.

Walter, E. & Pronzato, L. (1997), *Identification of Parametric Models from Experimental Data*, Springer, Berlin.

Waterman, P. C. & Truell, R. (1961), "Multiple scattering of waves," *J. Math. Phys.* **2**, 512–537.

Weaver, R. L. (1982), "On diffuse waves in solid media," *J. Acoust. Soc. Am.* **71**(6), 1608–1609.
Weaver, R. L. (1989a), "On the ensemble variance of reverberation room transmission functions, the effect of spectral rigidity," *J. Sound Vib.* **130**(3), 487–491.
Weaver, R. L. (1989b), "Spectral statistics in elastodynamics," *J. Acoust. Soc. Am.* **85**(3), 1005–1013.
Weaver, R. L. (1990a), "Anderson localization of ultrasound," *Wave Motion* **12**(2), 129–142.
Weaver, R. L. (1990b), "Diffusivity of ultrasound in polycrystals," *J. Mech. Phys. Solids* **38**, 55–86.
Weaver, R. L. (1997), "Multiple-scattering theory for mean responses in a plate with sprung masses," *J. Acoust. Soc. Am.* **101**(6), 3466–3474.
Weaver, R. L. (2005), "Information from seismic noise," *Science* **307**, 1568–1569.
Weaver, R. L. & Burkhardt, J. (1994), "Weak anderson localization and enhanced backscatter in reverberation rooms and quantum dots," *J. Acoust. Soc. Am.* **96**(5), 3186–3190.
Weaver, R. L. & Lobkis, O. I. (2000a), "Anderson localization in coupled reverberation rooms," *J. Sound Vib.* **321**(4), 1111–1134.
Weaver, R. L. & Lobkis, O. I. (2000b), "Enhanced backscattering and modal echo of reverberant elastic waves," *Phys. Rev. Lett.* **84**(21), 4942–4945.
Weaver, R. L. & Lobkis, O. I. (2001), "Ultrasonics without a source: Thermal fluctuation correlations at MHz frequencies," *Phys. Rev. Lett.* **87**(13), 134301.
Weaver, R. L. & Lobkis, O. I. (2003), "Elastic wave thermal fluctuations, ultrasonic waveforms by correlation of thermal phonons," *J. Acoust. Soc. Am.* **113**(5), 2611–2621.
Weaver, R. L. & Lobkis, O. I. (2005), "Fluctuations in diffuse field–field correlations and the emergence of the Green's function in open systems," *J. Acoust. Soc. Am.* **117**(6), 3432–3439.
Weaver, R. L. & Sornette, D. (1995), "The range of spectral correlations in psuedointegrable systems: GOE statistics in a rectangular membrane with a point scatterer," *Phys. Rev. E* **52**, 3341–3350.
Wegler, U. & Sens-Schönfelder, C. (2007), "Fault zone monitoring with passive image interferometry," *Geophys. J. Int.* **168**(3), 1029–1033.
Wheeler, C. T. (2005), "Curved boundary corrections to nodal line statistics in chaotic billiards," *J. Phys. A* **38**, 1491–1504.
Whitney, R. S. & Jacquod, P. (2005), "Microscopic theory for the quantum to classical crossover in chaotic transport," *Phys. Rev. Lett.* **94**(11), 116801.
Wigner, E. (1951), "On a class of analytic functions from the quantum theory of collisions," *Ann. Math.* **53**, 36–67.
Wigner, E. (1955), "Characteristic vectors of bordered matrices with infinite dimensions," *Ann. Math.* **62**, 548–564.
Wigner, E. (1957), "Characteristic vectors of bordered matrices with infinite dimensions II," *Ann. Math.* **65**, 203–207.
Wigner, E. (1958), "On the distribution of the roots of certain symmetric matrices," *Ann. Math.* **67**, 325–326.
Wigner, E. (1959), *Group Theory and Its Applications to the Quantum Mechanics of Atomic Spectra*, Academic, New York.
Wirzba, A. (1999), "Quantum mechanics and semiclassics of hyperbolic *n*-disk scattering systems," *Phys. Rep.* **309**(1–2), 1–116.
Wisniacki, D. A., Vergini, E., Benito, R. M. & Borondo, F. (2005), "Signatures of homoclinic motion in quantum chaos," *Phys. Rev. Lett.* **94**(5), 054101.
Wisniacki, D. A., Vergini, E., Benito, R. M. & Borondo, F. (2006), "Scarring by homoclinic and heteroclinic orbits," *Phys. Rev. Lett.* **97**(9), 094101.
Wolfson, M. A. & Tomsovic, S. (2001), "On the stability of long-range sound propagation through a structured ocean," *J. Acoust. Soc. Am.* **109**, 2693–.
Worcester, P. F., Cornuelle, B. D., Dzieciuch, M. A., Munk, W. H., Howe, B. M., A. Mercer, J., Spindel, R. C., Colosi, J. A., Metzger, K., Birdsall, T. & Baggeroer, A. B. (1999), "A test

of basin-scale acoustic thermometry using a large-aperture vertical array at 3250-km range in the eastern north pacific ocean," *J. Acoust. Soc. Am.* **105**, 3185–3201.

Wright, M. C. M. (2001), "Variance of deviations from the average mode count for rectangular wave guides," *Acoust. Res. Lett. Online* **2**(1), 19–24.

Wright, M. C. M. & Ham, C. J. (2007), "Periodic orbit theory in acoustics: Spectral fluctuations in circular and annular waveguides," *J. Acoust. Soc. Am.* **121**(4), 1865–1872.

Wright, M. C. M., Howls, C. J. & Welch, B. A. (2003a), "Periodic orbit calculations of modecount functions for elastic plates," *in Proceedings of the VIII International Conference on Recent Advances in Structural Dynamics*, no. 42.

Wright, M. C. M., Morfey, C. L. & Yoon, S. H. (2003b), "On the modal distribution for circular and annular ducts," *in 9th AIAA/CEAS Aeroacoustics Conference*, no. 2003-3141.

Yakubovich, V. A. & Starzhinskii, V. M. (1975), *Linear Differential Equations with Periodic Coefficients*, Wiley, New York.

Yang, Y., Ritzwoller, M. H., Levshin, A. L. & Shapiro, N. M. (2007), "Ambient noise Rayleigh wave tomography across Europe," *Geophys. J. Int.* **168**(1), 259–274.

Yao, H., van der Hilst, R. D. & de Hoop, M. V. (2006), "Surface-wave array tomography in SE Tibet from ambient seismic noise and two-station analysis: I. Phase velocity maps," *Geophys. J. Int.* **166**(2), 732–744.

Ye, L., Cody, G., Zhou, M., Sheng, P. & Norris, A. N. (1992), "Observation of bending wave localization and quasi mobility edge in two dimensions," *Phys. Rev. Lett.* **69**(21), 3080–3083.

Zelditch, S. (1987), "Uniform distribution of eigenfunctions on compact hyperbolic surfaces," *Duke Math. J.* **55**, 919–941.

Zelditch, S. & Zworski, M. (1996), "Ergodicity of eigenfunctions for ergodic billiards," *Comm. Math. Phys.* **175**, 673–682.

Zienkiewicz, O. C. & Taylor, R. L. (2000), *The Finite Element Method*, Vols. 1–3, 5th edn, Butterworth-Heinemann, Oxford, UK.

Index

Printed in the United States
By Bookmasters

Printed in the United States
By Bookmasters